A Laboratory Book of Computational Organic Chemistry

Warren J. Hehre

Wavefunction, Inc.
18401 Von Karman, Suite 370
Irvine, CA 92612

Alan J. Shusterman

Department of Chemistry
Reed College
3203 S. E. Woodstock Blvd.
Portland, OR 97202

W. Wayne Huang

Wavefunction, Inc.
18401 Von Karman, Suite 370
Irvine, CA 92612

ISBN 0-9643495-5-8

Printed in the United States of America

Acknowledgments

The present book owes much to our first attempt at a computational laboratory manual "Experiments in Computational Organic Chemistry", and to William Pietro and Lonnie Burke who joined us at that time. It is also indebted to the many readers who offered their comments and suggestions for improvement, as well as ideas for new directions. The members of Wavefunction and in particular, Andy Driessen, Bernard Deppmeier and Jeffrey Johnson, are sincerely thanked for their efforts to provide in SPARTAN all that is needed to complete the experiments in this book. Finally, Pamela Ohsan is thanked for her expertise and perseverance in preparing the final manuscript and far too many earlier drafts.

To the memory of

Robert Wheaton Taft
1922 - 1996

colleague and friend

Preface

Why does computer-based molecular modeling play an increasingly important role in chemical research? Is it that, at the same time that the cost of experimental laboratory science has skyrocketed, the cost of modeling has sharply decreased? Is it that computational methods, and the software and hardware needed to implement these methods, have matured to the point where useful results can be obtained for real systems in a practical time period? Could it be that Dirac's famous quote,

> *The underlying physical laws necessary for the mathematical theory of a large part of physics and the whole of chemistry are thus completely known, and the difficulty is only that the exact application of these laws leads to equations much too complicated to be soluble.*
>
> P.A.M. Dirac 1902-1984

made at a time when quantum mechanics was still in its infancy, is now only half true? Could it be that "the whole of chemistry" is now open to computation?

All of these factors contribute, but together they have a cumulative effect that is considerably greater than the sum of their separate contributions. Effective and accurate theoretical models, combined with powerful and usable software, and inexpensive, powerful computer hardware, have made molecular modeling an affordable and widely accessible tool for solving real chemical problems. The time is now at hand for modeling tools to be used on a par with experimental methods, as a legitimate and practical means for exploring chemistry. All that is needed is for mainstream chemists to be trained in the practical use of modeling techniques, and for them to adopt molecular modeling in the same way that they have adopted other research tools - NMR, GC-MS, X-ray crystallography - once deemed the exclusive province of highly trained "experts."

Molecular modeling offers two major benefits as a tool for exploration. First, modeling tools can be used to investigate a much wider variety of chemical species than are normally accessible to the experimental chemist. Different conformers of a flexible molecule, reactive intermediates, and even fleeting transition state structures, can all be easily studied using computational models. Moreover, the computational effort required to identify and characterize each of these species is essentially the same. This compares very favorably with experiment, where procedures for isolating and characterizing a molecule become increasingly expensive and difficult as the lifetime and/or concentration of the molecule decreases.

The main benefit of molecular modeling, though, is that it allows chemists to think more clearly about issues that are really central to chemistry - structure, stability, and reactivity - than would be possible without the use of a computer.

To see why this is so, one need only consider the basic tool conventionally used by organic chemists to describe molecular structure: two-dimensional line drawings. While an "expert" organic chemist can easily understand and produce usable figures, the figures themselves do not look at all like the molecules they are supposed to depict. Even worse, students of organic chemistry must spend considerable time mastering the creation and interpretation of these figures, and this turns out to be a major educational hurdle. Computer-generated models, by contrast, "look" and "behave" much more like "real molecules." Good models can be produced even when a student is unable to make an accurate drawing, and the resulting model is more than a symbol or representation of a molecule; it also conveys quantitative information (geometry, volume, contact area, symmetry, etc.) about the molecular structure. Thus, a chemistry student or research chemist, working with a computer, can explore "new areas of chemistry." Finally, computer models can also be constructed for molecules that cannot be represented by simple line drawings. Such molecules appear throughout organic chemistry, and include molecules containing delocalized charges (or spins), many unstable molecules, and, perhaps most important, reaction transition states connecting stable molecules.

The advantages of computer modeling over conventional representational tools are not limited to molecular structure. Structure is just the beginning. Models based on quantum mechanics can be easily and routinely used to calculate and display a myriad of chemical and physical properties, among them, stability, spectra and charge distribution. And, as already mentioned, these properties can be studied and compared for a much wider range of molecules than can be investigated experimentally. Therefore, computer modeling can provide students and chemists with a firm grasp of both sides of the structure-property relationship.

This book and its companion volume, **A Textbook of Computational Organic Chemistry**,[1] have been written to serve as joint texts for a one semester upper division lecture-laboratory course in computational organic chemistry. The two books should, however, be of general interest to anyone who wants to learn how to use molecular modeling both to "communicate chemistry" and ultimately to "do chemistry." The present volume takes a very practical approach and focuses almost entirely on the application of computer-based molecular modeling to problems in organic chemistry. An analogy might be made between "computational organic chemistry" and "organic spectroscopy". Most chemists think of the latter in terms of the identification and characterization of organic compounds using spectroscopic techniques, and not of the quantum mechanical description of the

---

1. A.J. Shusterman and W.J. Hehre, **A Textbook of Computational Organic Chemistry**, Wavefunction, Inc., Irvine, California, to be published, Winter, 1997.

interaction of "light" with matter or of the construction and repair of spectrometers. Readers who are interested in treatment of the theories underlying molecular models, or the algorithms used to construct these models, should consult one of the many texts that have been written on these topics.[2]

This book is a true laboratory manual; a tested procedure is provided for each experiment, but the results of these procedures are not reported in the text. You must do the experiments yourself in order to find out how they come out. There will be surprises, as there are with any "experimental science", and you must decide how these fit in (or do not fit in) to an overall picture.

The experiments have been kept very short (generally 2 to 4 pages), and fairly simple, in that each covers a single, narrowly-defined aspect of some chemical problem, and most can easily be completed inside of a "normal laboratory period" (2 to 3 hours).[3] Nevertheless, the experiments are based on "real" chemistry, much of which has been drawn directly from scientific journals, and they collectively illustrate how practical computations can provide insight into a wide range of phenomena. Some of the experiments offer "Optional" parts built on the material covered in the basic experiment, and which provide opportunities for extended study. In most cases, these should not be attempted until the basic material has been completed.

Naturally, the experiments in this book assume a certain knowledge of chemistry and some familiarity with the basic operations of a suitable molecular modeling program.[4] All of the experiments should be accessible to any reader who has completed a year of organic chemistry and physical chemistry (including quantum mechanics). However, a number of the experiments can be successfully understood and performed by a reader with far less training, such as a student in an introductory organic chemistry course. These experiments have been identified by a ❖ in the Table of Contents.

A series of short essays are interspersed among the experiments. A number of these deal with practical aspects of molecular modeling ("Finding Transition States"), while others illustrate "chemical thinking" in terms of models ("Reactive Intermediates and the Hammond Postulate").

---

2. A sampling might include: Atomic and molecular quantum mechanics. I.N. Levine, **Quantum Chemistry**, 4th Ed., Prentice Hall, Englewood Cliffs, New Jersey, 1991; D.A. McQuarrie, **Quantum Chemistry**, University Science Books, Mill Valley, California, 1983. Molecular orbital theory. J.A. Pople and D.A. Beveridge, **Approximate Molecular Orbital Theory**, McGraw Hill, New York, 1970; W.J. Hehre, L. Radom, P.v.R. Schleyer and J.A. Pople, **Ab Initio Molecular Orbital Theory**, Wiley, New York, 1986.
3. The time needed to complete any given experiment may vary considerably according to the nature of the experiment, the reader's preparation and execution of the stated procedure, and the computer hardware and modeling software.
4. A tutorial for the *SPARTAN* program, is provided as an appendix.

Table of Contents

Notes to the Reader

I) General Remarks

Experience in teaching molecular modeling to large numbers of practicing chemists and chemistry students suggests that many of the habits and "rules" for successful work in the experimental laboratory also apply to the computational laboratory. In particular, we recommend that you:

- Come to lab prepared. Experience suggests that, in most cases, a large portion of your time will be spent constructing models, setting up calculations, and analyzing results, and considerably less time will be spent waiting for calculations to finish. Therefore, planning your work in advance may save considerable time at the computer. Pre-lab preparation includes, but is not limited to: 1) reading experiments and background essays in advance, 2) making a personal outline of the experimental procedure (the outline might include the sequence of operations, special notes about molecule construction or optimization, a list of the molecules that need to be built, a list of the calculations that need to be performed on each molecule, etc.), 3) planning in advance how to build each molecule.

- Follow all instructions to the letter. All of the experiments have been tested, but procedural mistakes (the equivalent of spilling a sample on your lab bench) are inevitable. Remember that a computer will only do what you tell it to do.

- Scrutinize your results carefully. The most common error in molecular modeling is to accept without scrutiny whatever result the computer churns out. Again, the computer will only do what you tell it to do. The latter may or may not conform to what you thought you were requesting, or to what you need.

- Save your results (models). Never delete or destroy a model until you are certain that you have no further use for it (models that are built as part of a research project should never be completely destroyed). For example, if you obtain an unexpected result that you would like to discuss with your instructor or a colleague, save your model. It is almost impossible to piece together what happened during a calculation without having the actual model to inspect.[1]

II) Literature References

General references to the topics discussed in the experiments have been provided. While some of these come from the primary literature, most have been drawn from a number of widely-used advanced organic chemistry textbooks:

---

1. Note that successive *SPARTAN* computations on a given model erase most of the information about the previous model. Therefore, if you want to save a *SPARTAN* model for later discussion, you should not perform any further computations on this model.

Advanced Organic Chemistry, 3rd ed.
Part A: Structure and Mechanisms; Part B: Reactions and Synthesis
F.A. Carey and R.J. Sundberg
Plenum Press, New York, 1990

Physical Organic Chemistry, 2nd ed.
N. Isaacs
Wiley, New York, 1995

Mechanism and Theory in Organic Chemistry, 3rd ed.
T.H. Lowry and K.S. Richardson
Harper Collins, New York, 1987

Advanced Organic Chemistry, 4th ed.
J. March
Wiley, New York, 1992

These have been referred to simply as "Carey and Sundberg A", "Carey and Sundberg B", "Isaacs", "Lowry and Richardson" and "March", respectively, together with the appropriate page number.

III) Units

Geometry
Bond distances are given in Ångströms (Å) and bond angles and dihedral angles in degrees (°).

Energies, Heats of Formation and Strain Energies
Total energies and orbital energies from *ab initio* calculations are given in atomic units (also termed hartrees).

Heats of formation from semi-empirical calculations are given in kcal/mol, and orbital energies from semi-empirical calculations are given in electron volts (eV).

Strain energies from molecular mechanics calculations are given in kcal/mol.

1 hartree = 627.5 kcal/mol
1 eV = 23.06 kcal/mol
1 hartree = 27.21 eV

To convert to SI energy units (kJ/mol),

1 kcal/mol = 4.184 kJ/mol

Dipole Moments, Charges and Electrostatic Potentials
Dipole moments are given in debyes.
Atomic charges are given in electrons.
Electrostatic potentials are given in kcal/mol.

Vibrational Frequencies

Vibrational frequencies are given in wavenumbers (cm^{-1}).

IV) Graphical Displays

The graphical displays utilized in the experiments fall into one of two categories. The first are isovalue surfaces (or isosurfaces), which seek to display the sizes and shapes of specific molecular orbitals, total electron densities or spin densities. These will be referred to simply as molecular orbitals, electron densities and spin densities, respectively, e.g., "display the highest-occupied molecular orbital for...". The second are total electron density isosurfaces (isodensity surfaces) onto which the value of a specific quantity has been "color mapped", most commonly the value of a specific molecular orbital, the spin density or the electrostatic potential. These displays will be referred to as molecular orbital maps, spin density maps and electrostatic potential maps, respectively, e.g., "display the electrostatic potential map for...".

A more complete account of graphical displays is found in the essay: **Graphical Models and Graphical Modeling**.

V) Computational Methods

There has been no attempt to provide insight or rationale into the choice of a specific computational method for a particular experiment. In order to minimize computer time, semi-empirical rather than *ab initio* molecular orbital methods have been used wherever appropriate. In some cases, semi-empirical and *ab initio* methods are combined. For example, equilibrium and/or transition-state geometries obtained from semi-empirical methods have been used for energy and property calculations using *ab initio* techniques. This results in significant computer time savings over full optimization using *ab initio* methods, usually with very little loss of accuracy.[2]

The specific computational methods that are recommended for use herein are briefly outlined in the essay: **Computational Tools**, and have all been extensively documented and assessed elsewhere.[3,4] These include the AM1 and AM1-SM2 semi-empirical methods for gas and aqueous-phase calculations, and *ab initio*

2. For a discussion of this and other time-saving strategies see: W.J. Hehre, **Practical Strategies for Electronic Structure Calculations**, Wavefunction, Inc., Irvine, California, 1995.
3. *Ab initio* methods: W.J. Hehre, L. Radom, P.v.R. Schleyer and J.A. Pople, **Ab Initio Molecular Orbital Theory**, Wiley, New York, 1986; semi-empirical methods: T. Clark, **A Handbook of Computational Chemistry**, Wiley, New York, 1986; J.J.P. Stewart, in **Reviews in Computational Chemistry**, **1**, 45 (1989); J.J.P. Stewart, *J. Computer Aided Molecular Design*, **4**, 1 (1990); M.C. Zerner, in **Reviews in Computational Chemistry**, **2**, 313 (1991).
4. W.J. Hehre, **Critical Assessment of Modern Electronic Structure Methods**, Wavefunction, Inc., Irvine, California, to be published, Winter, 1997.

methods using the 3-21G, 3-21G$^{(*)}$ and 6-31G* basis sets. In general, these methods are "interchangeable" and, as time permits, it may be informative to carry out some of the experiments at higher levels of calculation than recommended. This will provide appreciation both for the relative prediction abilities of the different methods and for their relative computation costs.

VI) Modeling Programs

The experimental procedures are not specific to a particular computer program. They have all been tested using the *SPARTAN*[5] electronic structure program, and can be completed using this program. A tutorial describing the basic use of *SPARTAN* is included as an appendix to this book. Additional information about *SPARTAN* can be found in the on-line help or the *SPARTAN* User's Guide. Nearly all of the experiments in this book can also be performed using one of the versions of *SPARTAN* written for personal computers: MacSPARTAN,[6] MacSPARTAN *Plus*,[7] and PC SPARTAN.[8] Users of these and other suitable programs should consult their User's Guides.

---

5. *SPARTAN*, version 4.1, Wavefunction, Inc., Irvine, California.
6. MacSPARTAN, version 1.0, Wavefunction, Inc., Irvine, California.
7. MacSPARTAN *Plus*, version 1.0, Wavefunction, Inc., Irvine, California.
8. PC SPARTAN, version 1.0, Wavefunction, Inc., Irvine, California.

Notes to the Instructor

Courses in computational organic chemistry are still in their infancy, but experience in teaching these courses has already made it clear that they have much more in common with traditional courses in organic chemistry than they do with courses in quantum mechanics or computational chemistry. This is partly because the goal of computational organic chemistry is to create "modeling-literate", but not necessarily "modeling-expert", organic chemists. Therefore, a computational organic chemistry course is grounded in chemical phenomena, and not the arcane mathematics of chemical physics. Still more important, though, is the fact that computational organic chemistry is a "hands on" course. Therefore it shares most of the features commonly associated with conventional laboratory courses, such as organic chemistry. A successful computational organic chemistry student, like any other laboratory student, must master an unfamiliar set of skills, successfully follow experimental procedures, make and record observations, and make chemically plausible connections between these observations and the rest of chemistry.

Instructors can best assist their students by approaching the computational chemistry "laboratory" in the same way that they approach other laboratory courses. In particular, it is important that instructors make an effort to be present and interact with students during the laboratory period. The most obvious benefit of instructor supervision is that innocent errors can be caught and rectified before too much time is wasted. It is not at all uncommon for a student to submit a calculation that is much more complex and time-consuming than desired or needed, or even worse, to set up a calculation, but forget to submit it. Either error, unless caught by an observant instructor, can result in precious minutes and hours being wasted in front of an unresponsive computer screen.

The presence of an instructor in the laboratory may have other significant payoffs too. Experience suggests that the two most common problems facing novice model-users are: 1) molecule building, and 2) manipulating and examining computer-optimized models. In the first case, a student may be unsure how to select from a potentially bewildering array of building tools, and an instructor can speed up a student's work simply by providing key bits of advice. In the second, students, because of their inexperience, tend to be superficial in their observation and analysis of modeling results. An artful instructor can greatly enhance the value of an experiment by engaging students in conversation, and by helping them identify points of interest in their models.

The analogy between computation and other types of laboratory experimentation also applies to student laboratory reports. Although the minimalist approach, one in which a student records a few measurements on a standard form, allows for efficient grading, it does little to foster analytical skills, and it is precisely these skills which must be cultivated before students can make effective use of molecular modeling. A more comprehensive approach is recommended. Each experimental procedure in this book contains specific instructions regarding the recording of data, and several questions concerning the interpretation of modeling results, that can be used as the basis for a laboratory report. For example, the student's experimental procedure, observations, and answers, might be written up in a form that closely resembles the "Results and Discussion" sections found in typical journal articles. Brief answers to the text questions are not nearly as useful as a careful exposition of the specific observations and logic that leads the student to the given answer.

Finally, it should pointed out that answers to the text questions (and not all of them can be easily answered) are not provided. However, a CD-ROM containing all of the models and calculations and readable by the SPARTAN, MacSPARTAN, MacSPARTAN *Plus* and PC SPARTAN programs is available from Wavefunction.

I

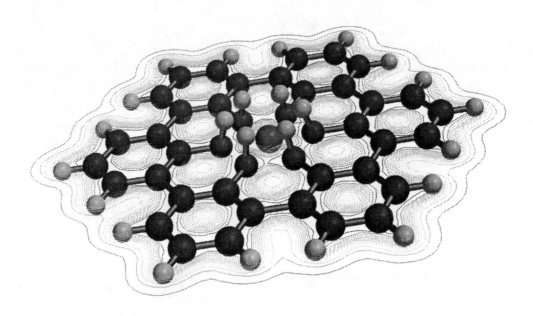

Computational
Tools

There are several tools available to chemists to describe molecular geometry and conformation. The simplest and most commonly employed among these is termed molecular mechanics. Here, the difference in "energy" of a molecule in a particular geometry from that of the same molecule in an "ideal geometry" is typically described in terms of a sum of contributions arising from distortions from "ideal" bond distances, bond angles and dihedral angles, together with contributions due to "non-bonded" interactions. As a consequence, molecular mechanics may not be used to calculate the relative energies of different (isomeric) molecules or of reaction energies, except in cases where the bonding is identical.

The molecular mechanics approach is very simple and often provides acceptable and sometimes very good descriptions of a molecule's equilibrium geometry with very little computational effort. Its major drawback is the need for explicit parameterization. Where sufficient experimental data do not exist, as for example in dealing with short-lived reactive intermediates, molecular mechanics techniques should not be expected to perform as well or as reliably as more sophisticated techniques. Where the task at hand is following reaction pathways connecting stable structures, and in particular, locating energy maxima (transition states) along these pathways, molecular mechanics is at a complete loss. There are no experimental structural data at all with which to build parameters. In these situations, there is little choice than to resort to quantum mechanical methods.

Ab Initio Molecular Orbital Models

Detailed description of quantum mechanical methods suitable for application to molecular systems ("molecular orbital" methods) is well beyond the scope of this book. The focus here is on the very broad issues, sufficient to distinguish quantum mechanical methods from molecular mechanics techniques.

Practical molecular orbital methods seek an approximate solution to the deceptively simple-looking differential equation formulated by Schrödinger in the 1920's.

$$\hat{H}\Psi = \varepsilon\Psi$$

In this equation, $\hat{H}$ is termed the "Hamiltonian operator"; it describes both the kinetic energies of the particles which make up the molecule, i.e., its nuclei and electrons, as well as the electrostatic interactions felt between individual particles. Nuclei, which are positively charged, repel other nuclei, and electrons, which are negatively charged, repel other electrons, but nuclei attract electrons. The quantity ε in the Schrödinger equation is the energy of the system, and Ψ is termed a "wavefunction". While the wavefunction has no particular physical meaning, its square times a small volume element corresponds to the probability of finding the system at a particular set of coordinates.

The Schrödinger equation has been solved exactly for the hydrogen atom (a one-electron system), where its solutions are actually quite familiar to chemists as s, p, d atomic orbitals, i.e.,

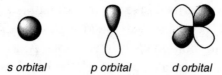

s orbital   p orbital   d orbital

They correspond to the ground and various excited states of the hydrogen atom.

Although the Schrödinger equation may easily be written down for many-electron atoms and for molecules, it cannot be solved. Approximations must be made. Practical *"ab initio"* molecular orbital methods start from the Schrödinger equation and then make three approximations:

1. Separation of nuclear and electron motions (the "Born-Oppenheimer approximation"). In effect, what this says is that "from the point of view of the electrons, the nuclei are stationary". This eliminates the mass dependence in what is now referred to as the electronic Schrödinger equation.[1]

2. Separation of electron motions (the "Hartree-Fock approximation"). What is actually done is to represent the many-electron wavefunction as a sum of products of one-electron wavefunctions, the spatial parts of which are termed "molecular orbitals". This reduces the problem of simultaneously accounting for the motions of several electrons to the much simpler problem of accounting for the motion of a single electron in a "field" made up of the nuclei and all the remaining electrons.

3. Representation of the individual molecular orbitals in terms of linear combinations of atom-centered basis functions or "atomic orbitals" (the "LCAO approximation"). This reduces the problem of finding the best functional form for the molecular orbitals to the much simpler problem of finding the best set of linear coefficients.

These three approximations lead to a series of equations termed the "Roothaan-Hall equations".[2]

Practical *ab initio* methods differ in the number and kind of atomic basis functions, and their cost increases as the fourth power of the number of basis functions (compared to a squared power dependence on the number of atoms for molecular

1. To see how mass enters into the picture and affects molecular properties, see: **Isotope Effects**.
2. For a detailed discussion, see: W.J. Hehre, L. Radom, P.v.R. Schleyer and J.A. Pople, *Ab Initio* **Molecular Orbital Theory**, Wiley, New York, 1986.

mechanics techniques).[3] The simplest *ab initio* methods utilize a "minimal basis set" of atomic orbitals, which includes only those functions required to hold all the electrons on an atom and to maintain spherical symmetry. Minimal basis set *ab initio* methods are often too restrictive to properly describe molecular properties, and "split-valence basis sets", which incorporate two sets of valence atomic orbitals, or "polarization basis sets", which, in addition, include d-type atomic orbitals, are often employed.

Ab initio molecular orbital methods using split-valence or polarization basis sets, have become a mainstay for routine and reliable descriptions of the structures, stabilities and other properties of organic molecules. They have also been applied with considerable success to the description of reaction pathways and to the elucidation of product distributions.

Semi-Empirical Molecular Orbital Models

The principal disadvantage of *ab initio* methods is their cost.[3] It is possible to introduce further approximations in order to significantly reduce cost while still retaining the underlying quantum mechanical formalism. "Semi-empirical" molecular orbital models follow in a straightforward way from *ab initio* models:

1. Elimination of overlap between functions on different atoms (the "NDDO approximation"). This is rather drastic but reduces the computation effort by more than an order of magnitude over *ab initio* techniques.

2. Restriction to a "minimal valence basis set" of atomic functions. Inner-shell (core) functions are not included explicitly, and because of this, the cost of doing a calculation involving a second-row element, e.g., silicon, is no more than that incurred for the corresponding first-row element, e.g., carbon.

3. Additional numerical approximations to further simplify the calculations, and more importantly introduction of adjustable parameters to reproduce specific experimental data. These distinguish among the various semi-empirical methods currently available. Choice of parameters, more than anything else, appears to be the key to formulating successful semi-empirical methods.

Which Models are Most Suitable?

Which of the available computational tools is the most appropriate? Molecular mechanics? Semi-empirical molecular orbital theory? *Ab initio* molecular orbital theory? The answer depends entirely on: 1) the problem at hand, 2) the level of confidence required in the results, and 3) available computational resources.

---

3. In practice for large systems, both *ab initio* and semi-empirical molecular orbital methods scale as the cube of size (number of basis functions) and molecular mechanics methods scale linearly with size (number of atoms).

With very few exceptions, only four calculation methods are used for the experiments in this book: AM1 and AM1-SM2 semi-empirical models and *ab initio* models with 3-21G (split-valence) and 6-31G* (polarization) basis sets. All four of these methods have become "practical standards" for calculations on organic systems, and while other methods may from time to time yield better results, none have as yet been as extensively tested and documented.

Very little use has been made of molecular mechanics. Molecular orbital methods, rooted in quantum mechanics, are to be preferred both because they are more generally applicable and because they are more reliable. Because of its low cost, molecular mechanics does have a role to play, in particular, for preliminary refinement of geometry and perhaps most importantly for conformational searching. However, the results of molecular mechanics calculations should be viewed with healthy skepticism.

AM1: This is certainly the most popular present-generation semi-empirical method for application to organic chemistry. It provides a good account of equilibrium structures, and usually reproduces transition-state geometries obtained from higher-level (*ab initio*) calculations. It is, however, less satisfactory in dealing with the geometries of molecules incorporating second-row (and heavier) elements. More importantly, AM1 and other presently-available semi-empirical models do not generally provide an acceptable account of reaction thermochemistry.

The success of AM1 in providing equilibrium (and transition state) geometries, together with its lack of success in providing reaction energetics, suggests its role in the experiments in this book (and more generally in research applications of computation). Specifically, AM1 will generally be employed to supply geometries, which in turn will be used for higher-level calculations of energetics.

AM1-SM2: This semi-empirical method is based on AM1 and is intended to be used in conjunction with AM1. Specifically, the difference between AM1-SM2 and AM1 heats of formation has been parameterized to reproduce experimental heats of (aqueous) solvation. The method generally performs well for neutral organic molecules, but often leads to sizable errors for charged species.

3-21G: This is among the simplest *ab initio* models which is successful in providing equilibrium (and transition state) geometries, as well as the energetics of reaction which do not involve bond formation and bond breaking. In particular, 3-21G generally provides an excellent account of the energetics of *isodesmic* reactions (see: **Isodesmic Reactions**). The 3-21G split-valence basis set, for hydrogen and first-row elements, needs to be supplemented by a set of d-type functions for second-row (and heavier) elements in order to yield satisfactory structures and energetics. The resulting method termed 3-21G$^{(*)}$, is used in conjunction with 3-21G, and is referred to by either name (3-21G or 3-21G$^{(*)}$).

6-31G*: This polarization basis set *ab initio* model, like 3-21G, is highly successful for the description of equilibrium (and transition state) geometries. It generally provides a superior account to 3-21G of reaction energetics, except where bonds are broken or formed, where neither method is particularly successful.

Relative Computation Times

Relative computation times for AM1 and AM1-SM2 semi-empirical and 3-21G and 6-31G* *ab initio* calculations on methylcyclohexane (a typical small organic molecule) and on lysergic acid (a typical large organic molecule) are given below. These are for single-point-energy calculations as well as for geometry optimizations and frequency evaluations, and are referenced to the time for a single-point-energy calculation on methylcyclohexane at the 3-21G *ab initio* level.

molecule and level of calculation	task		
	energy	geometry	frequency
equatorial methylcyclohexane (7 heavy atoms, C_s symmetry, 32 independent variables)			
AM1	.01	.08	.66
AM1-SM2	.03	4.5	92
3-21G	1	14	190
6-31G*	5.4	90	1100
lysergic acid (20 heavy atoms, C_1 symmetry, 102 independent variables)			
AM1	.05	1.9	11
AM1-SM2	.08	-	-
3-21G	17	600	-
6-31G*	120	-	-

There is a sizeable jump in going from semi-empirical and *ab initio* molecular orbital calculations, and a significant added cost of geometry optimization and especially frequency evaluation when compared to single-point-energy calculation. The AM1-SM2 semi-empirical model for treatment in water is comparable in cost to AM1 for single energy calculations, but significantly more expensive for both geometry optimization and for frequency evaluation. 6-31G* *ab initio* calculations are on the order of 5-7 times more costly than 3-21G calculations.

"Would you tell me, please, which way I ought to go from here?"

"That depends a good deal on where you want to get to," said the Cat.

"I don't much care where -" said Alice.

"Then it doesn't matter which way you go," said the Cat.

"-so long as I get *somewhere*," Alice added as an explanation.

"Oh, you're sure to do that," said the Cat, "if you only walk long enough."

Alice in Wonderland
Lewis Carroll

II

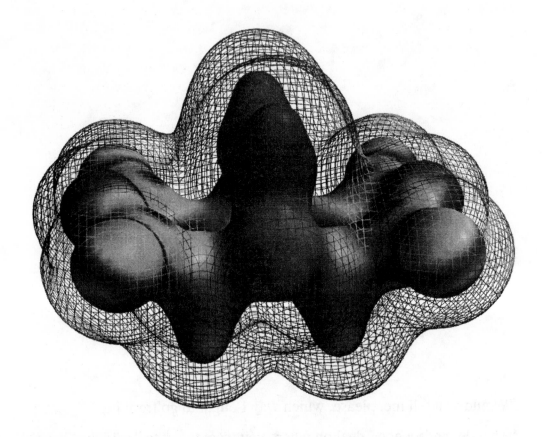

Finding Molecular
Geometries

$\mathbf{T}$he energy of a molecule depends on its geometry. Even small changes in structure can lead to quite large changes in total energy. Proper choice of molecular geometry is therefore quite important in carrying out computational studies. Experimental geometries, where they are available, would certainly be suitable. The trouble is, of course, that very few "high-quality" structures are available even for stable molecules. Experimental data for reactive or otherwise short-lived molecules are even more scarce, and no quantitative information at all is available for transition states. In the final analysis, there is really no viable alternative to obtaining geometries directly from calculation. Fortunately, this is not difficult, although it may be demanding in terms of computer time.

Finding and Verifying Molecular Geometries

Geometry optimization is an iterative process. The energy and energy "gradient" (first derivatives of the energy with respect to all geometrical coordinates) are calculated for the guess geometry, and this information is then used to project a new geometry. This process continues until the lowest energy or optimized geometry is reached. Several criteria must be satisfied before a geometry is accepted as optimized. First, successive geometry changes must not lower the total energy by more than a specified (small) value. Second, the energy gradient at the optimized geometry must closely approach zero. This insures that the optimization is nearing a flat region (the "bottom" of the energy surface). Third, successive iterations must not change any geometrical parameter by more than a specified (small) value.

In order for a geometry to correspond to an energy minimum, the curvature of the energy surface must be positive, i.e., the "minimum" must lie at the bottom of an energy well. The surface's curvature is defined by the eigenvalues of the "Hessian" (the matrix of second derivatives of the energy with respect to geometrical coordinates), all of which must be positive.[1]

Geometry optimization does not guarantee that the final structure has a lower energy than any other structure of the same molecular formula. All that it guarantees is a "local minimum", that is, a geometry the energy of which is lower than that of any similar geometry, but which may still not be the lowest energy geometry possible for the molecule. Finding the absolute or "global

1. The procedure for calculating the eigenvalues involves finding a set of geometrical coordinates ("normal coordinates") for which the Hessian will be diagonal, i.e., all off-diagonal elements will be zero. In this representation, all (diagonal) elements must be positive for the geometry to correspond to an energy minimum. The procedure involved in finding the normal coordinates is termed a "normal coordinate analysis", and is the same as that required for the calculation of vibrational frequencies, which relate directly to the square root of the elements of the (diagonal) Hessian. Positive Hessian elements yield real frequencies; negative Hessian elements yield imaginary frequencies.

minimum" requires repeated optimization starting with different initial geometries. Only when all local minima have been located is it possible to say with certainty that the lowest energy geometry has been identified.

In principle, structure optimization carried out in the absence of symmetry, i.e., in C_1 symmetry, must result in a local minimum. On the other hand, imposition of symmetry may result in a geometry which is not a local minimum. For example, optimization of ammonia constrained to a planar trigonal geometry (D_{3h} symmetry) will result in a structure which corresponds to an energy maximum in the direction of motion toward a puckered trigonal geometry (C_{3v} symmetry). This is the transition state for inversion at nitrogen in ammonia. (Indeed, transition states can sometimes be located simply by geometry optimization subject to an overall symmetry constraint, see: **Finding Transition States**.) The most conservative tactic is always to optimize geometry in the absence of symmetry. If this is not practical, it is always possible to verify that the structure located indeed corresponds to a local minimum by performing a normal-coordinate analysis on the final (optimized) structure. This analysis should yield all real frequencies.

How Good are Calculated Molecular Geometries?[2]

Both semi-empirical and *ab initio* molecular orbital models generally provide a satisfactory account of equilibrium geometries. A summary of mean absolute errors in heavy atom bond distances and skeletal bond angles is provided below.

Mean absolute errors in heavy atom bond distances (Å) and skeletal bond angles (degrees)		
model	heavy atom bond distance	skeletal bond angle
AM1 semi-empirical	0.050	3.3
3-21G(*) *ab initio*	0.016	1.7
6-31G* *ab initio*	0.020	1.4

Errors in bond lengths and angles from the *ab initio* methods are half the magnitude as errors from AM1, and are often inside the range of experimental errors. The slightly larger bond length error for 6-31G* calculations (relative to 3-21G(*) calculations) is a consequence of the fact that "limiting" Hartree-Fock bond distances are consistently larger than experimental values. Still, the errors resulting from semi-empirical AM1 calculations are small enough to make these techniques suitable for routine structure determination.

---

2. Extensive reviews of the performance of practical molecular orbital methods for structure determination are available: W.J. Hehre, L. Radom, P.v.R. Schleyer and J.A. Pople, **Ab Initio Molecular Orbital Theory**, Wiley, New York, 1986, chapter 6; W.J. Hehre, **Critical Assessment of Modern Electronic Structure Methods**, Wavefunction, Inc., Irvine, California, to be published, Winter 1997.

Property Calculations Using Approximate Molecular Geometries[3]

Given that semi-empirical models often provide geometries that are quite close to those obtained from *ab initio* methods, it is legitimate to ask whether or not structures from semi-empirical techniques may be used for energy and property calculations with *ab initio* methods. A favorable response would be of great significance as geometry optimization is a major "cost" in any computational investigation. In fact, the answer depends on what "property" is being calculated and the level of confidence required. Experience suggests that, except for "unusual molecules", use of semi-empirical structures has very little effect on relative energetics. Errors in reaction energies of 1 to 3 kcal/mol, which may arise from the use of "approximate geometries", must be balanced against the large savings in computer time. Other properties such as dipole moments, may show greater sensitivity to choice of structure, and use of appropriate geometries may lead to unacceptable errors.

Exact geometries must be used for frequency calculations. The reason for this is that frequencies are related to the first finite term in a Taylor series expansion of the energy (as a function of geometry). This is (assumed to be) the second-derivative term, which will not be true if the first-derivative term (the gradient) is not precisely zero. Frequencies not evaluated at precise geometries are meaningless.

---

3. This topic, among other practical issues concerned with carrying out calculations, is discussed in: W.J. Hehre, **Practical Strategies for Electronic Structure Calculations**, Wavefunction, Inc., Irvine, California, 1995.

Benzene or 1,3,5-Cyclohexatriene?

3-21G ab initio calculations are used to examine the energy and charge distribution in the hypothetical molecule 1,3,5-cyclohexatriene relative to the energy and charge distribution in benzene.

1,3,5-Cyclohexatriene with alternating double and single bonds,

is not a stable molecule, that is, it is not a minimum on the C_6H_6 potential energy surface.[1] Therefore, it cannot even be detected, let alone characterized. Nevertheless, 1,3,5-cyclohexatriene may be examined using computation, and its energy and other properties compared with those of benzene. This allows a number of interesting questions to be answered. Is geometrical reorganization (going from a structure with alternating single and double bonds to one in which all carbon-carbon bonds are equivalent) the full cause of aromatic stabilization, or are other factors involved? Is the distribution of charge in 1,3,5-cyclohexatriene localized as implied by its valence structure or delocalized as in benzene?

Procedure

Build benzene and optimize its geometry using 3-21G *ab initio* calculations. Next, build 1,3,5-cyclohexatriene as a planar molecule with alternating single and double bonds (constrained to 1.50 and 1.30 Å, respectively). You may either optimize its 3-21G geometry subject to the six bond length constraints, or calculate its 3-21G energy at this idealized geometry.

Compute the difference in energies between benzene and 1,3,5-cyclohexatriene (the "relaxation energy"). How does the "relaxation energy" compare with the "aromatic stabilization energy" of benzene (33 kcal/mol)? Does your result suggest that geometrical changes play a minor or major role in the overall stabilization of benzene?

Generate electrostatic potential maps for both benzene and 1,3,5-cyclohexatriene. Display the two on the same color scale. Describe the surface you obtain for benzene. Does it show a uniform distribution of charge, the region over the carbons (the π system) being negatively charged, the region around the hydrogens (the σ system) being positively charged? Describe the image you obtain for 1,3,5-cyclohexatriene. What do these maps tell you about the delocalization of charge in cyclohexatriene relative to that in benzene?

---

1.   S. Shaik, P.C. Hiberty, J.-M. Lefour and G. Ohanessian, *J. Am. Chem. Soc.*, **109**, 363 (1987). See also: **Stable Alternatives to Benzene**.

Optional

A number of molecules, among them biphenylene and tricyclobutabenzene, have been proposed as likely candidates for incorporating "localized" benzene rings.[2]

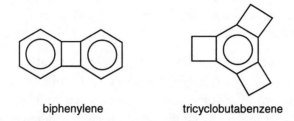

biphenylene   tricyclobutabenzene

Build each molecule and optimize their geometries using semi-empirical AM1 calculations. Do you see evidence for bond localization in either molecule (or in both molecules)? If you do, provide a rationalization.

2.   H.-B. Burgi, K.K. Baldridge, K. Hardcastle, N.L. Frank, P. Gantzel, J.S. Siegel and J. Ziller, *Angew. Chem. Int. Ed. Engl.*, **35**, 1454 (1995).

Is Thiophene Aromatic?

Semi-empirical AM1 calculations and ab initio 3-21G$^{(*)}$ calculations are used to investigate the energetics of stepwise hydrogenation of thiophene, first to a molecule incorporating one double bond and then to the fully saturated species. Differences in hydrogenation energies are interpreted in terms of aromatic stabilization.

Benzene possesses unusual thermodynamic stability relative to what might be expected for "1,3,5-cyclohexatriene", an unknown molecule (see: **Benzene or 1,3,5-Cyclohexatriene?**). A consequence of this is that hydrogenation of benzene to 1,3-cyclohexadiene is actually slightly endothermic, whereas the corresponding hydrogenation reactions taking 1,3-cyclohexadiene to cyclohexene and finally to cyclohexane, are both significantly exothermic.

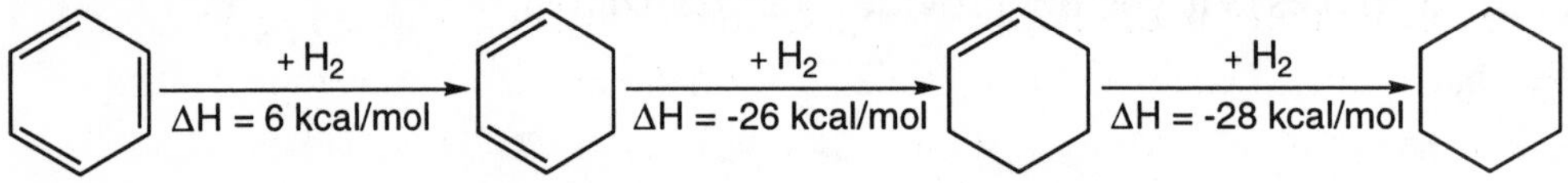

The usual interpretation is that the first addition of hydrogen to benzene "trades" the H-H bond in hydrogen and a carbon-carbon π bond for two carbon-hydrogen bonds, but in so doing destroys the aromaticity, whereas addition of the second hydrogen to 1,3-cyclohexadiene "trades" the same bonds but does not result in any loss of aromaticity. Therefore, the difference in the heats of hydrogenation of benzene and 1,3-cyclohexadiene (or cyclohexene) corresponds to the aromatic stabilization of benzene. From the data above, this difference is approximately 33 kcal/mol.[1]

Far less is known about the "aromaticity" of heterocycles.[2] For example, there is still controversy as to whether or not thiophene is aromatic. In this experiment, you will calculate the "aromaticity" of thiophene in an analogous fashion to that for benzene above, by comparing the energetics of hydrogen addition to thiophene and to the intermediate dihydrothiophene, i.e.,

These will be obtained from *ab initio* 3-21G$^{(*)}$ calculations, making use of equilibrium structures from semi-empirical AM1 calculations.

---

1. This is not the only, or necessarily the best, measure of the aromaticity of benzene. For discussions see: Carey and Sundberg A, p.499; March, p.40.
2. March, p.45; N.D. Epiotis, W.R. Cherry, F. Bernardi and W.J. Hehre, *J. Am. Chem. Soc.*, **98**, 4361 (1976).

Procedure

Build thiophene, dihydrothiophene and tetrahydrothiophene and optimize their geometries using AM1 semi-empirical calculations. Use the resulting AM1 geometries for single-point 3-21G$^{(*)}$ *ab initio* energy calculations. Work out the energetics of hydrogenation of thiophene and of dihydrothiophene. (You will require the 3-21G total energy for H_2; it is -1.1201 hartrees.) Determine the "aromatic stabilization" of thiophene as the difference between the first and second hydrogenation energies. Which compound enjoys the greater stabilization, benzene or thiophene?[3] How can you account for this difference using resonance theory?

Optional

Aromatic stabilization is typically accompanied by geometrical changes. For example, benzene does not incorporate three single and three double bonds, but rather six bonds of intermediate length. Compare AM1 CC single and double bond lengths for thiophene with those of *cis*-1,3-butadiene as a "non-aromatic standard". Does your comparison support the idea that thiophene exhibits aromaticity?

---

3. At the same level of calculation, hydrogenation of benzene is endothermic by 4 kcal/mol, hydrogenation of 1,3-cyclohexadiene exothermic by 42 kcal/mol and hydrogenation of cyclohexene exothermic by 42 kcal/mol.

Atomic Hybridization and Bond Lengths

Semi-empirical AM1 calculations and Natural Bond Orbital analysis are used to investigate the relationship between formal atomic hybridization and CH and CC bond distances in hydrocarbons.

CH bonds in hydrocarbons result from the sharing of two electrons, one in a hydrogen s orbital and one in a carbon sp^n hybrid orbital.[1] The CH bond distance appears to be correlated with the hybrid's p character, where % p is given by $100[n/(n+1)]$. Thus, experimental CH bond distances increase in the order: acetylene (1.061Å, 50% p), ethylene (1.085Å, 67% p), and ethane (1.102Å, 75% p). These data can be explained by the fact that p-type orbitals are larger and more diffuse than s-type orbitals. Increasing p character in a hybrid should have a corresponding effect on hybrid "size" and CH bond distance.

Similar trends are also observed in CC single bond distances, except that the bond distances appear to be even more sensitive to hybridization. For example, single bond distances fall in the order: propyne (1.459Å, 50% p), propene (1.501Å, 67% p), and propane (1.526Å, 75% p).

In this experiment, you will use semi-empirical AM1 calculations to determine CH and CC single bond distances in a variety of hydrocarbons, and Natural Bond Orbital (NBO) analysis[2] to calculate the hybridization at carbon. A good correlation between bond distance and % p character is expected for "normal" molecules, and deviations may be indicative of unusual bonding.

Procedure

CH bond distances: Build ethane, ethylene, and acetylene, and optimize their geometries using semi-empirical AM1 calculations. Record the energy and CH bond distance in each molecule. Calculate the % p character on carbon in each CH bonding orbital using the NBO method (these values will not agree exactly with the "ideal" values given above), and plot CH distance vs. % p character. Is there a good correlation between the two?

Build singlet methylene (CH_2), cyclopropane, and cyclobutane, and optimize their AM1 geometries. Record the energy, CH bond distance, and % p character (from an NBO analysis) in the CH bonding orbitals for each molecule. Plot these results on the previously constructed graph. Which molecules appear to be "normal"? Which CH bond distances deviate from the norm and in which direction? What factors might be responsible for these deviations?

---

1. Carey and Sundberg A, p.3; March, p.18; J.E. Huheey, **Inorganic Chemistry**, Harper and Row, 1983, p.239.
2. A.E. Reed, R.B. Weinstock and F. Weinhold, *J. Chem. Phys.*, **83**, 735 (1985).

CC Single Bond Distances: Build hydrocarbons **1-10**, and optimize their AM1 geometries.

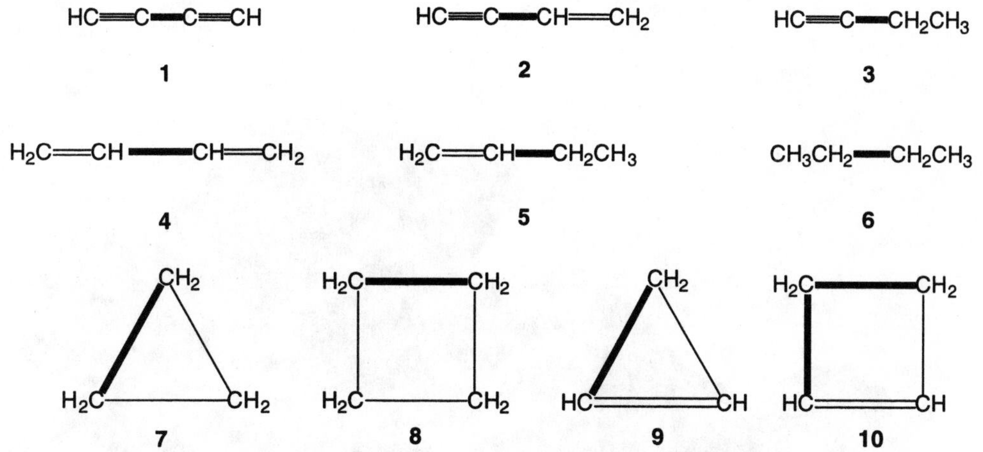

HC≡C—C≡CH

1

HC≡C—CH=CH$_2$

2

HC≡C—CH$_2$CH$_3$

3

H$_2$C=CH—CH=CH$_2$

4

H$_2$C=CH—CH$_2$CH$_3$

5

CH$_3$CH$_2$—CH$_2$CH$_3$

6

7

8

9

10

Record the CC bond distances for the "highlighted" bonds. Calculate the % p character on each carbon in the CC bond orbital for the highlighted bonds, and add the two % p values together to give a combined value for the bond. Plot CC bond distance vs. total % p character. Which molecules (bond distances) appear to follow a simple correlation? Which molecules (bond distances) deviate from this relationship, and in what direction? What factors might be responsible for these deviations?

Optional

Find three more examples of neutral hydrocarbons in which the CH bond distances deviate from the previously developed relationship between bond distance and % p character. Explain why these deviations occur. Also look for neutral hydrocarbons with "abnormal" CC bond distances.

III

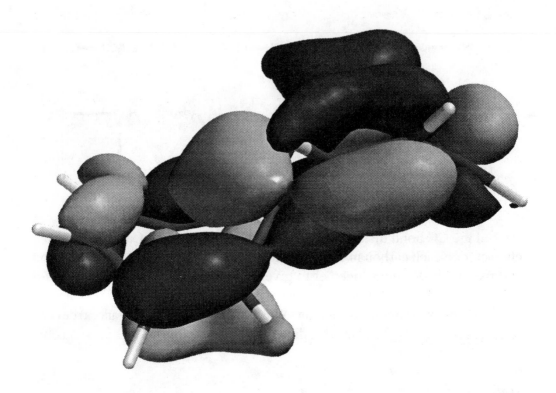

Isodesmic
Reactions

T here are many ways to compare molecular energies and to interpret energy differences. For example, the energy of hydrogen fluoride relative to separated hydrogen and fluorine atoms, i.e., ΔE for

$$H\text{-}F \Rightarrow H\cdot + F\cdot$$

may be interpreted as the bond energy of hydrogen fluoride. So too may the energy of dissociation of HF into a proton and fluoride anion, i.e.,

$$H\text{-}F \Rightarrow H^+ + F^-$$

The two reactions (and two bond dissociation energies) are obviously different. The energy of the first reaction (the "homolytic bond dissociation energy") is much smaller than the energy of the second reaction (the "heterolytic bond dissociation energy"), but it turns out that it is much more difficult to predict. The reason is that homolytic bond dissociation results in a change in the number of electron pairs, whereas the reactants and products in a heterolytic bond dissociation have the same number of electron pairs. Molecular orbital methods have greatest difficulty in accurately accounting for interactions among electrons, and their utility for thermochemical descriptions relies heavily on cancellation of errors for reactants and products which are "similar". Molecules with different numbers of electron pairs are more dissimilar than molecules with the same number of electron pairs.

Heterolytic bond dissociation does not maintain the total number of bonds, e.g., the HF σ bond in hydrogen fluoride is "converted" to a nonbonded lone pair on fluorine. Thus, while the reactant (HF) and the products (H^+ and F^-) are similar insofar as having the same number of electron pairs, from a chemical point of view they are quite different. There are, however, many important reactions which conserve both the total number of bonds and the total number of nonbonded lone pairs, e.g.,

$H_3C\text{-}CH_3 + H_2 \Rightarrow 2\ CH_4$ hydrogenation

$2\ CH_2{=}CH_2 \Rightarrow H_3C\text{-}CH_3 + HC{\equiv}CH$ disproportionation

$\overline{CH_2NHCH_2} \Rightarrow NH_2CH{=}CH_2$ structural isomerism

The first two reactions typify processes where the total number of σ bonds and total number of π bonds is strictly maintained. In the first reaction HH and CC bonds are exchanged for two CH bonds; in the second, two different CC π bonds are exchanged for a "double π bond" (an alkyne). The third reaction is typical of a process in which a σ bond is exchanged for a π bond. It is to be expected that molecular orbital methods will provide an even better account of these three and related types of reactions than they did for processes like heterolytic bond dissociation. Molecules that contain the same number of bonds are more similar than molecules with different numbers of bonds. This is indeed observed.

Even better results may be expected and are obtained for reactions in which reactants and products are even more closely related. For example, the reactants and products of so-called *isodesmic* ("equal bond") reactions have exactly the same numbers of each kind of formal chemical bond and each kind of formal nonbonded lone pair. Examples include the processes below.

$$H_3C-C\equiv CH + CH_4 \Rightarrow H_3C-CH_3 + HC\equiv CH \qquad \text{bond separation}$$

$$(CH_3)_3NH^+ + NH_3 \Rightarrow (CH_3)_3N + NH_4^+ \qquad \text{proton transfer}$$

In addition, all regio- and stereochemical comparisons are *isodesmic* reactions, as are conformation changes. Thus, *isodesmic* processes constitute a large class of reactions of considerable importance to organic chemistry.

All-and-all, energetic comparisons may be compartmented into one of several different categories, depending on the extent to which bonds and nonbonded lone pairs are maintained from reactant to product:

	type of process	examples
minimum cancellation of errors	no conservation of total number of electron pairs	homolytic bond dissociation
	conservation of total number of electron pairs; no conservation of total number of bonds	heterolytic bond dissociation
	conservation of total number of bonds and total number of nonbonded electron pairs; no conservation of number of each kind of bond, or number of each kind of nonbonded electron pair	hydrogenation, structural isomerism
maximum cancellation of errors	conservation of number of each kind of bond and number of each type of nonbonded electron pair (*isodesmic reactions*)	substituent effects, regio- and stereo-chemistry, conformation

21

It is to be expected (and it is observed) that the greater the similarity between reactants and products insofar as their bonding, the more accurate and reliable will be molecular orbital descriptions of energetics. The goal then is to express energetic comparisons in a way which takes greatest advantage of the similarity between reactants and products, if possible, using *isodesmic* reactions.

Wherever possible, the analysis of energy results in the experiments in this book rely on *isodesmic* reactions. When this is not possible, reactions have been written to relate molecules which are as similar as can be. In practice, what this generally means is forgoing absolute energies, e.g., bond dissociation energies, in favor of comparisons involving relative energies, e.g., substituent effects on bond dissociation energies. With a little bit of cleverness, it will usually be possible to formulate the problem of interest in a way that will allow good performance of practical molecular orbital methods.

Stable Alternatives to Benzene

Semi-empirical AM1 calculations and ab initio 3-21G calculations are used to examine a variety of plausible alternatives to benzene. A practical strategy for stabilizing Dewar benzene relative to benzene is explored.

Give a chemist the molecular formula C_6H_6 and the first thing, and perhaps the only thing, that will come to mind is benzene, **1**.

1

There are, however, several "reasonable" $(CH)_6$ structures, among them Dewar benzene, **2**, prismane, **3**, benzvalene, **4**, and 3,3'-bis (cyclopropene), **5**.

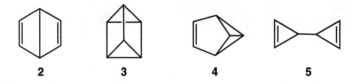

2   **3**   **4**   **5**

(Note that 1,3,5-cyclohexatriene has been excluded. While this is a relatively low-energy molecule, it is kinetically unstable with respect to benzene; see: **Benzene or 1,3,5-Cyclohexatriene?** While none of these alternatives are as thermally stable as benzene, all have been detected (and derivatives of some isolated and characterized), meaning that they are kinetically stable.

In this experiment, you will use semi-empirical AM1 calculations and *ab initio* 3-21G calculations first to see how high in energy (relative to benzene) these alternatives really are. You will then examine the Dewar benzene form of 1,2,4-tri-*tert*-butylbenzene, to see, as has been suggested, if steric effects may be used to reduce its energy relative to that of the benzene form.

Procedure

Build benzene, **1**, Dewar benzene, **2**, prismane, **3**, benzvalene, **4**, and 3,3'-bis (cyclopropene), **5**. Optimize their geometries using semi-empirical AM1 calculations and obtain and record the energy of each isomer relative to that of benzene. Calculate the vibrational frequencies of the three benzene alternatives. Are all indicated to be minimum energy forms? Recall, that this requires all of

---

1.   E.E. van Tamelen, *Angew. Chem. Int. Ed. Engl.*, **4**, 738 (1965).

the frequencies to be real (see: **Finding Molecular Geometries**). To get a better estimate of the relative stabilities, perform single-point energy 3-21G *ab initio* calculations (use the AM1 geometries). Which of the four alternatives is the best? Is this alternative stable enough that you might expect it to exist in detectable amounts in thermodynamic equilibrium with benzene?

Dewar benzene has actually been isolated, and found to revert only slowly to benzene (its half life is approximately 2 days at 25°C). This is remarkable given how similar the two geometries are, and the large thermodynamic driving force. Can you provide a rationalization for the apparent high kinetic stability of **2**? Hint: Consider one half of Dewar benzene as "cyclobutene" which undergoes ring opening to "1,3-butadiene".

A thermodynamically more stable Dewar benzene is **7**, formed from photolysis of 1,2,4-tri-*tert*-butylbenzene, **6**, and which reverts to its precursor only upon heating.[2]

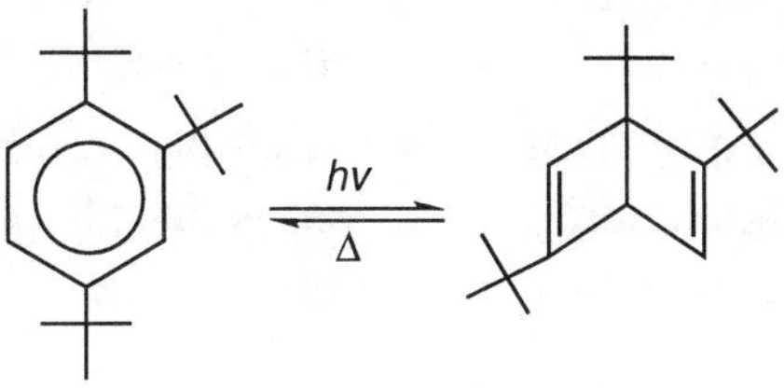

6   7

Optimize both **5** and **6** at the AM1 level. Compare their relative energies to those of benzene and Dewar benzene, i.e., write an *isodesmic* reaction (see: **Isodesmic Reactions**).

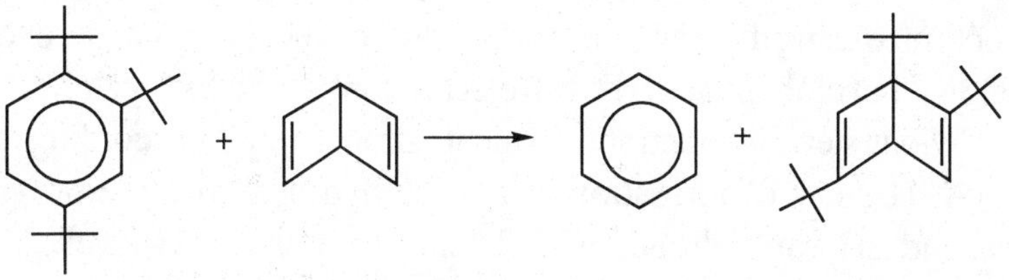

Does steric crowding in **6** lead to any obvious changes in the geometry of benzene? Look for evidence of localization of the benzene π system or for puckering of the ring. Do you see adverse effects of crowding in **7**? To what extent does steric crowding of the bulky *tert*-butyl groups affect the relative energies of benzene and Dewar benzene? Are your results consistent with experiment? Explain.

---

2.   Carey and Sundberg, p.604.

Ring Strain in Cycloalkanes

The semi-empirical AM1 method is used to calculate heats of hydrogenation of cycloalkanes which in turn are employed to obtain strain energies.

Small-ring cycloalkanes are thermodynamically less stable than the corresponding n-alkanes. This is due to a variety of factors, most notably distortion of CCC bond angles away from idealized tetrahedral values, and eclipsing interactions between CH bonds in cycloalkanes that are unable to assume fully staggered arrangements.[1] The destabilization of small-ring cycloalkanes, relative to open chain alkanes, is generally referred to as "ring strain".

In this experiment, you will employ semi-empirical AM1 calculations to evaluate the energetics of *isodesmic* reactions (see: **Isodesmic Reactions**).

$$\underset{CH_2 \rule{0.6em}{0.4pt} CH_2}{\overset{(CH_2)_n}{\bigcirc}} + 2\ CH_3CH_3 \longrightarrow \underset{CH_3 \qquad H_3C}{\overset{(CH_2)_n}{\bigcirc}} + CH_3CH_2 \text{-} CH_2CH_3$$

This reaction energy is equivalent to the difference in the hydrogenation energy of a small-ring cycloalkane and the corresponding hydrogenation energy of a "standard" acyclic alkane (in this case, n-butane), i.e.,

$$E\left[\underset{CH_2 \rule{0.6em}{0.4pt} CH_2}{\overset{(CH_2)_n}{\bigcirc}} \xrightarrow{H_2} \underset{CH_3 \quad H_3C}{\overset{(CH_2)_n}{\bigcirc}} \right] \text{-} E\left[CH_3CH_2 \text{-} CH_2CH_3 \xrightarrow{H_2} 2\ CH_3CH_3 \right],$$

and will allow you to assess the magnitude of ring strain as a function of ring size.

Procedure

Build and optimize using the AM1 method, cyclopropane, cyclobutane, cyclopentane, cyclohexane and cycloheptane. Try to select a conformer for cycloheptane which minimizes obvious steric interactions. Evaluate strain energies according to the above definition. (AM1 heats of formation of acyclic hydrocarbons are provided in the table below, and ΔH_f for hydrogen is -5.2 kcal/mol at the AM1 level.)

AM1 Heats of formation (kcal/mol)	
ethane	-17.4
propane	-24.3
n-butane	-31.1
n-pentane	-38.0
n-hexane	-44.8
n-heptane	-51.7

---

1. Carey and Sundberg A, p.157; March, p.150.

According to this measure, which rings are the most strained and which are the least strained? In particular, what is the relationship between strain energy and ring size? Examine the AM1 geometries of each of the cycloalkanes. Point out any significant deviations from tetrahedral CCC bond angles ("angle strain") and any eclipsing CH bonds ("torsional strain"). How do the CC bond distances in the cycloalkanes compare to those in the analogous n-alkanes? Try to explain why some bond distances, bond angles, and torsional angles are similar in cycloalkanes and alkanes, and why some are different.

Experimentally, cycloheptane is more strained than cyclohexane.[2] Do your calculations support this result? Try to explain why with reference to specific geometrical features.

---

2. Experimental strain energies (kcal/mol): cyclopropane, 27; cyclobutane, 26; cyclopentane, 7; cyclohexane, 0; cycloheptane, 6; Isaacs, p.320.

Bredt's Rule

Semi-empirical AM1 calculations and 3-21G ab initio calculations are used to verify Bredt's rule, "elimination to give a double bond in a bridged bicyclic system always leads away from the bridgehead".

Bredt observed that elimination of water from a bicyclic alcohol, such as **1**, gave an alkene in which the double bond did not involve the bridgehead position, **2**, rather than the bridgehead alkene, **3**.

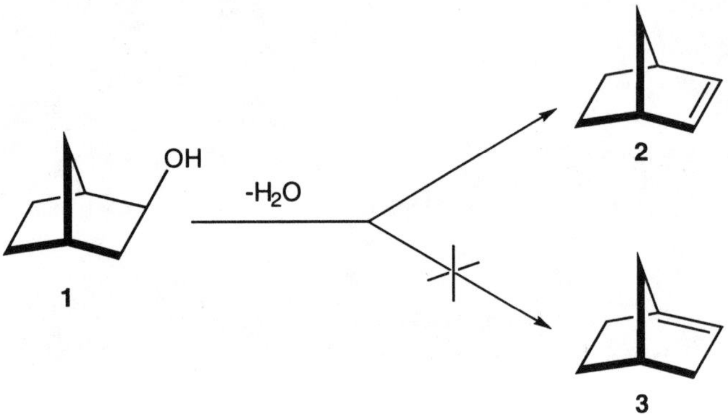

The preference is known to diminish with increasing ring size, which suggests that its origin is ring strain associated with incorporation of a double bond into a bridgehead position.[1] In this experiment, you will use semi-empirical and *ab initio* molecular orbital calculations to test this hypothesis by comparing the stabilities of several bridgehead and non-bridgehead alkenes.

Procedure

Build **2** and **3**, and optimize their geometries using semi-empirical AM1 calculations. Perform single-point 3-21G *ab initio* calculations using the AM1 equilibrium geometries. Identify the more stable isomer and work out the relative energy of the higher energy form. Do your results support Bredt's rule? Examine the geometries of **2** and **3**. Do you see any evidence of geometrical distortion that might explain their relative stabilities? (Hint: the C=C bond length in ethylene is 1.34Å and the molecule is planar.) Calculate and display the highest-occupied molecular orbital for **2** and **3**. What evidence do they provide for distortion of the alkene? Which alkene, if either, contains a distorted π orbital? Is this result consistent with your structure and energy results?

---

1. Carey and Sundberg A, p.161; March, p.160.

Repeat the calculations with the isomeric alkenes **4** and **5**, **6** and **7**, and **8** and **9**.

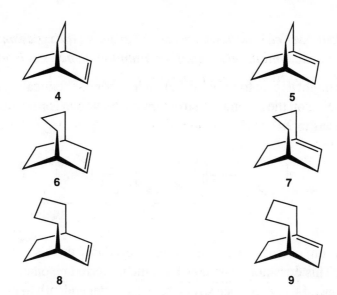

For the larger systems, try to choose conformers that minimize obvious steric interactions. Are the differences in energies and equilibrium structures for these compounds relative to those for **2** and **3** consistent with a steric origin for Bredt's rule? Based on your calculations, above what ring size would Bredt's rule not apply? Cite any experimental evidence for your speculation.

Optional

Electrostatic potential maps should be able to tell you whether or not bridgehead alkenes are also more reactive (kinetically less stable) than the corresponding non-bridgehead isomers. In particular, changes in potential on the π bond should signal changes in reactivity toward electrophiles. Pick one pair of isomers from the above systems (**2-9**) and, for each isomer, generate an electrostatic potential map. Simultaneously display the two on the same scale. According to this measure, is the π bond in the bridgehead alkene likely to be more or less reactive toward electrophiles than the π bond in the non-bridgehead system? Try to rationalize your result.

Octalene

Semi-empirical AM1 calculations are used to examine the conformational equilibrium in octalene, and to relate conformation to valence structure.

Naphthalene might appear to have a different number of chemically unique carbon atoms depending on the resonance structure that is drawn for it (three carbons in **1n**, five carbons in **2n**).

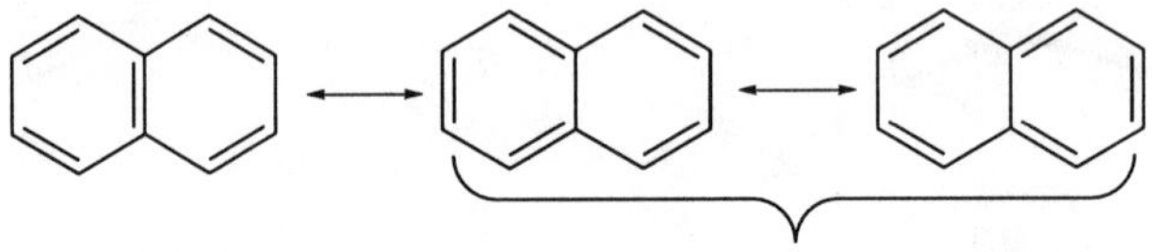

1n   **2n**

The ^{13}C spectrum of naphthalene, however, shows only three signals, even at low temperature. This does not mean that **1n** is the preferred resonance structure, but rather that none of the resonance structures exist independently – naphthalene is a resonance hybrid with the same symmetry as **1n**.

Octalene is the 8-membered ring analog of naphthalene. It too has different numbers of chemically unique carbons depending on the resonance structure that is drawn (four in **1o**, seven in **2o**). However, its NMR spectrum is completely different. Seven carbon signals are observed at room temperature, but these resolve into fourteen signals at −150°C![1] Clearly, octalene is not a resonance hybrid of **1o** and **2o** (this would require a four line spectrum), nor are the two structures labeled as "**2o**" resonance hybrids (this requires a seven line spectrum at all temperatures).

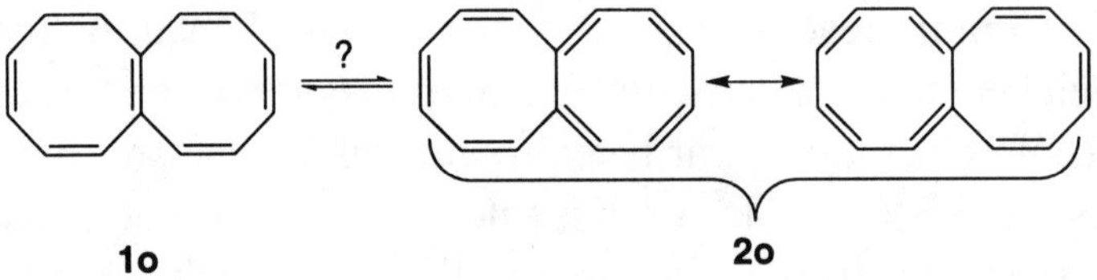

1o   **2o**

In this experiment, you will look for clues into the strange behavior of octalene by using semi-empirical AM1 calculations to examine the shapes and stabilities of 1,3,5,7-cyclooctatetraene, **1o**, and **2o**.

Procedure

1,3,5,7-Cyclooctatetraene: Build this molecule and optimize its geometry using AM1 semi-empirical calculations. Is the molecule planar? What are the CC bond distances in this molecule. How do you account for these observations (hint: does the molecule obey the Hückel 4N+2 rule for aromatic systems)?

---

1.   E. Vogel, *Angew. Chem. Int. Ed. Engl.,* **16**, 871, 872 (1977).

Octalene: Build **1o** and **2o**, and optimize their AM1 geometries. Does either molecule exhibit any internal symmetry? Calculate the vibrational frequencies of any symmetric molecule and determine whether it is a minimum-energy structure or a transition state (the latter will have one imaginary frequency; see: **Finding Molecular Geometries** and **Finding Transition States**). What are the CC bond distances in the two molecules? Are these consistent with octalene being a resonance hybrid or a structure with localized single and double bonds (for comparison, the CC bond distance in benzene is 1.40 Å)? Are the CCCC dihedral angles at the fused bond what you might have anticipated based on the corresponding dihedral angles in cyclooctatetraene? Try to account for the NMR spectrum of octalene. To do so, you must determine how many chemically unique carbons are found in **1o** and **2o**. Does this rule out either of these molecules as the preferred structure? Is this result consistent with the AM1 energies of the two molecules?

Optional

1. Were cyclooctatetraene to adopt a planar geometry, it would be a good candidate for an antiaromatic molecule. Build planar cyclooctatetraene and optimize using AM1 (you will need to start with a planar structure and maintain the symmetry plane). Does the resulting structure show evidence of delocalization, or does it incorporate localized single and double bonds? Use the same thermochemical analysis previously applied to thiophene (see: **Is Thiophene Aromatic?**) to decide whether planar cyclooctatetraene is stabilized or destabilized due to incorporation of eight π electrons into a ring.

2. The temperature-dependence of the octalene NMR spectrum implies a facile intramolecular rearrangement. This might involve a conformational change, a change in the positions of the single and double bonds, or both. One way to untangle this problem is to characterize the rearrangement according to whether it makes different carbons inside the same ring equivalent, or whether it equilibrates different carbons inside different rings. Which, if either, of these rearrangements requires a change in the location of single and double bonds and which allows the bonds to remain localized, i.e., might be purely a conformational change? A bond shift requires a delocalized transition state. In light of Hückel's rule, does this seem reasonable?

IV

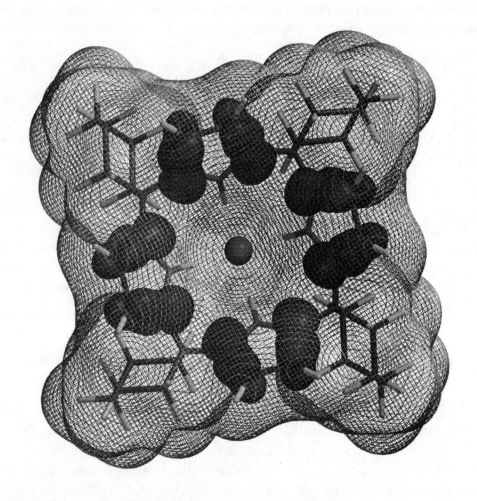

The Role
of Solvent

Quantum mechanical calculations are performed on isolated molecules (a "dilute gas"), whereas most chemistry is carried out in solvent. It is, therefore, relevant to inquire about the role of solvent in altering gas-phase structures and conformations, relative energetics, and chemical reactivity and selectivity.

The geometries of only a very few molecules are known with sufficient accuracy both in the gas and in solution to judge the possible role of solvent in altering structure. What comparative experimental data do exist for neutral molecules suggest that any changes are modest. This conclusion is supported by molecular orbital calculations in which solvent effects are taken into account. The geometries of charged species would be expected to change more, if for no other reason because the heats of solvation of ions are typically an order of magnitude greater than those for neutral molecules. Unfortunately, there is a nearly complete absence of experimental structure data for gas-phase ions. Calculations show a modest solvent effect on ion structure. Most apparent is a lengthening of bonds involving atoms which bear formal charges, e.g., the CO bonds in enolates. This is not unexpected; bond lengthening leads to a greater charge separation (polarity) which is, of course, favored by polar solvents.

In summary, solvent effects on the equilibrium structures of neutral molecules are small, and perhaps can be safely ignored. Solvent plays a more significant role in altering the geometries of ions, but the changes still appear to be relatively small.

Solvent is known to effect significant changes in conformation. In general, "polar conformations" will be stabilized to a greater extent than "non-polar conformations" by a (polar) solvent. For example, the relative populations of *anti* and *gauche* conformers of 1,2-dichloroethane shift with solvent

anti (non-polar)   gauche (polar)

In this case, the energy difference between the non-polar *anti* conformer and the polar *gauche* conformer, is approximately 1 kcal/mol in the gas phase and in non-polar solvents in favor of *anti*, but is reduced to nearly zero in polar media. While this may not seem very large, you should realize that a conformational energy difference of 1 kcal/mol leads to a 10:1 equilibrium preference for the lower-energy conformer at room temperature.

1.   E.L. Eliel and S.H. Wilen, **Stereochemistry of Organic Compounds**, Wiley, New York, 1994.

Another conformational change due to solvent is well illustrated by the structures of polypeptides and proteins. In the gas phase (and in non-polar solvents), polypeptides typically assume "stretched out" arrangements, presumably to minimize unfavorable steric interactions, while in water they adopt compact structures, presumably to minimize contacts between the hydrophobic regions of the polypeptide and the solvent. Indeed, it might be argued that the solvent has as much to do with the structures of proteins as their chemical makeup.

In summary, conformational equilibrium may be very sensitive to solvent. Polar conformers are favored in polar solvents and gross structures may adjust to optimize their contact with the solvent.

There is plentiful evidence to suggest the importance of solvent in altering gas-phase energetics. The general rule is that the more polar the molecule, the more it will be stabilized by solvent. Electrostatic considerations generally discourage the buildup of charge in isolated molecules, and a polar solvent will, therefore, usually act to reduce (gas-phase) energy differences between highly polar and less polar species. Charged molecules greatly exaggerate the situation. Solvation will normally be more effective for charge-localized ions than for delocalized species. A simple example of this concerns the site of protonation of phenol.[2] In the gas phase, carbon-protonated phenol **C** is estimated to be 15 kcal/mol more stable than oxygen-protonated phenol **O**,

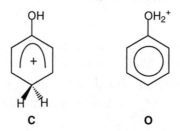

whereas in water protonation occurs predominantly at oxygen.

Closely related to this is the observation that solvent will typically reduce the effects of substituents in stabilizing charged molecules. For example, the gas-phase difference in base strengths between methylamine and ammonia. (methylamine is stronger by 9 kcal/mol) nearly disappears in water.[3]

In summary, solvent preferentially stabilizes structures with localized "charges", and therefore acts to moderate electrostatic factors which disfavor such structures in the gas phase.

---

2. D.J. DeFrees, R.T. McIver, Jr. and W.J. Hehre, *J. Am Chem. Soc.*, **99**, 3853 (1977).
3. E.M. Arnett, in **Proton Transfer Reactions**, E.F. Caldin and V. Gold, Ed., Wiley, New York, 1975.

Computational techniques available for treatment of molecules in solution fall into two general categories.[4] Explicit models treat the solvent as a set of distinct molecules. Operationally, the solute is placed in a "box" containing several hundred to several thousand solvent molecules, and the energy of the system is determined by averaging over all possible orientations of the individual molecules. While this is quite a reasonable approach, the need to average over a huge number of different solvent-solute configurations makes it extraordinarily expensive, and in practice solvent-solute and solvent-solvent interaction energies need to be calculated using empirical energy functions (molecular mechanics) rather than with quantum mechanics. This is the "weak link", as too little is known about solvent-phase structures and energetics to develop good empirical methodology. Even so, some very elegant work has appeared using such a technique.[5]

An entirely different approach treats the solvent as an electric field which interacts with the molecule.[6] This is a drastic simplification of the real situation, in which individual solvent molecules optimally organize themselves around the solute. Nevertheless, models have been formulated and parameterized to reproduce experimental heats of solvation. The experiments in this book make use of the AM1-SM2 parameterized solvation model.

4. Several excellent reviews of computational models for solvation have recently been collected in one place: **Structure and Reactivity in Aqueous Solution**, C.J. Cramer and D.G. Truhlar, ed., ACS Symposium Series No. 568, American Chemical Society, Washington DC, 1994.
5. W.J. Jorgensen, *Acc. Chem. Res.*, **22**, 184 (1989).
6. C.J. Cramer and D.G. Truhlar, *J. Am. Chem. Soc.*, **113**, 8305 (1991); R.W. Dixon, J.M. Leonard and W.J. Hehre, *Israel J. Chem.*, **33**, 427 (1993).

Solvent Effects on Tautomeric Equilibria

AM1 and AM1-SM2 semi-empirical calculations are used to investigate the tautomeric equilibrium involving 2-hydroxypyridine and 2-pyridone both in the gas phase and in water.

Many molecules, especially heterocyclic systems, exist as equilibrium mixtures of tautomers.[1] Aromatic stabilization, substituents, and solvent, among other factors, can influence which tautomer is favored thermodynamically. The tautomeric equilibrium for 2-hydroxypyridine/2-pyridone is typical.

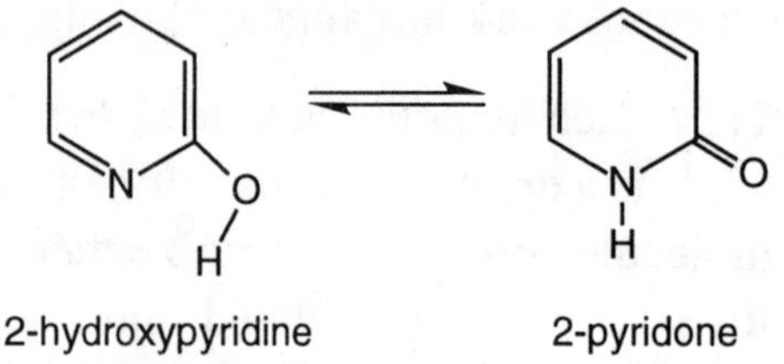

2-hydroxypyridine   2-pyridone

2-Hydroxypyridine is only slightly more stable than 2-pyridone in the gas phase ($\Delta E < 0.5$ kcal/mol). This behavior differs markedly from that of other phenol-type compounds, which strongly prefer the "enol" form, and may indicate that 2-pyridone enjoys some degree of aromatic stabilization.

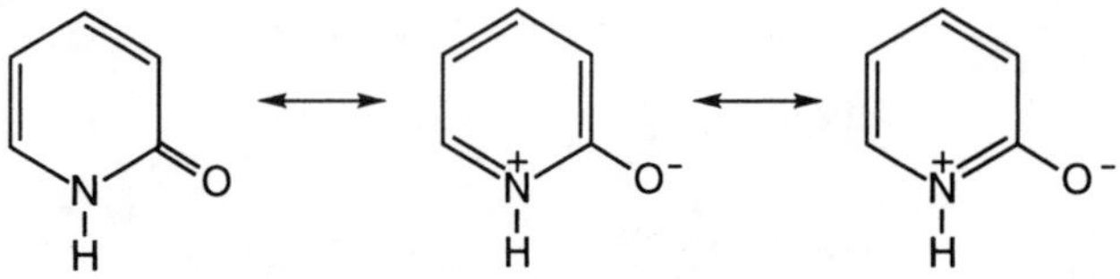

Interestingly, 2-pyridone is the more stable tautomer in polar solvents, again inconsistent with most "phenol" chemistry. It can, however, be explained if 2-pyridone benefits from aromaticity, and/or if 2-pyridone is significantly more polar than the "enol" tautomer (one expects a more polar molecule to be stabilized more strongly by polar solvents; see: **The Role of Solvent**).

In this experiment, AM1 and AM1-SM2 semi-empirical calculations are used to investigate the tautomeric equilibrium involving 2-hydroxypyridine and 2-pyridone both in the gas phase and in water.[2] The electric dipole moment will be employed as a measure of polarity as an attempt to rationalize the experimentally observed change in favored tautomer from gas phase to water.

---

1. March, p.69.
2. Isaacs, p. 205; M.W. Wong, K.B. Wiberg and M.J. Frisch, *J. Am. Chem. Soc.*, **114**, 1645 (1992).

Procedure

Build 2-hydroxypyridine and 2-pyridone and optimize their geometries using semi-empirical AM1 calculations. Choose a conformer for 2-hydroxypyridine which allows for internal hydrogen bonding. Identify the more stable tautomer and calculate the energy of the higher-energy tautomer relative to it. Do the calculations correctly indicate the fact that 2-hydroxypyridine is the lower energy tautomer, but that 2-pyridone is very close in energy? Which of the two tautomers has the larger electric dipole moment?

Perform single-point calculations on 2-hydroxypyridine and 2-pyridone using the AM1-SM2 model.[3] What is the effect of the solvent on the relative stabilities of the two tautomers? Note that the solvent dramatically increases the dipole moment of 2-pyridone, but has only a modest effect on the dipole moment of 2-hydroxypyridine. Why?

Optional

Perform AM1 (gas phase) and AM1-SM2 (aqueous phase) calculations on 4-hydroxypyridine and 4-pyridone.

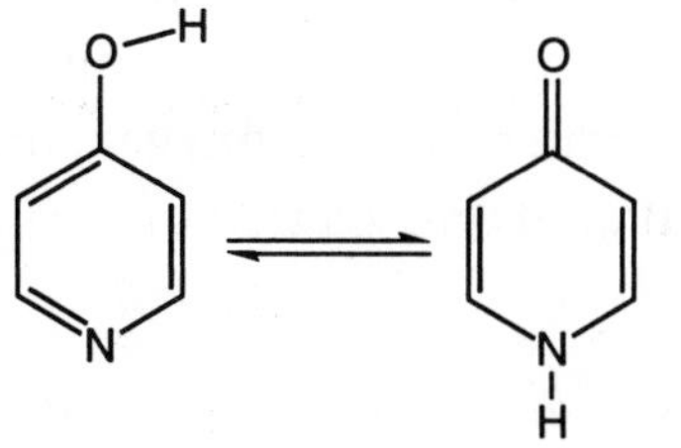

4-hydroxypyridine   4-pyridone

Point out any differences with the 2-hydroxypyridine/2-pyridone system and try to rationalize them using resonance theory.[4]

---

3. Because inclusion of solvent has very little effect on the geometries of neutral molecules, AM1 (gas phase) and AM1-SM2 (aqueous phase) geometries will be nearly identical.
4. March, p. 73.

Acid and Base Catalysis of Keto-Enol Tautomerism

Semi-empirical AM1 calculations and ab initio 3-21G calculations together with corrections for solvent using the AM1-SM2 method are used to investigate the energetics of both acid and base-catalyzed keto-enol tautomerism in acetaldehyde.

Carbonyl compounds with α hydrogens undergo rapid acid or base-catalyzed equilibration with their enol tautomers.[1]

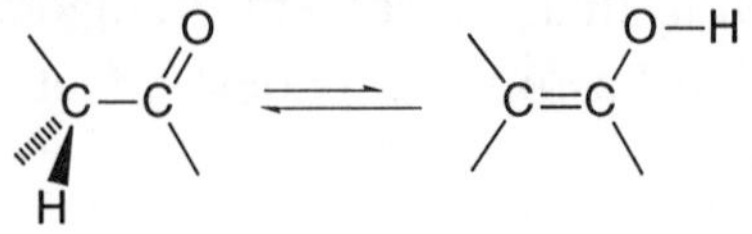

Although the equilibrium concentration of enol tautomers is typically very small and, because of this, enols are very difficult to isolate and characterize, they are believed to play an important role in the chemistry of carbonyl compounds.

The mechanism of base catalysis of keto-enol tautomerism in water involves removal of an α hydrogen by hydroxide, leading to formation of an enolate anion. This may then add water either at oxygen, resulting in the enol tautomer, or at the α carbon, resulting in reformation of the carbonyl compound.

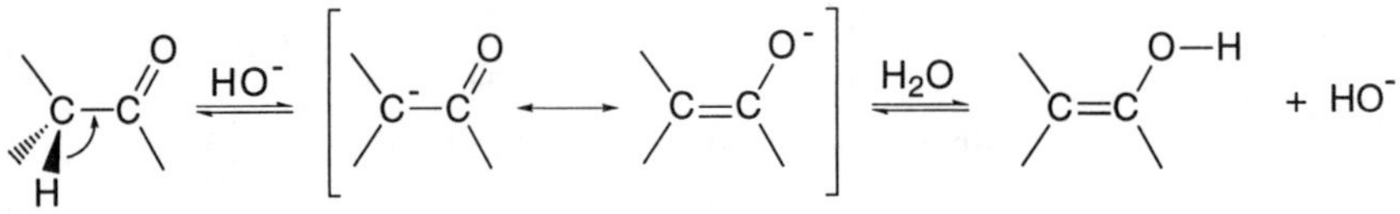

The mechanism of acid catalysis involves protonation at the carbonyl oxygen followed by deprotonation either at the α carbon, leading to the enol tautomer, or at oxygen, leading back to the original keto form.

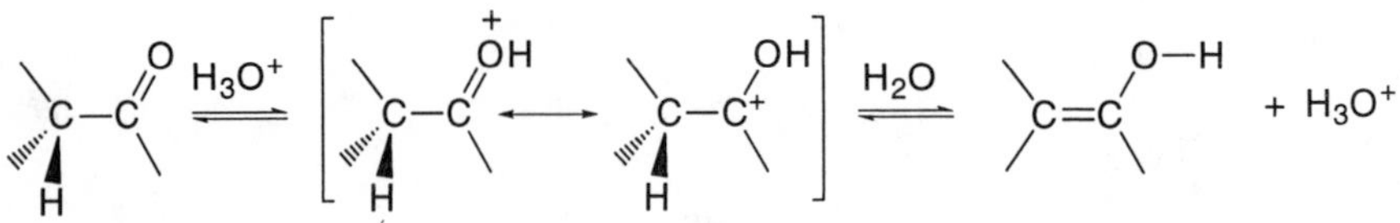

In this experiment, you will use both semi-empirical and *ab initio* molecular orbital calculations to examine the energetics of keto-enol tautomerism of acetaldehyde in water, both under conditions of base and acid catalysis.

---

1. March, p.69.

Procedure

Build all the molecules required to describe both acid and base-catalyzed tautomerism of acetaldehyde to vinyl alcohol, and optimize the geometry of each using AM1 semi-empirical calculations. Which valence structure for the enolate anion is the more appropriate, i.e., is the formal negative charge on carbon or on oxygen? (Compare bond lengths and atomic charges with those in acetaldehyde.) Which valence structure for protonated acetaldehyde is the more appropriate, i.e., is the formal positive on oxygen or on carbon?

Calculate the AM1-SM2 energy of each molecule at its AM1 geometry, and subtract it from the corresponding AM1 energy to get the aqueous solvation energy. Which molecules have the largest (magnitude) solvation energies, neutral molecules or ions? Which are larger, solvation energies for small ions, e.g., OH^- or for larger ions, e.g., CH_2CHO^-? Rationalize your results (see: **The Role of Solvent**).

Next, calculate the *ab initio* 3-21G energy of each molecule (use AM1 geometries). Add to these energies the previously obtained solvation energies. Work out the overall thermodynamics of tautomerism, acetaldehyde $\rightleftharpoons$ vinyl alcohol, using the 3-21G energies corrected for solvent. Which tautomer is more stable? Is your result supported by observation?

Work out the thermodynamic "barriers" to tautomerism for the acid and base-catalyzed routes. You need to assume that there are no significant barriers to the proton transfer reactions in solution, and that the intermediate ions (enolate in the case of base catalysis and protonated acetaldehyde in the case of acid catalysis) represent the "top" of the reaction energy profiles. Are the energy barriers you predict consistent with the observation that both acid and base-catalyzed reactions occur readily?

Optional

Is tautomerism always catalyzed, or can it occur via an unimolecular mechanism? The latter requires a single transition state in which hydrogen migrates from C to O. Build the structure shown below and use it to search for the AM1 transition state.

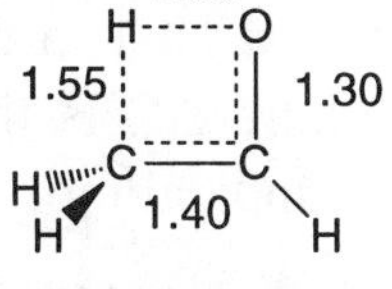

Then calculate the AM1-SM2 and 3-21G energies of the transition state, and use these to estimate the barrier for unimolecular rearrangement. Can this process compete with the catalyzed pathways? Under what conditions?

---

2. For the molecules involved in the base-catalyzed reaction you might instead perform single-point energy calculations with the 3-21+G method. This includes diffuse atomic orbitals on carbon and oxygen and provides a superior account of the energetics of reactions involving anions.

Hydrochloric Acid in the Gas Phase and in Water

AM1 and AM1-SM2 semi-empirical calculations together with atomic charges and electrostatic potential maps are used to point out differences in the dissociation of hydrochloric acid in the gas phase and in water.

Every chemist knows that hydrochloric acid is a strong acid and "completely dissociates" in aqueous solution. Does this mean that hydrogen chloride gas exists in dissociated form? Of course not. In fact, the gas-phase equilibrium constant for the dissociation of HCl lies so far toward HCl that no ions can be detected at all. The dissociation of HCl is driven by the stabilizing effect the solvent (usually water) has on the ions (see: **The Role of Solvent**). Acidity is therefore extremely sensitive to solvent.

In this experiment, semi-empirical AM1 and AM1-SM2 calculations will be used to investigate the effect of aqueous solution on the dissociation of HCl. Specifically, you will examine the energy of HCl as a function of interatomic distance in both the gas phase and aqueous solution. You will also examine the separation of charge as a function of HCl distance, both in terms of atomic charges and as shown by electrostatic potential maps. This will help clarify how the solvent promotes heterolytic bond cleavage.

Procedure

Perform single-point AM1 and AM1-SM2 semi-empirical calculations (two sets of calculations) on HCl at bond distances of 1.0, 1.1, 1.2, 1.3, 1.4, 1.5, 2.0, 2.5, 3.0, 4.0 and 5.0Å. Plot energy (both AM1 and AM1-SM2) versus HCl distance. How (if at all) do the two curves differ? In particular, how many minima appear on each curve? Do you see evidence for two distinct structures (bound and dissociated) for HCl in water? Extrapolate the curve which you computed for HCl dissociation in water to infinite interatomic separation. Use this energy and the aqueous energy of undissociated HCl to determine the pK_a for HCl(aq) according to $pK_a = \Delta E/1.36$, where ΔE is the energy difference between dissociated and undissociated acid in kcal/mol.

Calculate and plot the solvation energy of HCl (the difference in AM1-SM2 and AM1 heats of formation) as a function of distance. Is this largest for small or large HCl separations? Explain.

To rationalize the effect of solvent in promoting bond dissociation in HCl, calculate AM1 (gas phase) atomic charges at hydrogen, as well as electrostatic potential maps at three different bond distances: 1.3Å (near the equilibrium bond distance), 2.0Å and 3.0Å. According to these measures, at which of the three distances is HCl most polar? Looking at your previous energy data, what do you conclude about the relationship between polarity and solvation energy?

Repeat your atomic charge and electrostatic potential map calculations using the AM1-SM2 model (for aqueous solvent). What differences exist between the aqueous phase and the gas phase results. Are these in line with what you expect?

Optional

Although solvent can facilitate acid dissociation, it need not affect all acids in the same way. For example, toluene, $PhCH_3$, is a slightly stronger acid than H_2O in the gas phase, but is a considerably weaker acid in the aqueous phase (in fact, the pK_a of toluene cannot be measured).

A crude estimate of the difference between pK_a (toluene) and pK_a (H_2O) can be obtained by calculating $\Delta(pK_a) = \Delta E/1.36$ where ΔE is the AM1-SM2 energy for the reaction.

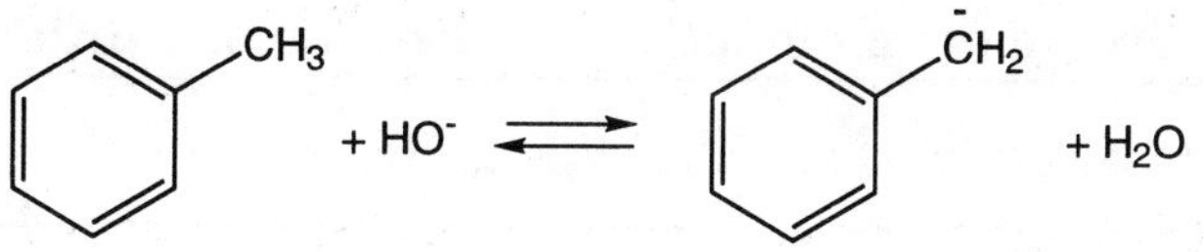

Build each molecule that appears above, optimize its AM1 geometry, and use this geometry to calculate its AM1-SM2 energy. Calculate ΔE for the gas phase (AM1) and aqueous phase (AM1-SM2) reactions. Are the results qualitatively consistent with experimental observations? What would you estimate the pK_a of toluene to be? Why is there such a large change in ΔE upon solvation? (Hint: calculate the solvation energy for each molecule and ion.)

Gas and Aqueous-Phase Basicities of Methylamines

AM1 and AM1-SM2 semi-empirical calculations and 6-31G* ab initio calculations are used to investigate the basicities of alkylamines in the gas phase, and to account for changes in basicities from the gas phase to water.

The basicity of different amines, B, can be defined relative to ammonia by means of the following *isodesmic* reaction: $BH^+ + NH_3 \rightarrow B + NH_4^+$ (see: **Isodesmic Reactions**). Experimentally, increasing methyl substitution increases amine basicity in the gas phase, but not necessarily in solution (see table). Trimethylamine is a stronger base than ammonia, but a weaker base than either methylamine or dimethylamine in water. Note also, that aqueous-phase basicity is much less sensitive to methyl substitution than gas- phase basicity.[1]

amine	ΔH_{rxn} (gas)	ΔH_{rxn} (H_2O)	aqueous phase pK_a
NH₃	0	0	9.26
CH₃NH₂	9	1.9	10.64
(CH₃)₂NH	16	2.0	10.73
(CH₃)₃N	19	0.7	9.79

Experimental gas and aqueous phase basicity of methylamines (kcal/mol, relative to ammonia) and aqueous phase pKa's.

In this experiment, you will first see if calculations can account quantitatively for relative gas-phase basicities in methylamines, and then investigate the role of the solvent in altering relative gas-phase basicities.

Procedure

Build ammonia, methylamine, dimethylamine, and trimethylamine, as well as their corresponding ammonium ions, and optimize their geometries using AM1 semi-empirical calculations. Follow each by a single-point energy calculation using the 6-31G* *ab initio* method (to provide a better description of relative gas-phase base strengths), and independently by a single-point calculation using the AM1-SM2 semi-empirical method (to account for the effects of solvent). Calculate heats of solution by subtracting AM1 heats of formation from the corresponding AM1-SM2 heats of formation. Finally, add these to the 6-31G*

1. E.M. Arnett, in **Proton Transfer Reactions**, E.F. Caldin and V. Gold, Ed., Wiley, New York, 1975; March, p. 269.

energies to produce aqueous phase energies. Are solvation energies larger for the neutral amines or for their corresponding protonated forms? Rationalize your finding (see: **The Role of Solvent**).

Calculate $\Delta H_{rxn}(gas)$ and $\Delta H_{rxn}(H_2O)$ for each amine, B, relative to ammonia using the 6-31G* data. Do the calculations reproduce the experimental gas-phase relative basicities? Do they reproduce the experimentally observed decrease in the overall range of basicity of the methylamines? Do they reproduce the observed reversal in methyl substituent effects in going from the gas phase into water?

Are the changes in relative base strengths from the gas phase to water due primarily to changes in solvation energy of the neutral amines, or primarily to changes in solvation energy of the corresponding ammonium ions? Calculate and display on the same color scale 6-31G* electrostatic potential maps for the amines and ammonium ions. Regions of maximum charge will be stabilized most effectively by the solvent. Are the regions more prominent in the amines or in the ammonium ions? What effect does methyl substitution appear to have on these regions? Is this effect consistent with your calculated heats of solvation?

Solvent Effects on S_N2 Reactions

Semi-empirical AM1 and AM1-SM2 calculations are used to construct reaction coordinate diagrams for gas and aqueous-phase S_N2 reactions.

Nucleophilic substitution (S_N2) is one of the most important synthetic reactions in organic chemistry. The reaction is stereospecific (inversion at C), and proceeds with a large variety of nucleophiles, Nu, and leaving groups, X.[1]

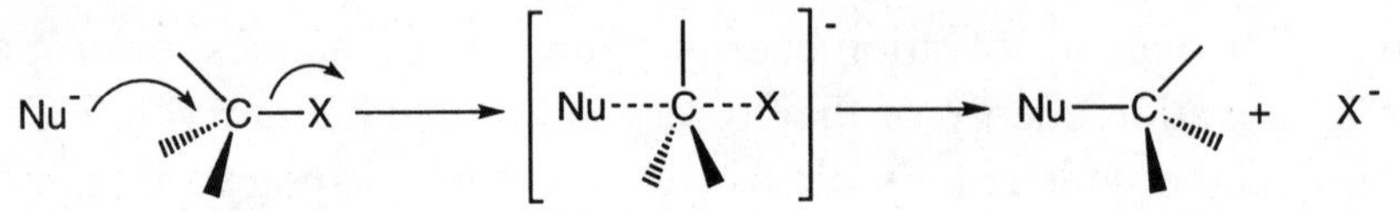

Although most textbooks draw the reaction coordinate with a barrier, and attribute this barrier to weakened bonds in the transition state, the fact is that gas-phase S_N2 reactions generally proceed with little or no barrier.[2] Thus, "barriers" observed in solution must be due to solvation. To understand, recognize that polar solvents are much more effective in stabilizing small, charge-localized ions (the nucleophile and the leaving group after it departs), than they are large, delocalized ions (the transition state). That is, solvent lowers the energy of the reactants and products more than it does the energy of the transition state, causing an increase in reaction barrier. (See: **The Role of Solvent.**)

In this experiment, you will use the AM1 method to construct potential energy diagrams for addition of Cl⁻ and OH⁻ as nucleophiles to methyl chloride. You will then construct potential energy diagrams for aqueous-phase reactions by recalculating the energy of each of the gas-phase species using the AM1-SM2 method.

Procedure

Build anionic complexes of Nu + CH_3Cl (Nu = Cl⁻, HO⁻) with NuC distances of 6.0, 5.0, 4.0, 3.0, 2.5, 2.0 and 1.5Å, and constraining the Nu-C-Cl angle to be 180°. Optimize (constrained optimization) the remaining geometrical coordinates in each structure using AM1 semi-empirical calculations. Record your data and plot heat of formation versus NuC distance, i.e., construct a reaction coordinate diagram. Also calculate the sum of the energies of the separated reactants.

1. Carey and Sundberg A., p. 261; Lowry and Richardson, p.327; Isaacs, p. 418; March, p.294.
2. D.K. Bohme and G.I. Mackay, *J. Am. Chem. Soc.*, **103**, 978 (1981); K. Morokuma, *ibid*, **104**, 3732 (1982).

Identify regions that contain minima or maxima, and perform additional calculations as needed in order to refine your diagram. Examine the structures corresponding to energy minima and/or maxima, and characterize them appropriately as "separated reactants", "complexes", "transition states", "products", and so on. What is the magnitude of the reaction barrier, i.e., the energy of the highest-energy structure relative to the energy of the separated reactants for each process? Which reaction is predicted to be more exothermic? Is there a relationship between overall thermodynamics and the size of the barrier? (See: **Reactive Intermediates and the Hammond Postulate**.)

For each of the AM1-optimized structures, calculate and record the AM1-SM2 energy. Again plot energy versus distance, i.e., construct an aqueous-phase reaction coordinate diagram. You may need to perform additional calculations to refine your diagram adequately. Repeat your previous analysis with the new reaction coordinate diagrams. Does the geometry around carbon in the transition state correspond to the standard picture of a trigonal bipyramidal carbon? Does solvent affect barrier height? Does solvent affect reaction exothermicity?

Optional

1. The solvation enthalpy of a structure is defined as the difference between its gas-phase and solution-phase heats of formation. Calculate the solvation enthalpy for each of the structures which you previously obtained. Are your findings consistent with the proposed relationship between solvation and ion size? AM1-SM2 solvation enthalpies (AM1 geometries) of the isolated molecules are: -0.6 (CH_3Cl), -5.8 (CH_3OH), -77.0 (Cl^-), -108.2 (HO^-) kcal/mol.

2. Sketch a "textbook" reaction coordinate diagram for the aqueous-phase reaction of $NH_3 + CH_3Cl \rightarrow CH_3NH_3^+ + Cl^-$. Based on the previously defined solvation model, predict how solvation energies will change along with this reaction coordinate. Use these predictions to sketch a qualitative reaction coordinate for the gas-phase reaction. **No calculations are required**.

V

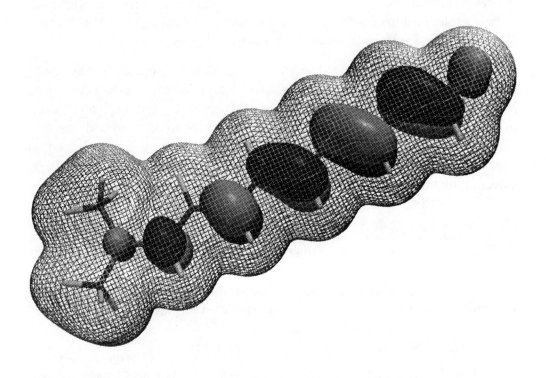

Charges on Atoms
in Molecules

harges are a part of the everyday language of organic chemistry, so much so that many chemists have come to accept them at face value. Charge distributions imply something about where the electrons reside in molecules, and this in turn implies something about the "chemistry" which molecules can undergo. For example, the obvious resonance structures for phenoxide,

suggest that the molecule's negative charge resides not only on oxygen, but also on the *ortho* and *para* (but not on the *meta*) ring carbons. This, in turn, suggests that the addition of an electrophile will occur at these sites.

Despite their obvious utility to chemists, there is actually no way either to measure atomic charges or to calculate them, at least not uniquely. The reason behind this surprising statement is quite simple. Take a careful look at a molecule from the point of view of quantum mechanics. It is made up of nuclei, each of which bears a positive charge equal to its atomic number, and electrons, each of which bears a charge of -1. It is reasonable to assume that the nuclei can be treated as point charges, i.e., they do not occupy appreciable space. On the other hand, electrons need to be viewed in terms of a distribution of negative charge, which although primarily concentrated in regions around the individual nuclei and in between nuclei that are close together, i.e., are bonded, extends throughout all space. The region of space occupied by a conventional space-filling or CPK model corresponds roughly to the molecule's van der Waals surface and encloses something on the order of 90-95% of the electrons in the entire distribution.

How then can charges be assigned to individual atoms? It is clearly necessary to account both for the nuclear charge and for the charge of any electrons associated with the particular atom. While the nuclear contribution to the total charge on an atom is easy to handle (as already mentioned, it is simply the atomic number), it is not at all obvious how to partition the total electron distribution by atoms. To see that this cannot be done uniquely, consider the heteronuclear diatomic molecule HF, sketched below.

The surrounding line roughly corresponds to a van der Waals surface and encloses a large fraction of the total electron density. The surface has been drawn to suggest that more electrons are associated with fluorine than with hydrogen. While this is qualitatively reasonable, how exactly is this surface to be divided between the two atoms? Are any of the divisions shown below better than the rest?

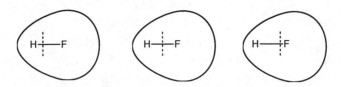

The answer to the first question is that it is not at all apparent how to divide the surface; the answer to the second question is clearly no! Atomic charge is not a molecular property, and it is not possible to provide a unique definition (or even a definition that will satisfy all). Electronic charge distributions can be calculated (and even measured using X-ray crystallography[1]), but it is not possible to uniquely partition them among the atomic centers.

Despite the fact that charges on atoms in molecules cannot be defined uniquely, methods have been developed to assign atomic charges.[2] The important point is that none of these methods is necessarily any better than any of the alternatives; while charges are able to provide insight into the properties and chemical reactivities of molecules, they are not themselves molecular properties, and cannot be treated in the same way as quantities such as the electron density, which is a molecular property. It is not possible to say that a particular set of charges obtained from a particular calculation using a particular charge analysis method (or for that matter from experiment) is "correct". In fact, there is no such thing as "correct" charges! The best that can be done is to say that the charges are consistent (or inconsistent) with specific molecular properties, such as the electric dipole moment, and with known patterns of reactivity.

Two of the most popular methods for obtaining atomic charges are the Mulliken partition method and charge calculation based on fits to electrostatic potentials.

Mulliken Method for Charge Partitioning[3]

The electron density function $\rho(\mathbf{r})$ for a molecule is given by the square of the wavefunction, and the probability of finding an electron at $\mathbf{r}$ is given by $\rho(\mathbf{r})d\mathbf{r}$, where $d\mathbf{r}$ is the small volume centered at $\mathbf{r}$. Summing over all space necessarily gives the total number of electrons contained in the molecule.

1. Nuclei are not seen in an X-ray experiment. Rather, they are assumed to be associated with very high concentrations of electrons.
2. For a review, see: S.M. Bachrach, in **Reviews in Computational Chemistry**, **5**, 171 (1994).
3. R.S. Mulliken, *J. Chem. Phys.*, **23**, 1833, 1841, 2338, 2343 (1955).

The same concept can be applied to molecular orbital wavefunctions, where the wavefunction is now expressed as a linear combination of atomic orbitals. The electron density function defined by the square of this wavefunction involves a summation that contains terms that are products of atomic orbitals. Some of these products involve orbitals located on the same atom, and the electron density associated with these terms can reasonably be assigned to this atom. However, some of the products involve orbitals on different atoms, and it is not at all obvious how to apportion this electron density (referred to as the "overlap population") between the two atoms. This was first recognized by Mulliken, who proposed that each atom be given half of the overlap population. This scheme, which allows all of the electron density to be assigned to specific atoms and generates a specific set of atomic charges, is referred to as "Mulliken Population Analysis". Other, physically more reasonable schemes for dividing the overlap population have also been developed, perhaps the most notable being "Natural Population Analysis".[4]

Charges Based on Fits to Electrostatic Potentials[5]

This alternative method of obtaining atomic charges is not a partitioning scheme as is the Mulliken procedure, but a scheme in which the electrostatic potential, the energy of interaction of a point positive charge with the nuclei and electrons of a molecule, is fit to atomic charges.[6] One can then think of the method as providing that set of atomic charges that best reproduces the "exact value" of the electrostatic potential.

Several steps are involved in its implementation after a molecular orbital wavefunction has been obtained:

1) Define a grid of points around the molecule. Choice of the grid obviously introduces ambiguity into the method. In practice, several thousand points make up the grid, and these are primarily located in a belt 2-5Å thick above the van der Waals surface.

2) Calculate the electrostatic potential at each of these points, using the nuclear charges and electronic wavefunction.

3) Fit the calculated potential to a potential based on atomic charges (treated as variables), subject to the constraint that the sum of atomic charges is equal to the total charge on the molecule.

---

4. A.E. Reed, R.B. Weinstock and F. Weinhold, *J. Chem. Phys.*, **83**, 735 (1985).
5. L.E. Chirlian and M. Francl, *J. Computational Chem.*, **8**, 894 (1987); C.M. Breneman and K.B. Wiberg, *ibid.*, **11**, 361 (1990).
6. For further discussion of the electrostatic potential, see: **Graphical Models and Graphical Modeling**.

Which Charge Calculation Method is Better?

This is not a simple question to answer as neither Mulliken partitioning methods nor charge fitting methods is unique. One must go on the relative strengths and weaknesses of the two.

The principal advantage of the Mulliken analysis is that it is computationally simple. It is one of those few procedures which can (almost) be carried out "on the back of an envelope". Mulliken charges are perhaps as good as electrostatic fit charges for qualitative assignments, but generally fail to provide a good account of electric dipole moments which indicate overall molecular polarity (see: **Atomic Charges and Dipole Moments**).

The electrostatic potential fitting procedure, while computationally much more expensive is the obvious method of choice if the reason for calculating charges in the first place is to use them in the construction of empirical energy functions, i.e., to reproduce intra and intermolecular electrostatic interactions. Electrostatic-fit charges typically provide a good account of dipole moments.

Mulliken and electrostatic-fit schemes lead to different atomic charges. Within each scheme, charges also depend on the underlying molecular orbital method and even change with basis set for *ab initio* methods. In the latter regard, Mulliken charges generally show larger variations with basis set than electrostatic-fit charges. Calculated charges for formaldehyde provide an example.

Atomic Charges for Formaldehyde (electrons) and Dipole Moments μ (debyes)		
	AM1	**3-21G**	**6-31G***
Mulliken	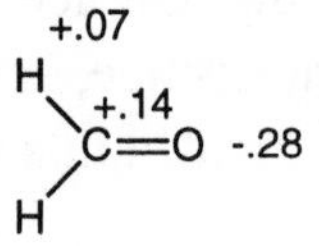 +.07 H +.14 C=O −.28 H	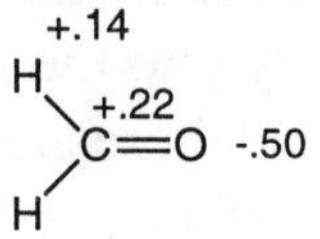 +.14 H +.22 C=O −.50 H	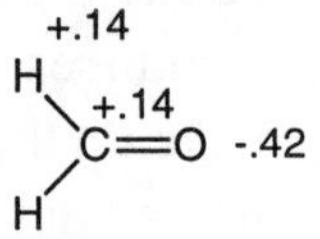 +.14 H +.14 C=O −.42 H
μ (Mulliken)	2.02	3.77	3.15
fits to electrostatic potentials	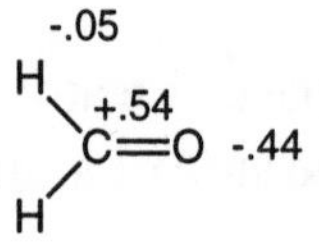 −.05 H +.54 C=O −.44 H	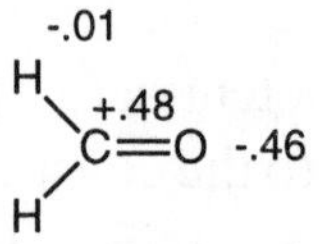 −.01 H +.48 C=O −.46 H	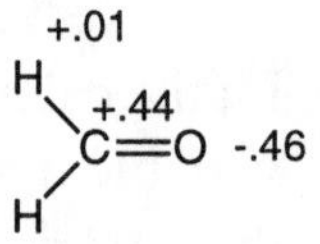 +.01 H +.44 C=O −.46 H
μ (electrostatic fit)	2.34	2.64	2.68
μ (exact fit)	2.32	2.66	2.67

Both Mulliken and electrostatic-fit methods show carbon as positive and oxygen as negative, although the magnitudes of the charges exhibit wide variation. The Mulliken analysis depicts the hydrogens as positively charged while the fitting scheme shows them as essentially neutral.[7]

Which set of charges is the more realistic? It is really impossible to say. Clearly the electrostatic-fit charges more closely reproduce the calculated dipole moment at each level of theory, although with the exception of AM1, the calculated dipole moment is significantly larger than the experimental value (2.34 debye). What is important to emphasize over and over is that the different methods give different charges. Charge comparisons among molecules should only be made with a single quantum mechanical method and charge calculation scheme.

---

7. It should be noted that the electrostatic potential for formaldehyde is strongly positive in the vicinity of the hydrogens. This lack of correspondence between electrostatic potential maps and electrostatic-fit charges is commonly observed, and suggests that if one quantity, e.g., the map, is successful for comparisons, then the other cannot be.

Atomic Charges and Dipole Moments

Ab initio 3-21G calculations are used to explore the influence of molecular geometry on dipole moments, and to examine relationships between atomic charges, bond dipoles and dipole moments.

The electric dipole moment is among the simplest experimentally measurable quantities related to the charge distribution in a molecule. It is often invoked to explain a variety of physical and chemical properties from boiling points to chemical reactivities. The origin of the dipole moment is easily understood by considering that a molecule is made up of a set of point charges, positively charged nuclei and negatively charged electrons,[1] or alternatively, "atoms" which bear either net positive or negative charge. In either representation, the scalar sum of these individual charges for a neutral molecule must be zero.[2] However, the vector sum leads instead to a simple "two charge" picture, and the notion that a molecule has a "positive end" and a "negative end".

x = positively charged nuclei

• = negatively charged electrons

vector sum

scalar sum=0

The magnitude of the resulting dipole moment vector is routinely measured and discussed. The direction and sign of the dipole moment vector may also be measured, but this is difficult and, in fact, is known experimentally only for a very few simple systems.

Chemists often view the dipole moment in a slightly different light, as arising from a collection of bond dipoles. Covalent bonding results from the sharing of electrons between atoms. If the two bonded atoms are in different chemical environments, then the electrons involved in the bond will not be shared equally. As a consequence, the bond will be polarized, with the two constituent atoms attaining slightly different net charges. The resulting bond dipole can be represented by a vector, the direction of which lies along the interatomic axis, and (by convention) points from relative positive to relative negative (the word "relative" here is important, since the two bonded atoms need not necessarily be charged or have opposite charges). The dipole moment for a molecule may then be viewed as the vector sum of the individual bond dipoles. This means that the dipole moment is related not only to bond polarity, but also to the overall shape of the molecule.

---

1. This is an alternative to the previous view of the electrons as forming a distribution spread out over all space. See: **Charges on Atoms in Molecules**.
2. If the scalar sum of the charges is not zero, then the vector sum will depend on the origin of the coordinate system. As a consequence, the dipole moment of a charged species is ill defined.

In this experiment, *ab initio* 3-21G calculations will be used to examine the charge distribution and dipole moment in carbon dioxide both in its linear (equilibrium) form and in a hypothetical bent geometry. The objective will be to clearly distinguish between polar bonds (present in both structures) and an overall dipole moment (present only in the hypothetical bent structure).

Procedure

Build (linear) carbon dioxide and optimize its geometry using *ab initio* 3-21G calculations. Obtain charges and use them to calculate the magnitude of each bond dipole by multiplying the difference in charge between carbon and oxygen by the bond length (number of electrons times distance in Å are acceptable units for the bond dipole). What is the direction of each bond dipole? What is the magnitude of the total dipole moment for linear carbon dioxide? Explain.

Construct carbon dioxide in which the OCO bond angle is constrained at 120°. Optimize its 3-21G geometry (subject to the bond angle constraint). Again examine both the charges and the electric dipole moment. As you did before, calculate the dipole moment for each bond based on atomic charges, and then, summing the bond dipoles, calculate a total dipole moment. How does it compare with that for linear CO_2? Rationalize the difference.

Optional

1. A general rule is that any molecule with a center of symmetry, i.e., for each atom located at coordinates x, y, z there will be another atom of the same kind located at coordinates -x, -y, -z, will have a zero dipole moment. Using this rule, decide which (if any) of the following molecules will have a zero dipole moment. **No calculations are required**.

2. How well do calculated charges reproduce dipole moments? Calculate both Mulliken and electrostatic-fit charges for CH_3F based on a 3-21G wavefunction (at the optimum 3-21G geometry) Obtain the exact dipole moment and calculate approximate dipole moments according to:

$$\mu \text{ (debyes)} = 2.54 \times \left[\left(\sum_i^{atoms} x_i q_i \right)^2 + \left(\sum_i^{atoms} y_i q_i \right)^2 + \left(\sum_i^{atoms} z_i q_i \right)^2 \right]^{1/2}$$

where x, y, z are Cartesian coordinates and q_i are atomic charges. Which set of charges is more successful in reproducing the exact dipole moment?

What Do Zwitterions Look Like?

AM1 and AM1-SM2 semi-empirical calculations are used to obtain atomic charges and electrostatic potential maps for a model polypeptide, zwitterionic hexaglycine, which in turn are examined to see to what extent the "positive" and "negative" centers are separated.

It is well known that peptides exist as zwitterions in aqueous solution at neutral pH. Indeed, even amino acids ("monopeptides") prefer zwitterionic forms in water, e.g., glycine exists as ^+H_3N-CH_2-CO_2^- and not H_2N-CH_2-CO_2H. Polar solvents such as water preferentially stabilize localized charges and therefore encourage "charged" tautomeric structures (see: **The Role of Solvent**). Of course, zwitterionic glycine is a neutral molecule, and it is interesting to ask to what extent it actually incorporates fully developed positively and negatively charged centers.

In this experiment, semi-empirical AM1 and AM1-SM2 calculations are used to obtain atomic charges and electrostatic potential maps for a model polypeptide, zwitterionic hexaglycine constructed as an α helix. These are then examined both for evidence of charge separation in the zwitterion, and to assess the role of solvent in promoting charge separation.

Procedure

Build hexaglycine as an α helix, and cap the ends with charged functional groups, i.e., NH_3^+ and CO_2^-. **Do not minimize.**[1] Perform single-point AM1 and AM1-SM2 semi-empirical calculations. Calculate the sum of electrostatic-fit atomic charges for the terminal NH_3^+ and CO_2^- groups for the model peptide. According to this measure, is the hexaglycine zwitterion accurately represented as a fully charged separated species, i.e., are the group charges +1 and -1? Which calculation (AM1 or AM1-SM2) shows the greater degree of charge separation? Explain your result.

Electrostatic potential maps provide another way to examine charge separation. Calculate and display maps corresponding to both AM1 and AM1-SM2 calculations. Do these show a molecule in which one end is strongly positively charged and the other is strongly negatively charged as implied by the zwitterionic structure? For which calculation does the charge separation appear to be greater? Is your result in line with the previous finding based on atomic charges?

---

1. Procedures which do not account explicitly for solvent will not necessarily yield the proper conformations for polypeptides in solution. For the purpose here (establishing charge distributions) it is better to use a "standard" helical conformation.

Optional

1. Perform single-point "gas-phase" (AM1) and "aqueous phase" (AM1-SM2) calculations on "neutral" hexaglycine, i.e., capped by non-charged groups NH_2 and CO_2H, also as an α helix. Compute the solvation energy (difference between the AM1-SM2 and AM1 heats of formation). It is larger or smaller than the solvation energy for zwitterionic hexaglycine? Does your result suggest that in the gas phase, amino acids and peptides might have different tautomeric structures than observed in water?

2. Does increased "spacing" between "acid" and "base" sites on zwitterionic polypeptides lead to increased charge separation? Perform single-point AM1 calculations on triglycine and decaglycine constructed as α helices. Compare calculated atomic charges and electrostatic potential maps to those previously obtained for hexaglycine at the same level. Are there significant differences?

Solvent Effects on Electron Distributions in Push-Pull Polyenes

Semi-empirical AM1 calculations are used to examine the charge distribution in a polyene substituted with π acceptor and π donor groups (a "push-pull" polyene). Dipole moments and electrostatic potential maps are used to assess the degree of charge transfer between these groups, and AM1-SM2 calculations to examine charge transfer in a polar medium such as water.

Polyenes substituted with π-electron donor and π-electron acceptor groups form the basis for many new materials, including organic conductors, batteries, and non-linear optical materials. The electron distribution in these molecules is described by a resonance hybrid of a "neutral" structure and a "charge transfer" structure, i.e., the donor and acceptor groups "push" and "pull" electron density, respectively.

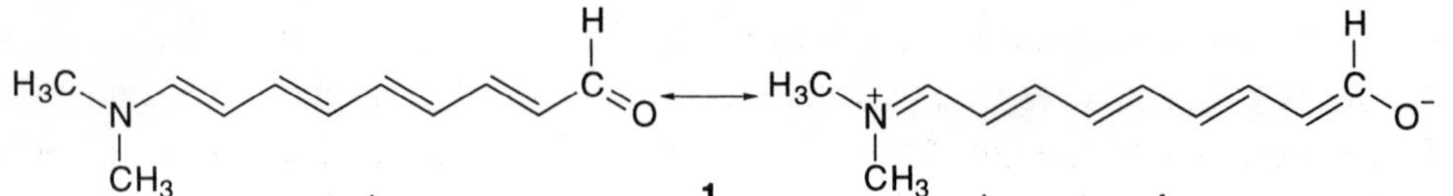

The degree of charge transfer between the terminal groups has been shown to depend on external electromagnetic fields (this is the basis for using these compounds as non-linear optical materials).[1] In this experiment, you will explore whether charge transfer also depends on the surrounding medium. A nonpolar medium, and the gas phase, should favor an electron distribution that minimizes charge separation. A polar medium will stabilize charged sites, and should favor an electron distribution in which there is more charge development. (See: **The Role of Solvent**.) Specifically, you will use AM1 semi-empirical calculations to examine the electron distribution in **1** in the gas phase, and then AM1-SM2 calculations to predict changes in this distribution in water. The degree of charge transfer will be determined by calculating and comparing dipole moments, and by comparing electrostatic potential maps.

Procedure

Build **1** as a planar molecule, and optimize its geometry using the AM1 semi-empirical method. Note the dipole moment. If the donor and acceptor groups did not interact, one could predict the dipole moment of **1** by adding together the AM1 dipole moments for its component amines and aldehydes, e.g.,

1.   C.B. Gorman and S.R. Marder, *Chem. Materl.*, **7**, 215 (1995).

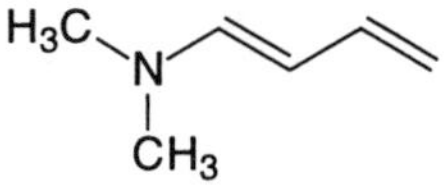

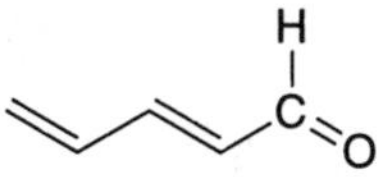

2.35 debye   +   3.55 debye   = 5.90 debye

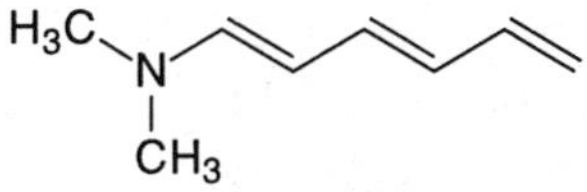

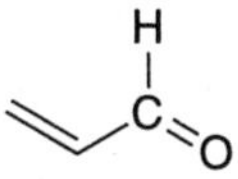

2.87 debye   +   3.06 debye   = 5.93 debye

How does the dipole moment of **1** compare to these predictions? Is this consistent with the type of charge transfer interaction proposed above?

Perform a single-point energy calculation on **1** using the AM1-SM2 semi-empirical model (AM1 geometry). Note the dipole moment. Is it larger or smaller than the gas-phase (AM1) dipole moment? What does this suggest about the degree of charge transfer and the effect of solvent? Calculate and display the electrostatic potential maps for the gas-phase (AM1) and aqueous- phase (AM1-SM2) models on the same color scale. Which molecule shows more charge transfer between the terminal groups? Are these maps consistent with the conclusions based on dipole moments?

VI

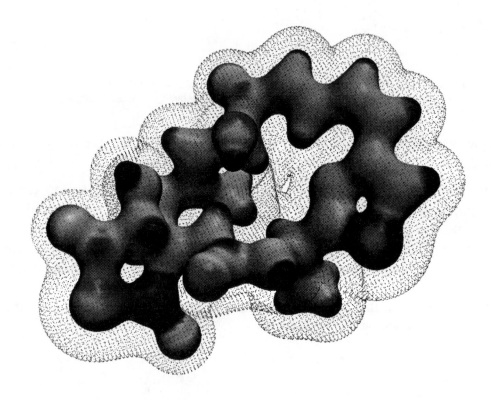

Interpreting
Conformational
Preferences

Rotation about single bonds is periodic, retracing itself every 360°. Therefore, any function that seeks to describe the energy of internal rotation must also repeat itself every 360°. In fact, it is possible to write a general energy function, $E^{torsion}$, as a combination of simpler functions, V_n, each of which repeats n times in a 360° interval. For example, we might write the following truncated Fourier series,

$$E^{torsion}(\omega) = V_1(\omega) + V_2(\omega) + V_3(\omega)$$

where ω is the torsion angle, and V_1, V_2, and V_3 are independent functions of ω that repeat every 360°, 180°, and 120°, respectively. These functions are referred to as one-fold, two-fold, and three-fold potentials.[1]

The different n-fold potentials are useful because each can be associated with a particular chemical phenomenon. For example, a one-fold potential describes the different energies of *anti* and eclipsed conformers of dimethylperoxide, while a two-fold potential describes the different energies of planar and perpendicular conformers of benzyl cation. Three-fold potentials, which are more familiar to chemists, describe internal rotation in molecules like ethane.

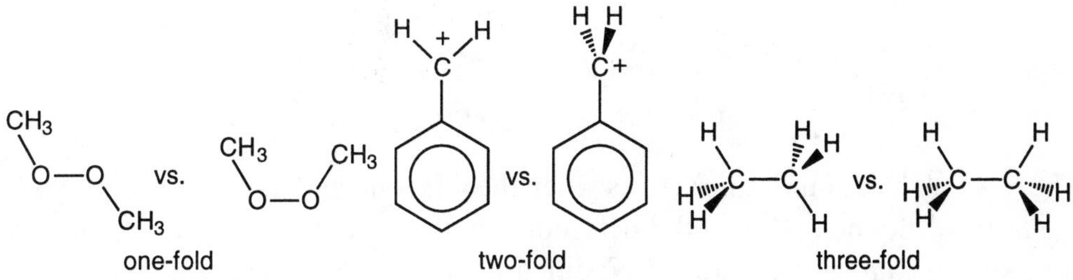

one-fold   two-fold   three-fold

While rotation in a symmetric molecule might be described using only one potential or a combination of two potentials, less symmetric molecules require more complex combinations of potentials. This is illustrated by fluoromethylamine.

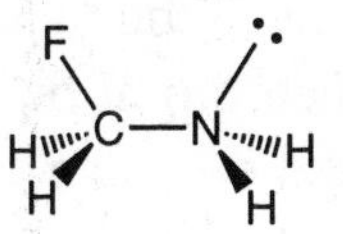

FCN: = 0°

---

1. A general formula for V_n is $k_n(1-\cos(n[\omega - \omega_{eq}]))$. For a description of the use of Fourier series in analyzing internal rotation, see: L. Radom, W.J. Hehre and J.A. Pople, *J. Am. Chem. Soc.*, **94**, 2371 (1972).

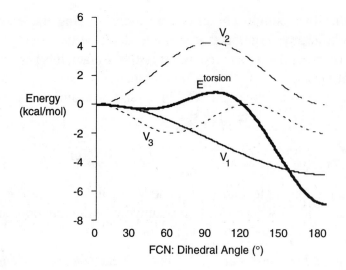

The heavy solid line in the figure describes $E^{torsion}$ for rotation about the CN bond. Note that the curve is relatively complicated, and clearly does not correspond to a simple one-fold, two-fold, or three-fold potential. There are two distinct minima.

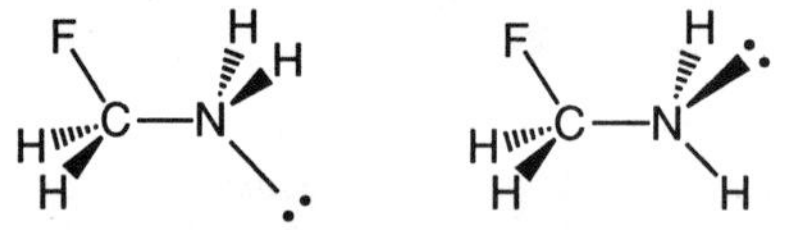

FCN: = 180°   FCN: ≈ 45°

The lower (global minimum) arises when the CF bond and the nitrogen lone pair are *anti*, while the higher and much more shallow minimum is almost, but not precisely, a *gauche* structure (FCN: dihedral angle ~45°). Also, note that one of the two energy maxima (FCN: dihedral angle ~115°) does not exactly correspond to an eclipsed structure.

The unusual behavior of $E^{torsion}$ becomes clearer when it is resolved into one-fold, two-fold, and three-fold components (in this case, the sum of these components provides a virtually perfect fit of $E^{torsion}$). The one-fold term reflects a clear and very strong preference for the CF bond and the nitrogen lone pair to be *anti* and not eclipsed. This preference might be electrostatic since the *anti* structure arranges the dipoles associated with the CF bond and nitrogen lone pair in opposite directions.

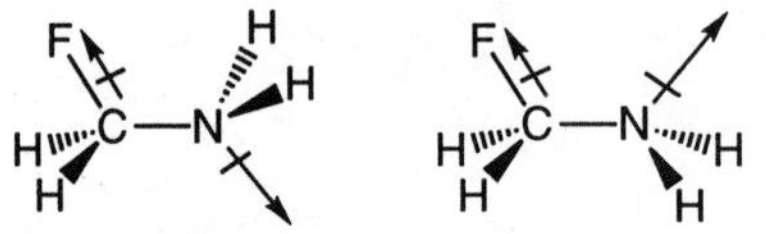

FCN: = 180°   FCN: = 0°
dipoles subtract   dipoles add

The three-fold term is also easy to explain. It reflects the preference for staggered over eclipsed structures. This terms contributes much less to $E^{torsion}$ than either the one-fold or two-fold terms, consistent with the low steric demands of the CH_2F and NH_2 groups.

What is most interesting perhaps, and what might not have been anticipated without this type of analysis, is the large contribution made by the two-fold potential. This potential reflects a strong preference for a planar arrangement of FCN:, and can be attributed to a stabilizing π interaction between the high-energy lone pair orbital on NH_2 and the low-energy CF σ^* orbital (see: **π Interactions Involving σ Electrons**). This interaction involves only two electrons and is stabilizing when the FCN: unit is planar, but vanishes when the unit is twisted into a perpendicular geometry, i.e.,

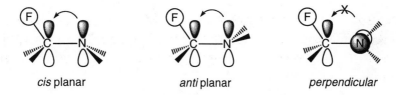

cis planar   anti planar   perpendicular

Most of the peculiar features of $E^{torsion}$ can now be attributed to either V_1 or V_2. V_1 accounts for the *anti* geometry being the global minimum. V_2, however, is responsible for the position of the maximum and the shift in the higher-energy minimum to smaller dihedral angles.

It is important to note that the terms that contribute to $E^{torsion}$ are completely independent of each other, and each may be treated as one part of a larger picture. Thus, the observation that electron donation from the nitrogen lone pair into the CF σ^* orbital is optimal when the two groups are planar is independent of the observation that the *cis* coplanar structure is destabilized, relative to the *anti* structure, by dipole-dipole interactions.

Such analyses, as provided above for fluoromethylamine, are not yet routine, but there is no longer any reason why they should not be.[2] What they do point out is that conformational preferences, even for simple systems, may arise from a combination of factors, and that molecular modeling may help to unravel these factors.

---

2. Operationally, Fourier component analysis requires just two steps, energy calculations on conformers with different torsion angles and least-squares fitting of the energy to a truncated Fourier series. The entire process can be automated.

Conformations of Alkanes

Semi-empirical AM1 calculations are used to construct energy profiles for rotation in n-butane and in 2-methylhexane. The latter provides information about the relationship between rotation barrier and degree of substitution.

Acyclic alkanes prefer conformations in which CH and CC bonds are staggered. Steric interactions between alkyl groups, however, can affect both the geometry and energy of these conformers. For example, the *anti* conformer of *n*-butane is known experimentally to be 0.8 kcal/mol more stable than the *gauche* conformer, a fact that can be rationalized if CH_3/CH_3 *gauche* interactions are larger than CH_3/H and H/H *gauche* interactions.[1]

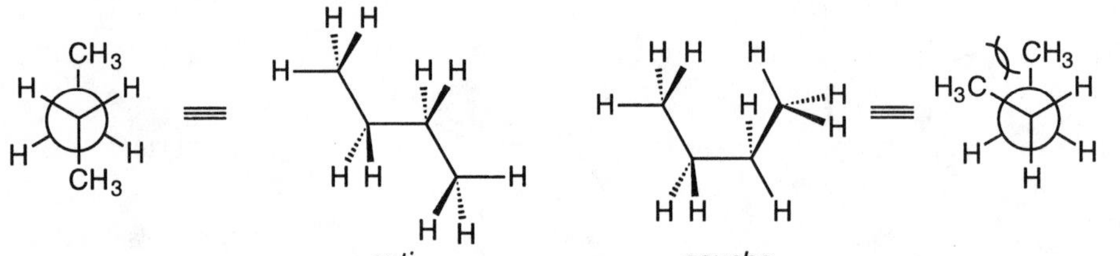

anti   gauche

The conformational behavior of larger alkanes is more difficult to predict, both because there are more degrees of conformational freedom, and because steric interactions between atoms separated by several bonds can be difficult to anticipate. 2-Methylhexane, for example, can experience internal rotation about any of its six CC bonds, and there may be as many as 3^6 staggered conformations.

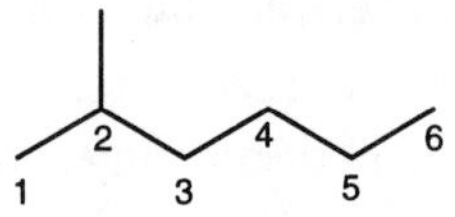

2-methylhexane

In this experiment, you will use AM1 calculations to map the energy for internal rotation about the C_2-C_3 bond in *n*-butane, and about the C_2-C_3 and C_3-C_4 bonds in 2-methylhexane. While the latter do not constitute a complete map of the energy surface for this molecule, they still provide interesting insight into the importance of "long range" steric interactions, and the difficulty of analyzing the conformational preferences of complex molecules.

Procedure

n-**Butane**: Build completely staggered conformers of *n*-butane in which the CCCC dihedral angle is 0°, 30°, 60°, ..., 180° (constrain this angle). Optimize the geometry of each constrained molecule using the AM1 method. Tabulate the final heat of formation for each conformer, and plot the heat of formation versus CCCC dihedral angle.

---

1. Carey and Sundberg A, p. 121, March, p. 122.

Which structures correspond to minima, and which correspond to maxima? One way a molecule might relieve unfavorable steric interactions is to expand its CCC bond angle; how do the CCC bond angles vary with CCCC dihedral angle? Make a plot. What can you infer from your data about the relative magnitude of CH_3/CH_3, CH_3/H, and H/H *gauche* interactions?

2-Methylhexane: Build completely staggered conformers of 2-methylhexane in which the $C_1C_2C_3C_4$ dihedral angle is 0°, 60°, 120°, ..., 300° (constrain this angle). Optimize the geometry of each constrained molecule using the AM1 method, and plot the heat of formation versus dihedral angle. Repeat this procedure using staggered conformers in which the $C_2C_3C_4C_5$ dihedral angle is constrained to be 0°, 60°, 120°, ..., 300°, and plot the heat of formation versus dihedral angle. Examine your plots and determine whether they have the expected symmetry (what does an unsymmetric plot indicate and how should it be corrected?). Identify the most stable conformer on each energy plot, and characterize its structure, i.e., is the carbon skeleton perfectly staggered or do some CCCC dihedral angles deviate from ±60° or 180°? Which bond, the trisubstituted C_2-C_3 bond, or the disubstituted C_3-C_4 bond, has a higher barrier to complete rotation? Try to rationalize your observations in terms of steric interactions.

Optional

Use AM1 calculations to map energy profiles for rotation about the C_2-C_3 and C_3-C_4 bonds in 2,2-dimethylpentane. How many minimum-energy conformers are there for each rotation? What is the barrier to complete 360° internal rotation about each bond? How do you account for these results?

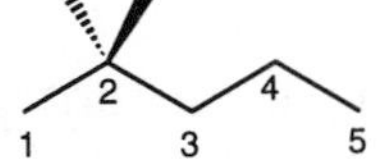

2,2-dimethylpentane

Conformations of Alkylcyclohexanes

Semi-empirical AM1 calculations are used to assign and interpret conformational preferences in alkylcyclohexanes.

Alkylcyclohexanes prefer a conformer in which the alkyl group is *equatorial*.[1] The *equatorial* conformer of methylcyclohexane is 1.7 kcal/mol lower in energy than the *axial* conformer, and the *equatorial-axial* difference increases to 5.4 kcal/mol in *tert*-butylcyclohexane.

R *equatorial*   R *axial*

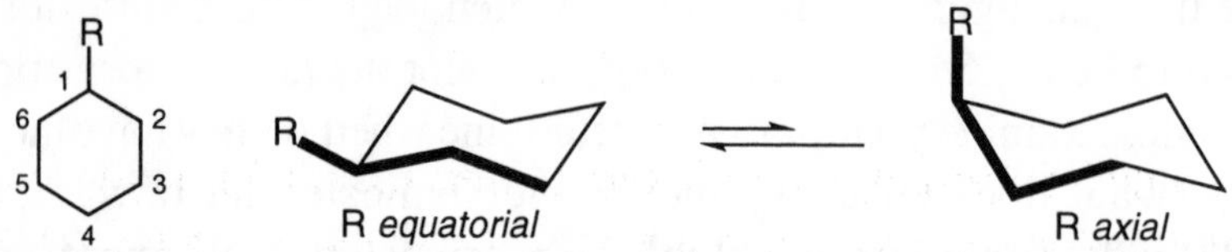

This preference can be attributed to steric interactions between the substituent and the ring (*gauche* interactions). The alkyl group is *anti* to C-3 and C-5 in the *equatorial* conformer, and *gauche* in the *axial* conformer. Liken this to *n*-butane, where steric interactions disfavor the *gauche* conformer by 0.8 kcal/mol relative to the *anti* conformer.

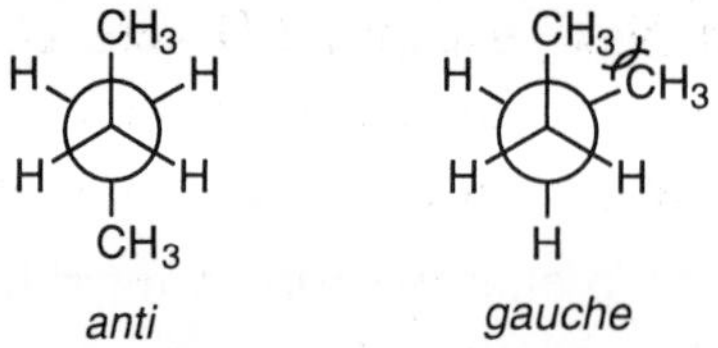

anti   *gauche*

Conformational preferences in disubstituted cyclohexanes can be predicted by adding together substituent effects, and significant deviations from additivity are indicative of interactions between substituents. For example, direct steric interaction between alkyl substituents in a 1,3-*diaxial* conformer might give an especially large preference for the diequatorial conformer.

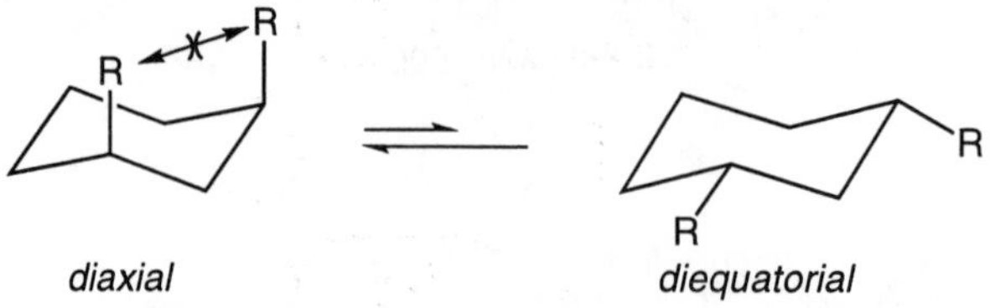

diaxial   *diequatorial*

In this experiment, you will first examine conformational preferences in methylcyclohexane and *tert*-butylcyclohexane, and then investigate additivity of substituent effects in *trans*-1,2- and *cis*-1,3-dimethylcyclohexanes.

Procedure

Build *equatorial* and *axial* methylcyclohexane, optimize using the AM1 semi-empirical method, and calculate and record the *equatorial-axial* energy

---

1. Carey and Sundberg A, p. 130; Lowry and Richardson, p. 138; March, p. 143.

differences. Repeat for *equatorial* and *axial tert*-butylcyclohexane. How do your results compare with the experimental energy differences? Examine the optimized models. What geometric features support or reject the explanation given above? (Cite specific changes in distances, angles, or dihedral angles.)

Build *diequatorial* and *diaxial* conformers of both *trans*-1,2-dimethylcyclohexane and *cis*-1,3-dimethylcyclohexane, and optimize using AM1. What would the conformer energy differences be if the two methyl groups did not interact? Account for any discrepancies. (Cite specific geometric features to support your argument.)

Optional

1. *Cis*-2-methyl-5-*tert*-butyl-1,3-dioxane prefers the conformation in which the methyl group is *equatorial* and the bulky *tert*-butyl group is *axial*.[2]

   Build both conformers, and optimize their geometries using AM1 semi-empirical calculations. Next, use the AM1 geometries to calculate 3-21G *ab initio* energies. What are the relative conformer energies obtained using AM1? 3-21G? Which method provides a better model for understanding the experimental results? How do you account for the unusual preference for an *axial tert*-butyl group? (Cite specific geometric features of your models that support your argument.)

2. Poly(tetrahydrofuranyl)cyclohexanes prefer conformations in which the methylene groups are *axial*.[3] For example, the X-ray crystal structure of **1** shows four methylene groups *axial* and two oxygens *axial*.

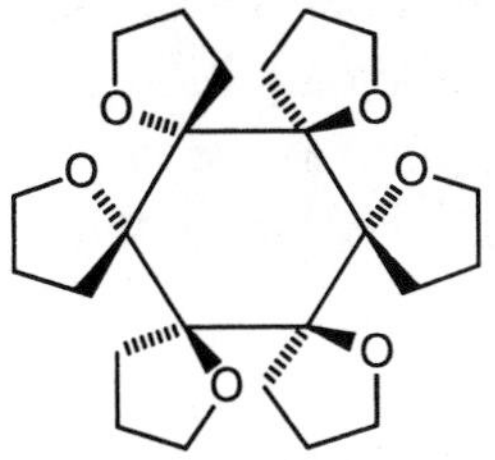

1

   Why is this unexpected? Hint: consider the relative sizes of methylene and oxygen. What is the reason for the observed conformational preference in **1**? (Draw both chair conformers and consider nonbonded interactions between the furan rings.) **No calculations are necessary**.

---

2. Carey and Sundberg A, p. 145; E.L. Eliel, M.C. Knoeber, *J. Am. Chem. Soc.* **90**, 3444 (1968).
3. L.A. Paquette, M. Stepanian, B.M. Branan, S.D. Edmonson, C.B. Bauer and R.D. Rogers, *J. Am. Chem. Soc.*, **118**, 4504 (1996).

Conformations of 1,3-Butadiene. A Price to Pay for Diels-Alder Cycloaddition

Ab initio 3-21G calculations are used to map the energy surface for internal rotation in 1,3-butadiene, and to determine the structure of its "cis" conformer. The "cis"-anti energy difference is compared to the known increase in activation energy for Diels-Alder cycloadditions involving 1,3-butadiene relative to the corresponding reactions involving cyclopentadiene.

The conformations of conjugated dienes, such as 1,3-butadiene, are governed by a combination of electronic and steric interactions. The preferred *anti* arrangement minimizes steric interactions while allowing the two π bonds to be coplanar and maximize conjugation.[1] The geometry of the higher-energy conformer is open to question. Is the carbon backbone planar in order to maximize conjugation, or is it twisted slightly in order to relieve steric congestion?

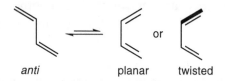

anti   planar   twisted

In this experiment, you will use 3-21G *ab initio* calculations to examine the energy for rotation about the CC single bond in 1,3-butadiene in order to establish the geometries of the two minimum-energy conformers, and their equilibrium ratio. You will also examine the effect of this equilibrium on the reactivity of 1,3-butadiene in the Diels-Alder reaction. It has been estimated that the barrier for cycloaddition of butadiene and acrylonitrile is approximately 10 kcal/mol higher than the barrier for cycloaddition of cyclopentadiene and acrylonitrile.[2] Cyclopentadiene is "locked" into a *cis* conformation, and the extra energy required for butadiene addition might be due to butadiene's *cis-anti* energy difference.

Procedure

Build conformers of 1,3-butadiene with the CCCC dihedral angles of 0°, 30°, 60°, ..., 180°. Optimize each using a 3-21G *ab initio* model (constrain the dihedral angle), record your data and plot energy relative to the most stable conformer versus dihedral angle.

1.   Carey and Sundberg A, p. 128. For a review of experimental and theoretical work, see: A.J.P. Devaquet, R.E. Townshend and W.J. Hehre, *J. Am. Chem. Soc.*, **98**, 4068 (1976).
2.   K.N. Houk, *Pure and Appl. Chem.*, **61**, 643 (1989).

What are the structures of the minimum-energy conformers? Is the *cis* conformer planar or twisted? How does the barrier separating the *cis* and *anti* conformers compare to that in an alkene (2.0 kcal/mol in propene, 1.7 kcal/mol in 1-butene)? What is the *cis-anti* energy difference? Can this energy difference account in full for the difference in activation energies for Diels-Alder cycloaddition of butadiene and cyclopentadiene? If your results suggest that it cannot, what other factors might affect the barrier? (See: **Optional** part 2.)

Optional

1. Isoprene and 2-methoxybutadiene are two commonly used Diels-Alder dienes. Use 3-21G *ab initio* calculations to identify their preferred conformers. How do the conformational preferences of these dienes compare to butadiene? What effect does the substituent have on the *"cis-anti"* energy difference, and how will this affect Diels-Alder reactivity? How else might the substituents affect Diels- Alder reactivity? (Make specific predictions, and provide specific reasons to support your predictions.)

isoprene   2-methoxybutadiene

2. Consider the importance of electronic effects on Diels-Alder reactivity differences between 1,3-butadiene and cyclopentadiene. Generate and compare electrostatic potential maps for *cis*-1,3-butadiene and for cyclopentadiene as obtained from 3-21G calculations. The more negative the electrostatic potential on the diene, the more an electrophile (the dienophile) will be attracted to the π system.[3] Display the maps on the same color scale, and measure the minimum (most negative) values of the electrostatic potential. Relate any differences in electrostatic potentials that you observe to known differences in reactivity of the two dienes.

3. Acrolein, like butadiene, has two conformers, *anti* and *cis* (or nearly *cis*).

trans   cis

Both are able to participate in the role of a dienophile in Diels-Alder cycloadditions, but there is evidence that the predominant pathway involves the *cis* conformer. Explain.

---

3. S.D. Kahn, C.F. Pau, L.E. Overman and W.J. Hehre, *J. Am. Chem. Soc.*, **108**, 7381 (1986), See also: **Graphical Models and Graphical Modeling**.

Conformations of Hydrazine. Keeping Lone Pairs Out of Each Other's Way

6-31G* ab initio calculations are used to map the energy surface for internal rotation in hydrazine, and to determine the structures of its minimum-energy conformers. The shape of the energy surface is compared to trends in the energies of nonbonding orbitals, and the energy difference between these orbitals. An optional experiment shows how the conformations of larger hydrazines might be determined experimentally using observed orbital-energy splittings and AM1 semi-empirical calculations.

The same guidelines that tell us that alkanes prefer staggered conformations can also be applied to the conformational analysis of molecules containing heteroatoms. Thus, hydrazines, R_2NNR_2, prefer conformations in which the lone pairs and bonds are staggered.

Hydrazine has two "staggered" conformers, one with the nitrogen lone pairs (:) *anti* and the other with lone pairs *gauche*.[1]

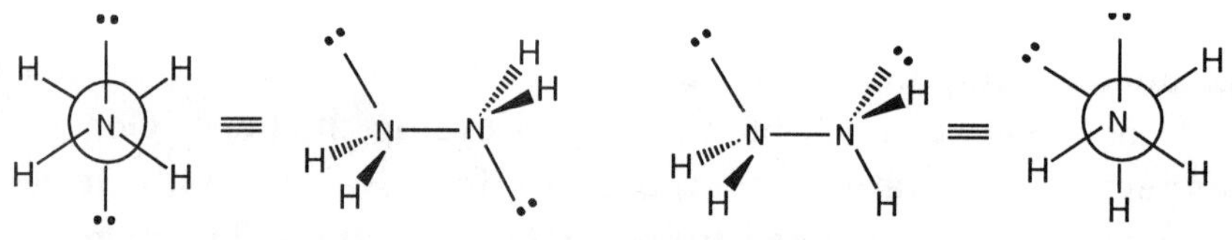

lone pairs *anti*   lone pairs *gauche*

Interactions between the two lone pair orbitals gives rise to two "nonbonding" molecular orbitals of different energy. The energies of these orbitals, as well as the energy gap between them, depends on the hydrazine conformation. The weakest interaction, and the smallest energy gap, should occur when the two lone pairs are roughly orthogonal, i.e., in the *gauche* conformer. Therefore, accepting the usual suggestion that lone pair-lone pair interactions are more important than lone pair-bond and bond-bond interactions, it is to be expected that the *gauche* conformer will be the preferred conformer.

In this experiment, you will use *ab initio* 6-31G* calculations to examine both the energy surface for internal rotation in hydrazine, and the behavior of the two nonbonding orbitals. The goal will be to establish how these orbitals affect conformational preferences.

---

1.   S.F. Nelsen in **Molecular Structure and Energetics**, vol. 3, J.F. Liebman and A. Greenberg, Eds., VCH Publishers, Deerfield Beach, Fl., 1986, p.1.

Procedure

Construct hydrazine in a conformation in which the nitrogen lone pairs are *gauche* and optimize using 6-31G* *ab initio* calculations. Make seven copies of the optimized structure and, without changing any other geometrical parameters, set the :NN: dihedral angle to 0°, 30°, 60°, ..., 180°, respectively.[2] Calculate the energy of each using the 6-31G* *ab initio* model (**do not optimize the geometry**), and determine the energy of each molecule relative to the lowest-energy molecule. Record your data, plot relative conformer energy versus :NN: dihedral angle, and identify any minimum-energy conformers. (You may need to perform additional calculations to refine your plot.) What are their structures? How does the barrier for internal rotation in hydrazine compare to 6-31G* barriers in ethane (2.9 kcal/mol) and methylamine (2.0 kcal/mol)? What, if anything, does this tell you about the importance of lone pair-lone pair interactions relative to bond-bond and bond-lone pair interactions?

Examine the energies of the two highest-energy occupied orbitals in each molecule for each conformer, then calculate the energy difference between these orbitals. This energy gap should be small if the lone pairs interact weakly, but should increase if there is a strong lone pair-lone pair interaction. What structures on the conformational energy surface correspond to the weakest and strongest lone pair-lone pair interactions? Are these interactions stronger in the *anti* or *gauche* conformers? To what extent can you account for the shape of the energy surface, and the conformational preferences of hydrazine, in terms of lone pair-lone pair interactions?

Calculate and display the two highest-energy occupied orbitals in the molecule where lone pair-lone pair interactions appear to be weakest, and in the molecule where the interactions appear to be strongest. Are the shapes of these orbitals consistent with their energy splitting? Explain. Be sure to cite specific aspects of the orbital shapes that support your conclusion.

Optional

1. Using the 6-31G* energy data, perform a Fourier component analysis on hydrazine (see: **Interpreting Conformational Preferences**). Plot the one, two and three-fold energy contributions on the same graph (and on the same scale) as the total energy. Provide interpretations for the individual terms. Which, if any, of the terms dominate the overall energy profile? Comment on the relative importance of the term describing lone-pair-lone-pair interactions and that favoring staggered as opposed to eclipsed arrangements.

---

2. You cannot "see" the nitrogen lone pairs in order to define the :NN: dihedral angle. What you can do is define "points" midway between the hydrogens for each of the NH_2 groups, and then to use these points to define the :NN: dihedral angle.

2. As shown below, the energies of the two highest-occupied molecular orbitals in hydrazine relate to their first and second ionization potentials, which may be measured using photoelectron spectroscopy.[1]

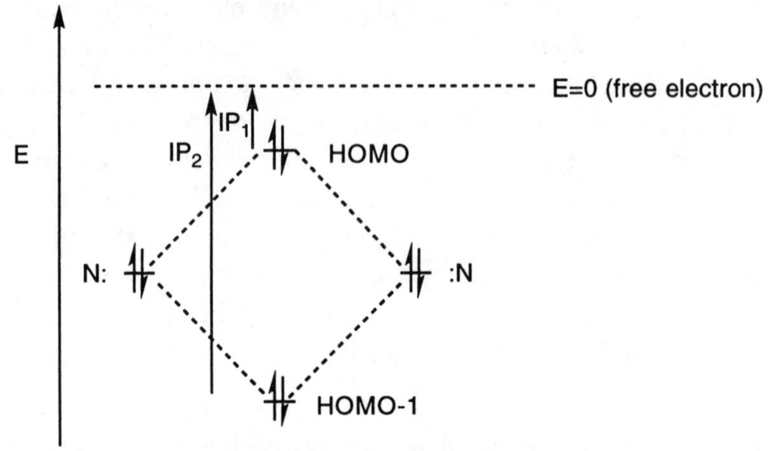

Differences in ionization potentials ($IP_2 - IP_1$) should relate to orbital energy splittings, and hence provide information about conformation. You can test this by examining a number of "rigid" hydrazines for which experimental ionization potentials are available.

Build, and optimize using AM1 semi-empirical calculations, hydrazines **1 - 6**.

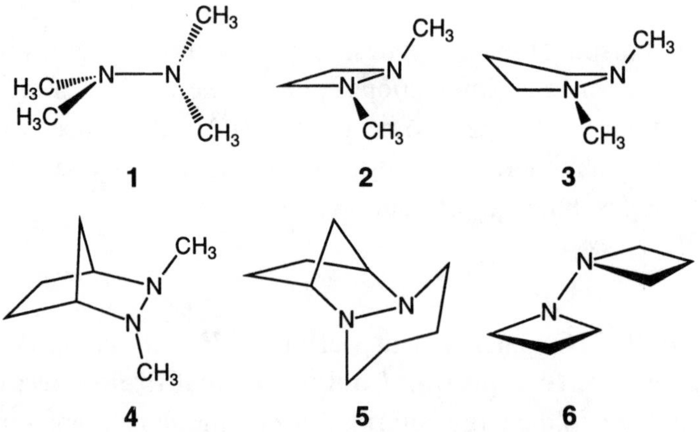

Position the lone pairs in **1** and **6** approximately orthogonally (90°) to start.[3] Plot the difference in orbital energies vs. $IP_2 - IP_1$. Is there a relationship? Plot the difference in orbital energies vs. :NN: dihedral angle. Does it parallel the plot previously constructed for hydrazine?

---

3. This angle cannot, of course, be measured as the exact positions of the nitrogen lone pairs are not known. The next best thing is to obtain an approximate angle based on the bisectors of the respective NR_2 groups.

3. You might anticipate that there will be a relationship between orbital energy splitting and proton affinity of hydrazines. All other factors being equal, it is reasonable to expect that the hydrazine with the highest-energy lone pair combination (largest splitting) will have the greatest proton affinity. Calculate proton affinities of hydrazines **1-5** using the AM1 model. You will need to combine your data for neutral molecules with new calculations for the protonated systems. You don't need to calculate absolute proton affinities but rather proton affinities relative to a standard. Do you observe a correlation? Try to account for any large deviations.

VII

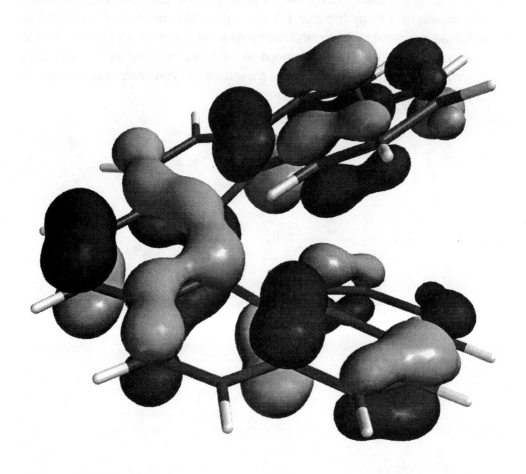

π Interactions Involving
σ Electrons

C hemists are accustomed to partitioning electrons into two sets, σ and π, and attributing different properties to each set. The π electrons are assumed to interact strongly via resonance, and are deemed responsible for most of a molecule's "chemistry", while σ electrons are assumed to be localized between specific pairs of atoms. An example of this follows from the usual interpretation of the resonance structures for 4-methylbenzaldehyde.

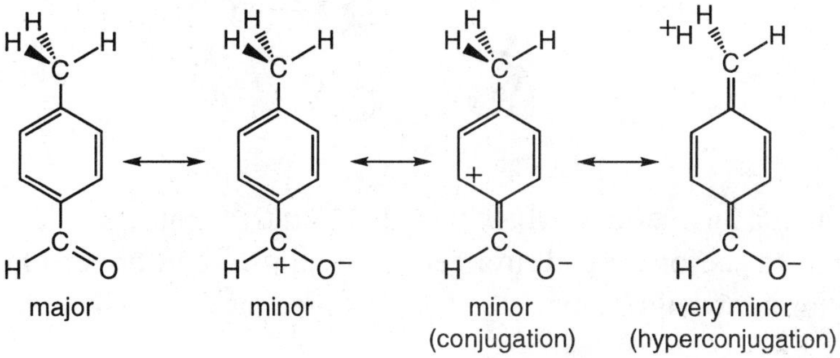

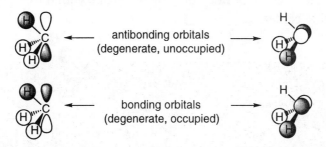

| major | minor | minor
(conjugation) | very minor
(hyperconjugation) |

Resonance interactions are assumed to occur between the π electrons on benzene and on the formyl group (conjugation), but interactions between the ring and the σ electrons of the methyl group (hyperconjugation) are usually ignored.

Of course, from the point of view of quantum mechanics, there is no difference between σ and π electrons, and molecular orbital theory takes an entirely different view of electronic structure. σ and π molecular orbitals are both fully delocalized, and π interactions can involve both types of electrons. A few examples of this are provided below.

The molecular orbitals of the methyl group include the four "π"orbitals shown below[1] (see: **Visualizing Molecular Orbitals**).

← antibonding orbitals →
(degenerate, unoccupied)

← bonding orbitals →
(degenerate, occupied)

The orbitals can be viewed as σ orbitals because they describe CH bonding interactions; the filled orbitals are CH σ orbitals and the empty orbitals are σ*. However, the orbitals are also π orbitals in that they have a nodal plane that contains the CX axis in CH_3X, and they provide a mechanism for methyl groups to engage in CX π bonding.

---

1. The term "π" designates an orbital with a single nodal plane, to be distinguished from a σ orbital which does not have a nodal plane.

Consider the π interactions between filled methyl orbitals in ethane. Four new filled π orbitals are formed, two of which are CC "π bonding" and two of which are CC "π antibonding". While these tend to cancel, i.e., there is no change in the CC bond order, the interaction is energetically destabilizing.[2] This can affect the molecule's choice of conformation. To see how, examine one of ethane's "π antibonding" orbitals for the staggered and eclipsed forms.

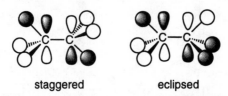

staggered   eclipsed

The antibonding interactions, which include HH interactions, are clearly weaker in the staggered conformer and stronger in the eclipsed conformer. On this basis alone, the former would be expected to be favored, as of course it is.

Similar considerations can account for the conformational preferences of propene, $CH_3CH=CH_2$, and acetaldehyde, $CH_3CH=O$. In each case, the molecule prefers a conformer in which "π antibonding" between the methyl and the π orbital on the CC or CO bond is minimized. This turns out to be the conformer in which one of the methyl hydrogens eclipses the double bond (see: **Conformational Isomerism in Alkenes**).

π interactions between filled and unfilled orbitals, which can stabilize a molecule, are most influential when there is a good energy match between the methyl and the π orbital on the group to which it is attached. This will occur when the π orbital is particularly electron rich (high-energy filled orbital) or particularly electron poor (low-energy unfilled orbital). An example of the latter is the interaction of the filled "π orbital" of a methyl group with the empty C 2p orbital on a carbocation center, i.e.,

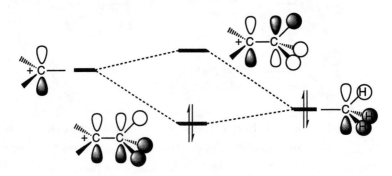

---

2. As a rule, antibonding orbitals are destabilized more than bonding orbitals are stabilized. Thus, if both sets of orbitals are filled, there is net destabilization.

The filled orbital is stabilized by π bonding between the two carbons. This bonding also transfers electron density from the methyl group to the carbocation, and leaves the methyl group with a small positive charge.

Substituted methyl groups also include filled and empty π type orbitals in their descriptions, although the substituent(s) may break the plane of symmetry. Interactions with π orbitals on groups that are directly bonded may lead to a number of interesting effects, among them the "anomeric effect" (see: **The Anomeric Effect**). In this case, the attached group (X) contains a π-type lone pair which may be able to interact favorably with a π* orbital on the substituted methyl group. If the substituent (Y) is electronegative, then the π* orbital will largely be concentrated along the CY bond and be polarized significantly toward C. Thus, the interaction depends on aligning the lone pair and the CY bond.

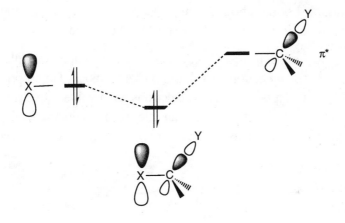

Conformational Isomerism in Alkenes

Ab initio 3-21G calculations are used to identify the preferred conformers of 1-butene and methyl vinyl ether.

Propene adopts a conformation in which one of the CH bonds on the methyl group eclipses the CC double bond.[1]

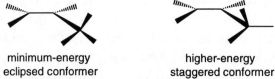

| minimum-energy eclipsed conformer | higher-energy staggered conformer |

The barrier to rotation through an unstable conformation in which a CH bond staggers the double bond is about 2 kcal/mol, comparable in magnitude to barriers for single bond rotation in ethane (2.9 kcal/mol) and propane (3.4 kcal/mol).

This preference for eclipsing arises because one of the valence molecular orbitals of a methyl group (see: **π Interactions Involving σ Electrons**) is of appropriate symmetry to interact with the CC π bond.[2] As four electrons are involved (two from the methyl group and two from the π orbital), such an interaction is energetically destablizing and needs to be minimized. This is best accomplished in an eclipsed geometry, i.e.,

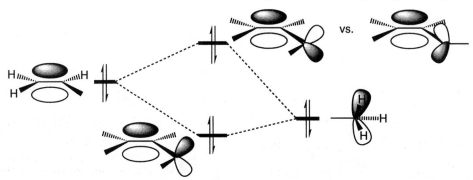

The qualitative argument suggests that the CC single bond in a molecule like 1-butene or the CO single bond in a molecule like methyl vinyl ether will also prefer to eclipse the CC double bond. However, it is not obvious which conformers will be global minima. In this experiment, you will use 3-21G *ab initio* calculations to survey the conformational energy profiles for 1-butene and methyl vinyl ether and try to account for the noted preferences.

Procedure

1-Butene: Build conformers for 1-butene in which the CCCC dihedral angle is fixed approximately at 0°, 30°, 60°, ..., 180°, and in which the CH bonds in the

1.   Carey and Sundberg A, p. 126.
2.   W.J. Hehre and L. Salem, *J. Chem. Soc., Chem. Commun.*, 754 (1973).

ethyl group are staggered. Optimize each conformer using the 3-21G *ab initio* model (constrained optimization), record your data and plot energy versus CCCC dihedral angle. Identify all of the minimum-energy conformers. Are they eclipsed like propene? Which conformation is the most stable? What are the barriers to rotation? Can you account for your results in terms of steric interactions alone, or do other factors appear to be operating?

Methyl vinyl ether: Build conformers for methyl vinyl ether in which the CCOC dihedral angle is fixed approximately at 0°, 30°, 60°, ..., 180°. Optimize each using the 3-21G *ab initio* model (constrained optimization), record your data, plot energy versus CCOC dihedral angle, and identify all minimum-energy conformers. Compare conformational preferences noted here to those in propene and in 1-butene. Examine the highest-occupied molecular orbitals for the minimum-energy conformers, and record their energies. Are these results consistent with the orbital interaction arguments advanced for propene? Explain.

Optional

1. What is the preferred conformation of *cis*-2-butene? Do CH bonds on both methyl groups eclipse the double bond, or do one or both "twist" to relieve the unfavorable nonbonded HH interaction?

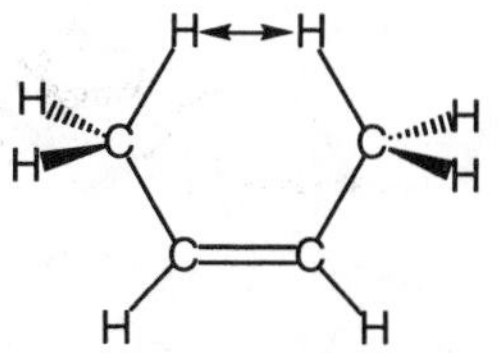

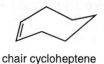

cis-2-butene

    Optimize *cis*-2-butene using the 3-21G method starting from the conformer you think is likely to be preferred (do not impose symmetry).

2. The conformational preferences of cycloalkenes are determined by the interplay of different factors: the staggering of single bonds, the eclipsing of single and double bonds, and steric interactions between nonbonded groups. Build cycloheptene in the chair conformer (shown below), and optimize its geometry using the 3-21G *ab initio* model. Does the calculated structure suggest that certain conformational factors are more important than others? Explain. Are your conclusions biased by ring size, i.e., would you expect to see similar behavior in cyclohexene or cyclooctene? (Support your arguments by calculating the structures of these molecules.)

chair cycloheptene

The Mechanism of Cyclohexane Ring Inversion

Semi-empirical AM1 calculations and ab initio 3-21G calculations are used to construct a mechanism for ring inversion in cyclohexane.

Cyclohexane preferentially exists in a chair conformation in which all six carbons are identical, and in which the twelve hydrogens are split into two chemically distinct sets: *equatorial* and *axial*.[1] Ring inversion, a rapid process at room temperature, causes the two sets of hydrogens to exchange roles, and NMR analysis of the hydrogen signal at various temperatures indicates that the barrier to ring inversion lies in between 10.1 and 10.5 kcal/mol.

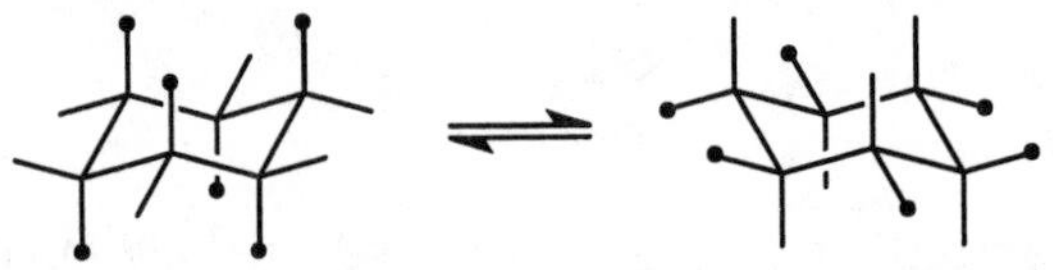

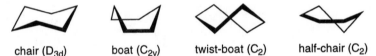

It is believed that inversion is a stepwise process involving several intermediates and transition states. Some of the possible structures on the inversion pathway are shown below (with symmetry point groups given in parentheses).

chair (D_{3d})   boat (C_{2v})   twist-boat (C_2)   half-chair (C_2)

In this experiment, you will use AM1 semi-empirical calculations to investigate the properties of these species, and to construct a mechanism for ring inversion. The chair and half-chair conformers are known to be minimum-energy and transition state structures, respectively, and you will use the calculated vibrational frequencies of the boat and twist-boat conformers to determine their properties. A minimum-energy structure must have only real frequencies, while a transition state must have exactly one imaginary frequency (see: **Finding Transition States**).

Procedure

Build chair, boat, and twist-boat conformers of cyclohexane (preferably with the symmetries listed above), and optimize their geometries using AM1 semi-empirical models.[2,3] To get a better estimate of the relative conformer energies,

---

1. Carey and Sundberg A., p. 130; Lowry and Richardson, p.138; March, p.143.
2. A "boat" conformer can be built by first building and minimizing a model of bicyclo[2.2.1]heptane (norbornane), a molecule with C_{2v} symmetry, and then deleting the one carbon bridge.
3. A "twist-boat" conformer can be built by adding *equatorial* methyl groups onto C_1 and C_4 of "boat" cyclohexane,[2] minimizing the model, deleting the methyl groups and again minimizing. The resultant structure should have C_2 symmetry.

perform single-point energy 3-21G *ab initio* calculations. Identify the most stable conformer, and compute all of the conformer energies relative to this conformer.

Calculate and display the vibrational frequencies of the chair, boat and twist-boat conformers. Identify each conformer as a minimum-energy structure or transition state.

Build a half-chair conformer of cyclohexane, and use this model to search for the closest AM1 transition state.[4] Calculate and display its vibrational frequencies to verify that it is a transition state. Perform a single-point 3-21G calculation on half-chair cyclohexane to get a better estimate of its energy relative to that of the most stable conformer.

Now that you have identified the transition states that might be involved in the overall mechanism of ring inversion, inspect the "imaginary" vibrational mode in each transition state. This motion should indicate which minimum-energy structures are connected by each transition state. Use this information to determine the mechanism of ring inversion. Finally, construct a reaction coordinate diagram for ring inversion. Plot energy against conformation, and clearly indicate the sequence of conformational changes required for ring inversion. What is the calculated barrier to ring inversion? How does this value compare with the experimentally observed value?

Optional

Use AM1 models to construct a mechanism, and estimate the barrier, for interconversion of *axial* and *equatorial* conformers of chair methylcyclohexane.[5] Write your mechanism using a reaction coordinate diagram on which you clearly indicate the energy of each intermediate and transition state relative to the chair conformer of *equatorial* methylcyclohexane. How does the methyl group change the conformational preferences of the cyclohexane ring? How does the methyl group affect the barrier for ring inversion? Does it "lock" the ring, i.e., prevent or significantly slow ring inversion?[5]

---

4. A reasonable "half-chair" starting structure can be obtained by using the linear synchronous transit method to find a structure midway between the chair and twist-boat conformers.
5. You will need to construct several isomers of each ring conformer in order to construct a complete inversion mechanism. The simplest way to build the necessary conformers is to build the appropriate conformer of cyclohexane and then add a methyl group to the desired position.

The Anomeric Effect

Semi-empirical AM1 calculations are used to compare the conformational preferences of substituted cyclohexanes and tetrahydropyrans. These differences are rationalized in terms of the anomeric effect.

Glucose exists in aqueous solution as a mixture of two diastereomeric cyclic hemiacetals, known as anomers.

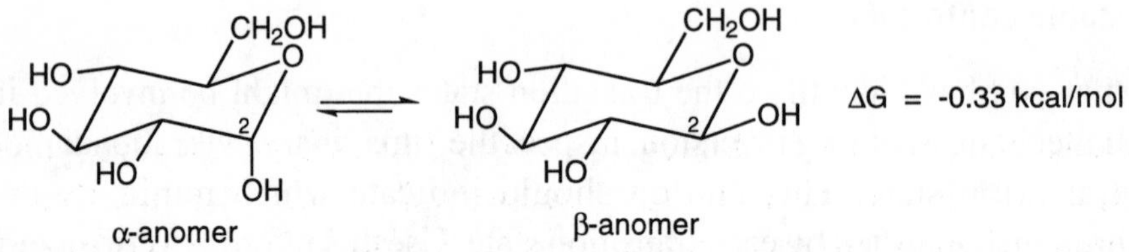

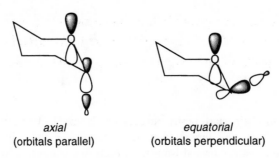

The OH group at C-2 is *axial* in the minor anomer, α-glucose, and is *equatorial* in the major anomer, β-glucose. This agrees with the general observation that substituents on six-membered rings prefer *equatorial* positions. However, the *axial* or α anomer is actually preferred in many other glucose derivatives, e.g.,

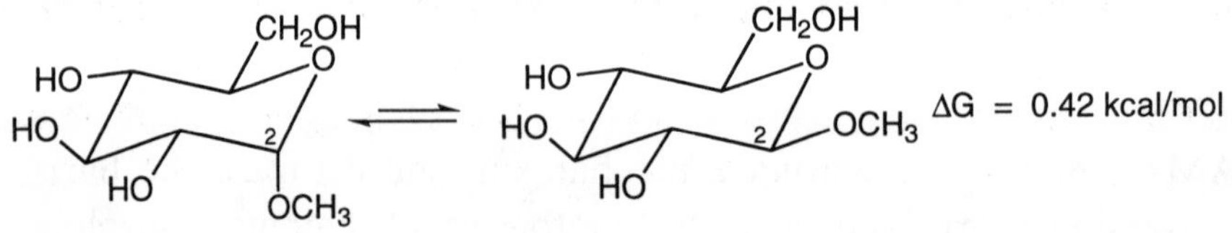

This unusual preference for *axial* substitution is due to the "anomeric effect",[1] and may be rationalized in terms of a geometry-dependent interaction between the frontier molecular orbitals (see: **π Interactions Involving σ Electrons**). In the case of glucose-like molecules, this involves a stabilizing interaction between the higher energy (π-type) lone pair on the ring oxygen and the σ* orbital localized on the (C-2)-O bond. This is much more favorable when the CO bond is *axial* (the two orbitals are nearly parallel) than when it is *equatorial* (the two orbitals are approximately perpendicular), i.e.,

axial
(orbitals parallel)

equatorial
(orbitals perpendicular)

---

1. Carey and Sundberg A, p. 146; March, p. 128.

In this experiment, you will use semi-empirical AM1 calculations to compare conformational preferences of chlorocyclohexane and 2-chlorotetrahydropyran.

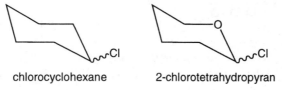

chlorocyclohexane   2-chlorotetrahydropyran

The latter serves as a model for glucose-like molecules and exhibits an anomeric effect (the *axial* conformer is preferred by 1.8 kcal/mol).

Procedure

Build *axial* and *equatorial* chlorocyclohexane and 2-chlorotetrahydropyran, and optimize their geometries using AM1 semi-empirical calculations. Determine the *equatorial-axial* energy difference for each molecule. Which conformer is more stable for each system? Are your results consistent with experiment? (Experiments show that *equatorial* chlorocyclohexane is preferred by 0.5 kcal/mol.) What assumptions must be made in order to compare experimental ΔG values with AM1 ΔH values?

Are the C-O and C-Cl bond lengths in the *axial* and *equatorial* conformers of 2-chlorotetrahydropyran abnormal? (Take the AM1 C-O bond length in tetrahydropyran, 1.423Å, and your AM1 C-Cl bond lengths in chlorocyclohexane, as "normal" values.) Given that interaction between the oxygen lone pair and the CCl σ^* orbital should result in π bonding (primarily in the *axial* conformation) you might expect bond length changes. Are your results consistent with these expectations?

Optional

1. The anomeric effect has also been rationalized in terms of dipole-dipole interactions, i.e., the favored conformer should have the lower dipole moment. Are the dipole moments for *axial* and *equatorial* 2-chlorotetrahydrofuran consistent with this hypothesis?

2. As expected, a methoxy group in the 2 position in tetrahydropyran prefers to be *axial*, while a methyl group prefers to be *equatorial*. Surprisingly, a methylamino group favors the *equatorial* position.[2] Explain.

2.   P. Deslongchamps, **Stereoelectronic Effects in Organic Chemistry**, Pergamon Press, Oxford, 1983.

Bond Angles in AH_3 Molecules. Walsh's Rules

Ab initio 6-31G* calculations are used to obtain bond angles in AH_3 molecules in order to test Walsh's rules for predicting equilibrium geometries.

Examination of key molecular orbitals may often provide insight into molecular geometry. For example, Walsh showed that the geometry of AH_3, where A is a main-group element, is correlated with the energy of the highest-energy occupied orbital (HOMO).[1] The HOMO for planar and pyramidal NH_3 are shown below, and it can be seen that NH_3 prefers the geometry with the lower HOMO energy (the orbital energy can be inferred from the presence or absence of AH and HH bonding interactions; see: **π Interactions Involving σ Electrons**). BH_3, by comparison, contains two fewer electrons and has a different HOMO (not shown). This orbital has a lower energy in the planar geometry and this is the preferred form for BH_3.

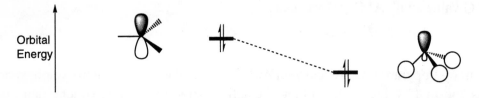

In this experiment, you will use 6-31G* *ab initio* calculations to establish the orbital energy ordering in ammonia at its optimal (pyramidal) and planar geometries. You will also use Walsh's analysis, supported by 6-31G* calculations, to interpret the variation in bond angle in an isoelectronic series (CH_3^-, NH_3, H_3O^+), and to explain the calculated bond angles in odd-electron species, such as $NH_3^{\cdot+}$ and $PH_3^{\cdot+}$.

Procedure

Ammonia: Build ammonia and optimize its geometry using 6-31G* *ab initio* calculations. Next, build ammonia as a planar (D_{3h}) species and optimize its 6-31G* geometry subject to this symmetry constraint. Record the energy of each molecule, and calculate the energy difference. Which form is more stable? Record the HNH bond angle. Calculate and display the four highest-energy occupied molecular orbitals for each molecule, and record their orbital energies. Use your results to assess Walsh's rule. Does the HOMO energy change in the same way as the total energy? Is the HOMO energy change larger than the other orbital

---

1.   A.D. Walsh, *Prog. Stereochem.*, **1**, 1 (1954). More precisely, Walsh argued that the change in geometry from BH_3 to NH_3 should be in the direction planar to pyramidal.

energy changes? Do the energies of the three underlying molecular orbitals change significantly with puckering of the nitrogen center? If so, how might the observed changes affect the simple picture based on changes in HOMO energy alone?

Methyl anion, hydronium ion: Build CH_3^- and H_3O^+ and optimize their 6-31G* geometries. Record their energies and bond angles. Build planar (D_{3h}) structures for each molecule and optimize their 6-31G* geometries subject to this symmetry constraint. Record the energy of each molecule, and calculate the pyramidal-planar energy difference. (Note: include the results for NH_3 when you answer the following questions.) What is the relationship between energy difference and bond angle? Record the energies of the four highest-energy occupied and the lowest-unoccupied molecular orbitals. Perturbation theory suggests that pyramidalization stabilizes the planar HOMO by mixing it with the planar LUMO, and that the degree of stabilization is inversely related to the HOMO-LUMO energy gap.[2] Is this argument supported by your calculated orbital energies? Can this argument account for the shapes of the HOMO and LUMO in the pyramidal geometry?

Ammonia and phosphine radical cations: The 6-31G* HOMO-LUMO energy gaps in planar NH_3 and PH_3 are 0.611 and 0.465 a.u., respectively. Build models of $NH_3^{\cdot+}$ and $PH_3^{\cdot+}$, and optimize their 6-31G* geometries.[3] Record the energy and bond angle in each molecule. Use Walsh's analysis, perturbation theory (see above), and the HOMO-LUMO energy gaps to explain the structures of the radical cations.

---

2. T.A. Albright, J.K. Burdett and M.-H. Whangbo, **Orbital Interactions in Chemistry**, Wiley, New York, 1985, p. 140.
3. These molecules are doublets (spin multiplicity = 2) and need to be handled using so-called unrestricted Hartree-Fock (UHF) models.

Configurational Stability of Amines

Semi-empirical AM1 calculations and ab initio 6-31G* calculations are used to explore the configurational stability of substituted amines. The trends are interpreted in terms of π interaction between the substituent and nitrogen and steric constraints imposed on cyclic amines.

Amines with three different nitrogen substituents, e.g., methylethylamine, can exist in either of two enantiomeric configurations. These enantiomers cannot be separated for most amines, however, because the two isomers rapidly invert their configuration via a planar transition state. Therefore, these amines exist mainly as racemic mixtures.[1]

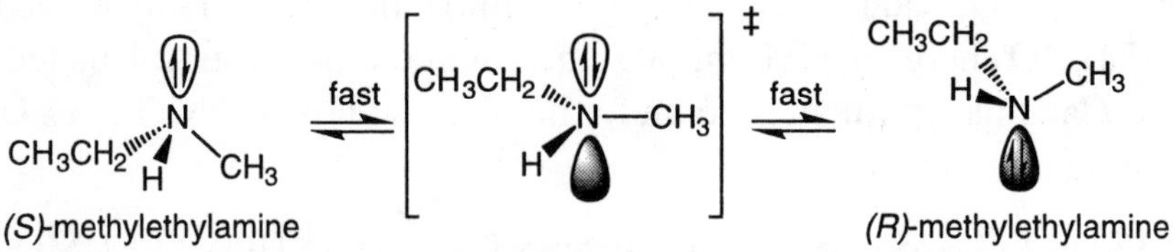

(S)-methylethylamine   (R)-methylethylamine

Although amine inversion is generally rapid, substituents can and do influence inversion rates, and in some cases inversion can be made sufficiently slow that the different enantiomers can be observed and even isolated. One way in which substituents might affect inversion rates is electronic. Inversion changes the hybridization of the nitrogen lone pair orbital from a low-energy sp^3 hybrid to a higher-energy p orbital. The energy of the molecule seems to parallel the energy of the lone pair, and substituents that affect the energy of this orbital might be expected to affect the energy of the molecule in an analogous fashion. An extreme case of this type of orbital interaction is seen in "amines" with π acceptor substituents, such as C≡N, C(=O)R, and NO_2. These "amines" are almost always planar or nearly so, presumably because the substituent's low-energy π* orbital stabilizes the nitrogen lone pair more in the planar geometry than in the pyramidal geometry. The same reasoning suggests that π donor substituents, such as NR_2, OR, and halogen, will raise the inversion barrier. Interaction between the donor substituent's filled π orbital and the nitrogen lone pair will be greatest in the planar geometry, and will destabilize this geometry relative to the pyramidal (see: **A Molecular Orbital Description of Substituent Effects**).

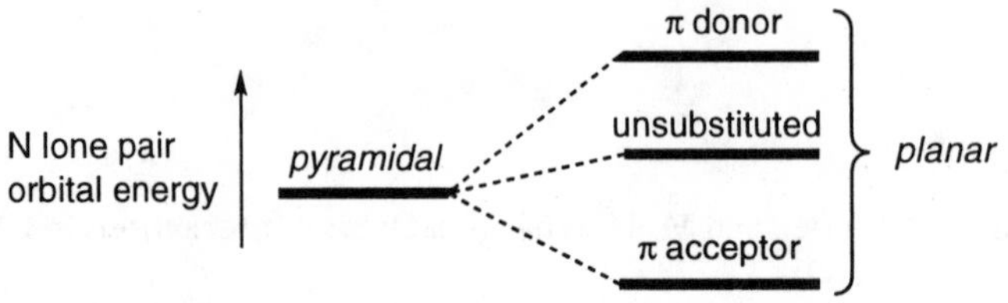

---

1.   March, p. 98.

Another type of substituent effect is seen in the inversion of cyclic amines. The barrier depends on ring size, and is largest in aziridines. For example, the inversion barrier in 1,2,2-trimethylaziridine is 20 kcal/mol, while the barrier in *N*-methylpyrrolidine is only 7 kcal/mol.[2] This behavior can be explained by the fact that inversion requires expansion of the C_{ring}-N-C_{ring} bond angle, and this becomes increasingly difficult in smaller rings.

1,2,2-trimethylaziridine   *N*-methylpyrrolidine

In this experiment, you will use semi-empirical AM1 calculations to explore the relationship between orbital energy and amine configuration, to examine the effect of different substituents on amine inversion barriers, and to examine the effect of ring size on amine inversion barriers.

Procedure

Orbital energy and inversion barrier: Build pyramidal and planar ammonia, and optimize their geometries using AM1 semi-empirical calculations (note: the planar geometry is not a true energy minimum, therefore, it is necessary to build this molecule with planar (D_{3h}) symmetry, and optimize its geometry subject to this symmetry constraint). Determine the inversion barrier. How does this compare to the experimental barrier of 5.9 kcal/mol? Perform single-point energy calculations on planar and pyramidal ammonia using the 6-31G* *ab initio* method (AM1 geometries). Is there a significant change in inversion barrier from that obtained using AM1 calculations?

Ammonia has four valence orbitals (three orbitals describing the NH bonds and one nonbonding orbital). Obtain the energies of these orbitals for both pyramidal and planar forms of ammonia. Make two "stacks" of orbital energies (AM1 or 6-31G*) side-by-side on the same graph, i.e.,

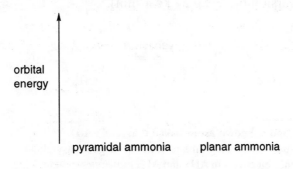

2.   J.B. Lambert, W.L. Oliver, *J. Am. Chem. Soc.*, **91**, 7774 (1969).

Find orbitals that have the same (or similar) shapes at each geometry, and connect these points on your graph so that your graph contains four line segments (note: the easiest way to match orbital shapes is to match orbital surfaces for each geometry).[3] Which orbital energy is most sensitive to bond angle? Is this orbital a bonding, nonbonding, or antibonding orbital? Use these results to critically evaluate the argument given above that inversion barriers are defined by changes in the lone pair orbital energy.

Electronic substituent effects: Build pyramidal and planar trimethylamine, $N(CH_3)_3$, trifluoroamine, NF_3, and tricyanoamine, $N(CN)_3$, and optimize their geometries using AM1 semi-empirical calculations (note: the planar geometries may or may not be a true energy minimum, therefore, build these molecules with planar (D_{3h}) symmetry and optimize their geometries subject to this symmetry constraint). Calculate the pyramidal-planar energy difference for each system (include your previous results for ammonia). Also, obtain the bond angles at nitrogen in the pyramidal geometry. Is this bond angle correlated with the size of the inversion barrier? Perform single-point 6-31G* energy calculations on planar and pyramidal forms of $N(CH_3)_3$, NF_3 and $N(CN)_3$ (AM1 geometries) and compute barriers to inversion. Are any of your previous results altered?

Obtain the energy of the nitrogen "nonbonding" orbital in each molecule at each geometry (note: this orbital may or may not be the highest-energy occupied orbital, therefore, you may need to examine several orbitals). Also, note the shape of the "nonbonding" orbital in the planar geometry of each molecule. Can the trends in orbital energy and shape be accounted for using the π donor/acceptor arguments given above? Explain your reasoning, and cite specific orbital properties that support your argument. Are the calculated substituent effects on inversion barrier consistent with the orbital model?

Effect of ring size: A good rule-of-thumb states that conformational isomers that are separated by a barrier of at least 20-25 kcal/mol can be separated at room temperature. Evidence for this is provided by 1-chloro-2-methylaziridine, which can be separated into two configurationally stable diastereomers at room temperature (inversion barrier ~ 27 kcal/mol).[4]

---

3. This type of graph is known as a "Walsh diagram"; A.D. Walsh, *Prog. Stereochem.*, **1**, 1 (1954). Walsh used these diagrams to rationalize geometry changes as a function of the number of valence electrons in AH_2 and AH_3 molecules, where A is a main-group element. See: **Bond Angles in AH3 Compounds. Walsh's Rules**.
4. S.J. Brois, *J. Am. Chem. Soc.*, **90**, 508 (1968).

Build both aziridine isomers and optimize their geometries using AM1 semi-empirical calculations. Next, build a model of "flat" aziridine, i.e., one in which the N-Cl bond lies in the plane of the ring, and use this model to search for the closest AM1 transition state. Verify that your structure corresponds to a true transition state by calculating and displaying its vibrational modes (there should be one imaginary frequency corresponding to nitrogen inversion; see: **Finding Transition States**). Use the AM1 energies of these molecules to construct an energy surface for the inversion reaction. What is the calculated barrier for this process? Which isomer is more stable? How does the CNC bond angle vary during inversion?

Use a similar procedure to construct an AM1 energy surface for inversion of chlorodimethylamine (an acyclic "analogue" of 1-chloro-2-methylaziridine). What is the calculated barrier for this process? Based on this result, do you think chloroethylmethylamine, $ClN(CH_3)CH_2CH_3$, could exist as an enantiomerically pure species at room temperature? What is the change in CNC bond angle during the inversion of chlorodimethylamine? What does this suggest about the role of ring strain in controlling the barrier for aziridine inversion?

Perform single-point 6-31G* energy calculations on the planar and pyramidal forms of the two aziridine isomers and on chlorodimethylamine (use AM1 geometries), and obtain inversion barriers. Do any of your qualitative conclusions (based on the AM1 calculations) change?

Stereodynamics of Dimethylisopropylamine

Semi-empirical AM1 calculations are used to establish the energy requirements for nitrogen inversion and internal rotation about the CN bond in dimethylisopropylamine. The AM1 results are then used to interpret the results of dynamic NMR studies.

The ^{1}H NMR spectrum of dimethylisopropylamine, $(CH_3)_2NCH(CH_3)_2$, at 170K shows signals for only two types of methyl groups.[1] The spectrum shows four broad methyl signals at 93K, however, indicating that conformational isomerism is relatively slow under these conditions, and that the preferred conformers are **GA** and **AG** (the **GG** conformer, which is 0.7 kcal/mol higher in energy, contains only two types of methyl groups and cannot be responsible for the observed spectrum).

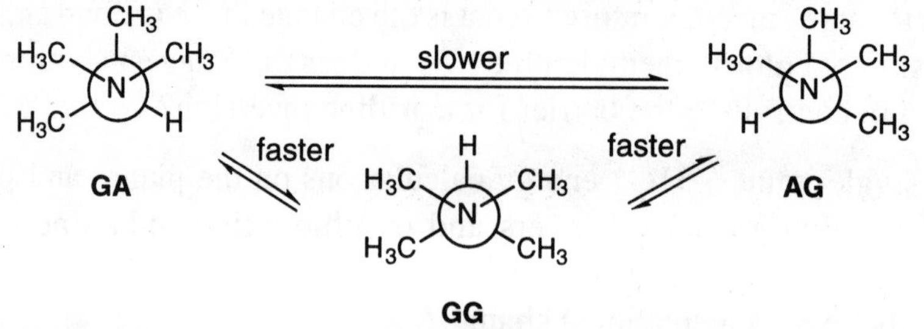

Analysis of the NMR spectra obtained at temperatures between 170K and 93K indicates that **GA/GG** and **AG/GG** interconversion is more rapid ($\Delta G^{\ddagger} = 4.5$ kcal/mol), than **GA/AG** interconversion ($\Delta G^{\ddagger} = 5.2$ kcal/mol). The NMR spectra are silent, however, regarding the mechanism of **GA/AG** interconversion. Two (limiting) processes can be imagined: one involving inversion at nitrogen via transition state **inv**, and one involving rotation about the CN bond via transition state **rot**.

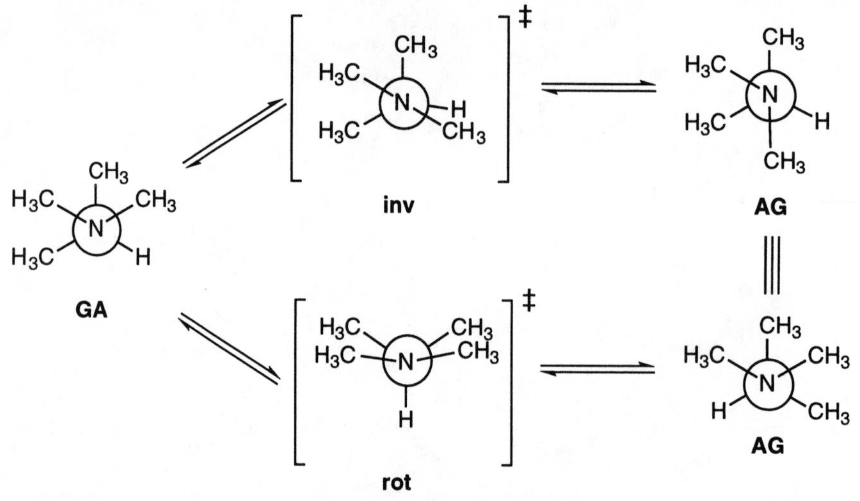

---

1. J.H. Brown and C.H. Bushweller, *J. Am. Chem. Soc.*, **114**, 8153 (1992).

If the inversion and rotation processes have significantly different energy requirements, i.e., there is a large difference in the two transition state energies, then it should be possible to assign **GA/AG** interconversion unambiguously to one mechanism or the other. In this experiment, you will use AM1 semi-empirical calculations to establish the conformational preferences of the amine, and to compare energies of the two transition states, **inv** and **rot**.

Procedure

Build models of **GA** and **GG** and optimize their geometries using AM1 semi-empirical calculations. Are your results in reasonable agreement with the experimental relative conformer energies?

Build models of the transition states, **inv** and **rot**, and the transition state for **GA/GG** interconversion, and use these models to search for the AM1 transition states. Verify that each model is a true transition state by calculating and displaying its vibrational frequencies (a transition state should have one imaginary frequency; see: **Finding Transition States**). Use the AM1 energies of these transition states to calculate AM1 barriers for **GA/GG** interconversion and for the two **GA/AG** pathways. Which, if any, of these results are in reasonable agreement with experiment? Which pathway do you think is responsible for **GA/AG** interconversion, inversion or rotation?

Racemization of Helicenes

AM1 semi-empirical calculations are used to reveal the mode of racemization in [5] and [6]helicene.

Helicenes are beautiful molecules made up of successively fused benzene rings so as to form a spiral, e.g., for [5]helicene and [6]helicene.[1]

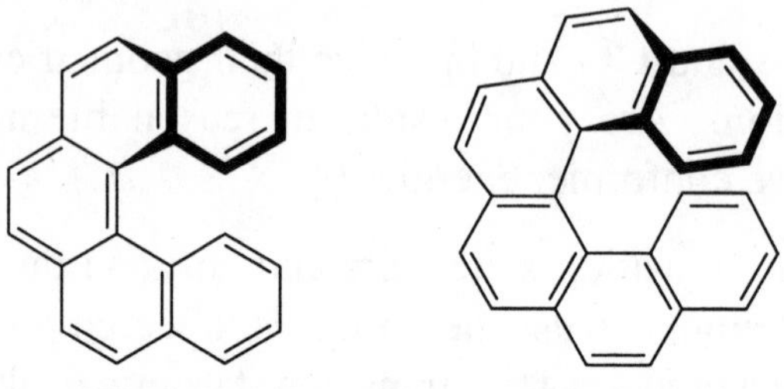

[5]helicene   [6]helicene

In principle, helicenes should exist in distinct (isolable) "left" and "right-handed" forms. This was first confirmed for [6]helicene, and a barrier to racemization of 36 kcal/mol ($\Delta G^{\ddagger}$) measured. [5]Helicene is now known to racemize with a significantly smaller barrier ($\Delta G^{\ddagger} = 24$ kcal/mol), but interestingly, barriers to racemization in larger helicenes are not much greater than that in [6]helicene. Even barriers of this magnitude for racemization in helicenes seem remarkable upon inspection of molecular models. The "obvious" transition state for racemization, with all rings in one plane, appears to be far too crowded to be a serious candidate.

In this experiment, AM1 semi-empirical calculations are first used to determine the structures of [5] and [6]helicenes, and then to attempt to locate the pathway for racemization.

Procedure

Build [5]helicene and optimize its geometry using AM1 semi-empirical calculations. Examine the calculated geometry. Does "coiling" result in significant bond length distortions or in localization of bonds? Compare your geometry to that of phenanthene.

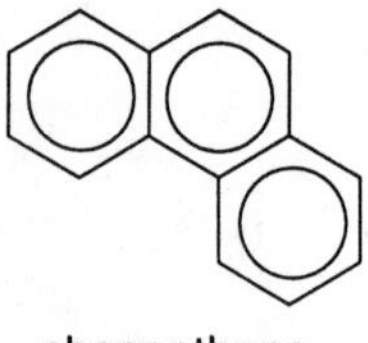

phenanthene

---

1.   R.H. Janke, G. Haufe, E.U. Würthwein and J.H. Borkent, *J. Am. Chem. Soc.*, **118**, 6031 (1996). For a general discussion, see: E.L. Eliel and S.H. Wilen, **Stereochemistry of Organic Compounds**, Wiley, New York, 1994, p. 1163.

Calculate electrostatic potential maps for [5]helicene and phenanthrene, and compare them on the same color scale. Is there any indication that [5]helicene might be more susceptible to electrophilic attack? If so, at which carbon(s)?

Repeat this procedure with [6]helicene. Are the structure and properties of this molecule more or less perturbed (relative to phenanthrene) than those of [5]helicene?

One possible mode of racemization in helicenes that does not necessitate a "planar" transition state is shown schematically below.

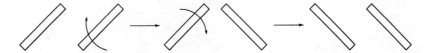

Here the transition state should possess a plane of symmetry (as opposed to a C_2 axis in the case of the helicenes). Construct such a structure for [5]helicene,[3] i.e.,

and use this structure to search for the AM1 transition state. Calculate the vibrational frequencies of this molecule and verify that it is indeed a transition state and that the reaction coordinate indeed smoothly connects left and right-handed [5]helicene (see: **Finding Transition States**). Calculate the racemization barrier. Is it in line with the experimental value?

Perform the same analysis with [6]helicene. Compare the two systems.

---

3. Building hints: Build the hypothetical molecule below,

minimize with molecular mechanics (it should posses C_s symmetry) and delete the methylene bridge. Do not minimize.

VIII

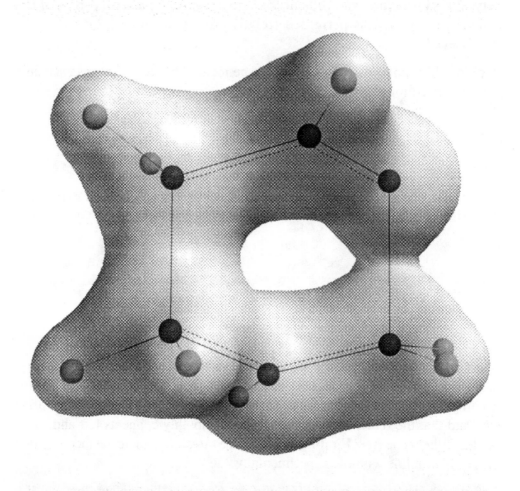

Chemical Reactivity and Selectivity in Conformationally Mobile Molecules

T he chemical reactivity of a molecule is often determined by its conformation and stereochemistry. For example, base-induced (E2) elimination of HBr from an alkyl bromide proceeds most rapidly when there is an *anti* arrangement of H and Br. If the molecule can easily adopt the desired conformation, it will be more reactive than a molecule that cannot. This situation is illustrated by the elimination rates of the two diastereomers of 4-*tert*-butylcyclohexyl bromide. The *cis* and *trans* isomers (shown below) are both conformationally mobile, equilibrating between two chair cyclohexane conformers, but the *cis* isomer undergoes elimination about 500 times more rapidly than the *trans* isomer.

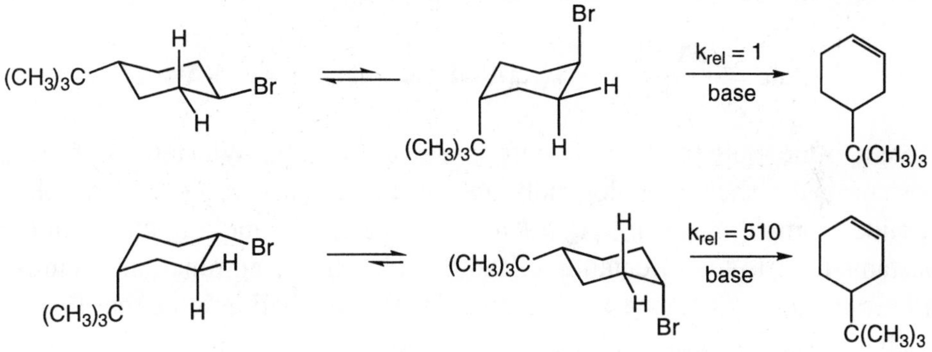

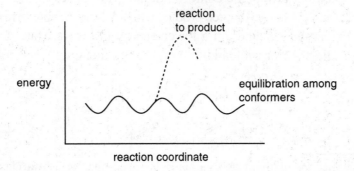

This difference in reactivity can be explained by noting that the preferred conformer for the *cis* isomer contains an *anti* H/Br relationship, but the preferred conformer for the *trans* isomer does not. Therefore, the *trans* isomer must either undergo an unfavorable conformational change before it can react, or it must eliminate HBr by a less favorable pathway.

The influence of conformation on chemical reactivity depends on the relative rates at which various conformers equilibrate and react. Since, in most cases, conformational isomerization is much more rapid than chemical reaction, it is reasonable to assume (for now) that the various conformers are always in equilibrium.

The behavior of this type of system is described by the "Winstein-Holness equation", and the "Curtin-Hammett principle", both of which can be best understood by referring to the following reaction sequence.[1]

$$P_A \xleftarrow{\ k_A\ } A \underset{k_{BA}}{\overset{k_{AB}}{\rightleftharpoons}} B \xrightarrow{\ k_B\ } P_B$$

This describes a reactant molecule with two reactive conformations, **A** and **B**, each of which yields a unique product, P_A and P_B, by means of an irreversible reaction. The reactivity of the starting material is described by the Winstein-Holness equation, which says that the overall reaction rate is a weighted average of the two reaction rate constants, k_A and k_B, i.e.,

$$\text{rate} = \frac{d[A+B]}{dt} = k_{WH}[A+B] = (n_A k_A + n_B k_B)[A+B]$$

where n_i is the mole fraction of conformer i, and k_{WH}, the Winstein-Holness rate constant, is the phenomenologically observed rate constant. For the cyclohexyl bromide example, $k_{WH} = n_{axBr}k_{axBr} + n_{eqBr}k_{eqBr}$, where the mole fractions and rate constants refer to the orientation of the Br atom. Since elimination requires an *anti* arrangement of H and Br, i.e., an axial Br, we shall assume that $k_{eqBr} \approx 0$, which gives $k_{WH} \approx n_{axBr}k_{axBr}$. It is, therefore, possible to account for the reactivity of the two isomers in terms of the variation in n_{axBr}; this quantity is large for the more reactive *cis* isomer, but is small for the less reactive *trans* isomer.

When different conformers yield different products, the product distribution, $[P_A]/[P_B]$, can be predicted using the Curtin-Hammett principle. This states that $[P_A]/[P_B]$ depends on the relative energies of the transition states leading to these products, $G(A^\ddagger)$ and $G(B^\ddagger)$.

$$\frac{[P_A]}{[P_B]} = (k_A/k_B)K = \exp[-(G(A^\ddagger) - G(B^\ddagger))/RT]$$

This is an extremely important principle because it shows that the product distribution need not reflect the conformational preferences of the reactant. It is even possible for P_A to be the favored product when **A** is not the preferred conformer, as long as P_A can be reached via a lower energy transition state. Consider, for example, the Diels-Alder cycloaddition of 1,3-butadiene with acrylonitrile.

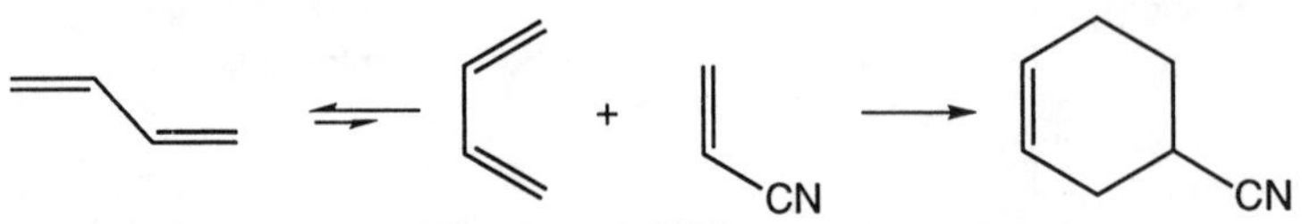

---

1. A more detailed description of the Winstein-Holness equation, the Curtin-Hammett principle, their range of application, and examples of systems that obey these relationships, can be found in J. I. Seeman, *Chem. Rev.,* **83**, *83* (1983).

Even though the diene exists primarily in an *anti* conformation (see: **Conformations of 1,3-Butadiene. A Price to Pay for Diels-Alder Cycloaddition**), the reaction is concerted and must and does proceed exclusively via the higher-energy *cis* conformer.

Another example of the Curtin-Hammett principle is provided by the reaction of tropanes **X** and **Y** with isotopically labelled methyl iodide, CH_3I. The favored conformer of the reactant is **X**, but the major product is P_Y, indicating that the transition state leading from **Y** to P_Y ($Y^{\ddagger}$) is lower in energy than the transition state leading from **X** to P_X ($X^{\ddagger}$). In this case, the steric repulsion that makes **X** the preferred conformer, also makes $Y^{\ddagger}$ the preferred transition state.

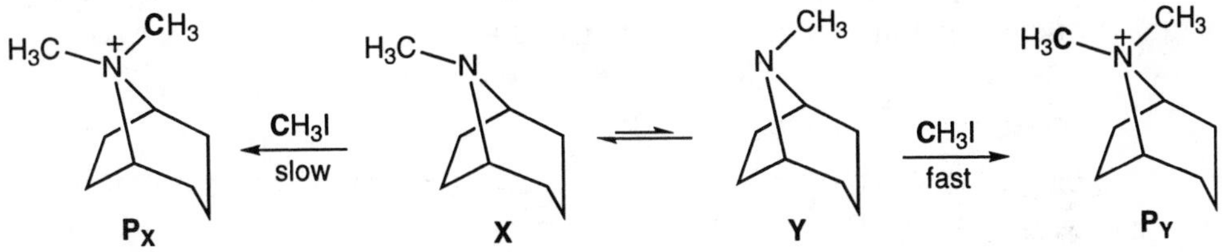

In this case, unlike the Diels-Alder reaction above, both conformers are able to react.

Stereochemistry of Free Radicals

Semi-empirical AM1 calculations are used to establish the geometry of carbon-centered free radicals, and to assess the effect of ring strain on free radical stability.

Diacylperoxides decompose upon heating to yield carbon-centered free radicals and carbon dioxide. Decomposition of the *cis-cis* and *trans-trans* diacylperoxides (R = *tert*-butyl) shown below might be expected to give different free radicals and lead to different mixtures of the two cyclohexyl bromides, but, in fact, the same mixture is obtained regardless of the starting material. This result is consistent with rapid equilibration of two pyramidal free radicals, or with a single planar (or nearly planar) free radical intermediate.

trans : cis = 2:1

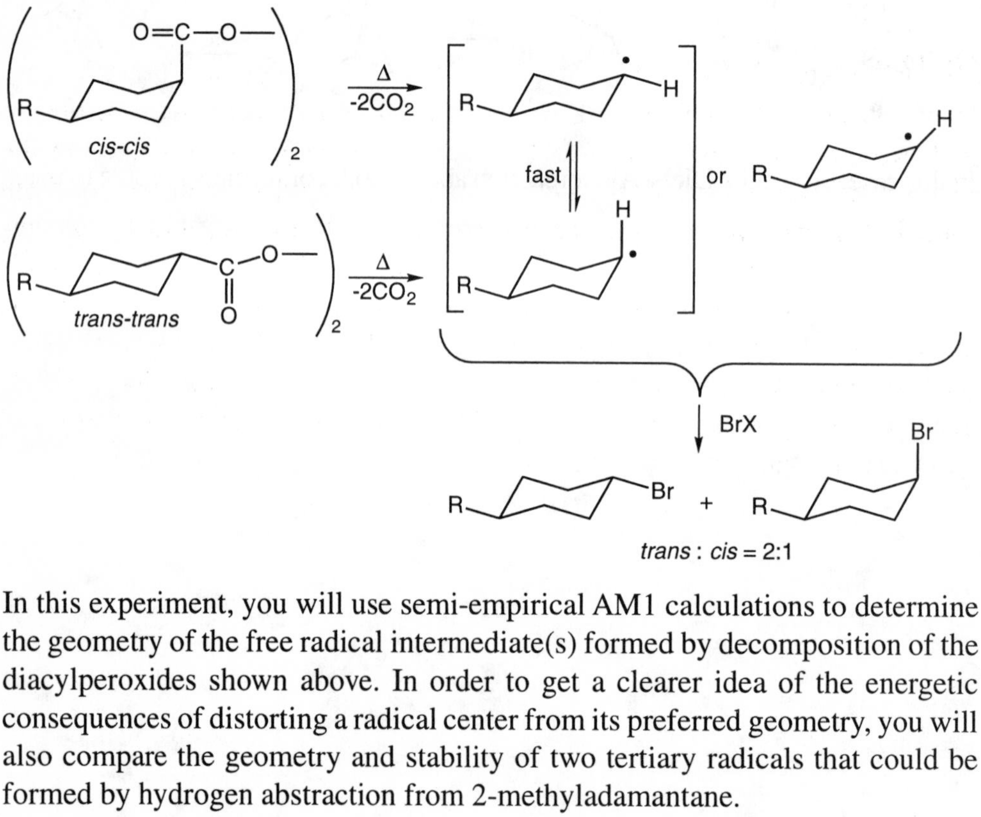

In this experiment, you will use semi-empirical AM1 calculations to determine the geometry of the free radical intermediate(s) formed by decomposition of the diacylperoxides shown above. In order to get a clearer idea of the energetic consequences of distorting a radical center from its preferred geometry, you will also compare the geometry and stability of two tertiary radicals that could be formed by hydrogen abstraction from 2-methyladamantane.

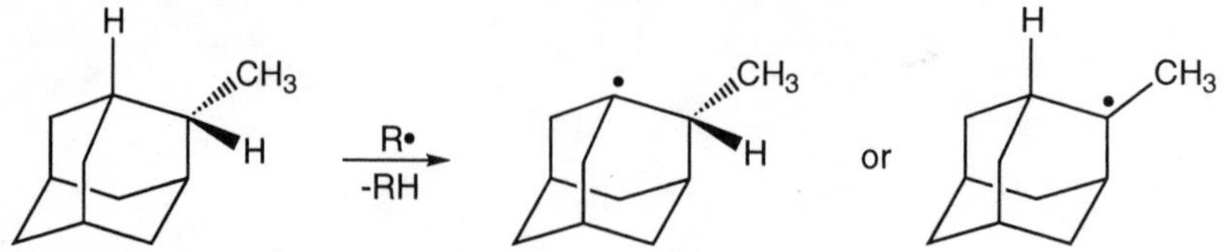

The adamantyl ring system is normally regarded as rigid, and different free radical geometries (pyramidal vs. planar) would favor one or the other free radical.

Procedure

4-*tert*-Butylcyclohexyl radical: Build *equatorial tert*-butylcyclohexane, and then remove a hydrogen from C-4 to create models of both *cis* and *trans* radicals. Optimize their geometries using AM1 semi-empirical calculations. Record each radical's energy, and compare their energies and structures. Would you describe the geometry of these radicals as pyramidal or planar? How do your results explain the formation of a single product mixture independent of starting material?

Adamantyl radicals: Build 2-methyladamantane, and then remove a hydrogen to create models of the two possible tertiary radicals. Optimize their geometries using AM1 semi-empirical calculations. Compare the energies and structures of the two radicals. Which radical is more stable, and why? Do the radicals have the same ring shape as the parent hydrocarbon?

Optional

Carbocations prefer a planar geometry at carbon. Why? (See: **Bond Angles in AH₃ Molecules. Walsh's Rules.**)This preference should be reflected in the relative stability of the tertiary carbocations formed by hydride abstraction from 2-methyladamantane, i.e., the carbocation analogs to the two tertiary radicals considered above. Build models of these carbocations and optimize their geometries using AM1 semi-empirical calculations. Record each cation's energy, and compare their energies and structures. Which cation is more stable? Based on these results, which intermediate has a stronger preference for a planar geometry, a free radical or a carbocation? Cite specific changes in energy and geometry that support your conclusion.

Configurational Stability of Sulfoxides. Racemization without Inversion?

Semi-empirical AM1 calculations and ab initio 3-21G$^{(*)}$ calculations are used to establish the energy requirements for sulfur inversion and sigmatropic rearrangement in a chiral allylic sulfoxide. These results are compared to the experimental racemization barrier, and are used to suggest a racemization mechanism.

Sulfoxides are configurationally stable, a fact that can be attributed to a large barrier for sulfur inversion. For example, the experimental racemization barrier ($\Delta G^{\ddagger}$) of phenyl *p*-tolyl sulfoxide is 38.6 kcal/mol; this presumably represents the energy difference between the pyramidal sulfoxide and a planar transition state.

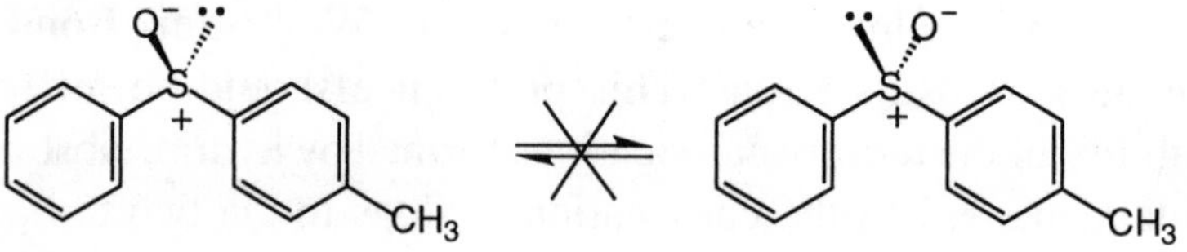

Racemization of allylic sulfoxides, on the other hand, occurs more easily. For example, racemization of allyl *p*-tolyl sulfoxide occurs with an activation barrier of only 24.7 kcal/mol, far less than the presumed pyramidal inversion barrier. What is the cause for this change in behavior between two systems which at first glance would seem to be very similar? One possible explanation is that the allyl group (relative to phenyl) might facilitate inversion at sulfur. An interesting alternative explanation is that racemization in allyl sulfoxides proceeds by means of sequential [2,3] sigmatropic rearrangements and an achiral intermediate.

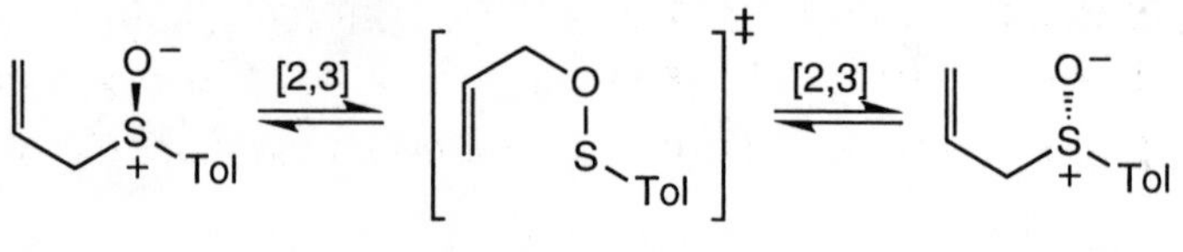

In this experiment, you will use semi-empirical AM1 calculations and *ab initio* 3-21G$^{(*)}$ calculations to construct energy surfaces for the pyramidal inversion and sigmatropic rearrangement pathways of allyl methyl sulfoxide, $CH_3S(O)CH_2CH=CH_2$, in order to evaluate their relative importance as racemization mechanisms.

Procedure

Inversion pathway: Build allyl methyl sulfoxide, and optimize its geometry using AM1 semi-empirical calculations. Next, construct a model of the inversion transition state by replacing the pyramidal sulfoxide group with a planar one. Use this model to search for the AM1 transition state, and verify that the new model is a true transition state by calculating and displaying its vibrational frequencies (a transition state should have one and only one imaginary frequency; see: **Finding Transition States**). Subtract the transition state energy from that of allyl methyl sulfoxide to obtain the AM1 barrier for inversion at sulfur. (A better estimate of the barrier may be obtained by performing single-point energy 3-21G$^{(*)}$ calculations using the optimized AM1 geometries for both allyl methyl sulfoxide and the transition state.) Is the calculated barrier more consistent with the experimental racemization barrier of allyl p-tolyl sulfoxide or phenyl p-tolyl sulfoxide?

Rearrangement pathway: Build a model of the transition state for sigmatropic rearrangement of allyl methyl sulfoxide, by building the cyclic molecule shown below, and optimizing its geometry using AM1 calculations (constrain the two bond distances to values shown in the diagram).

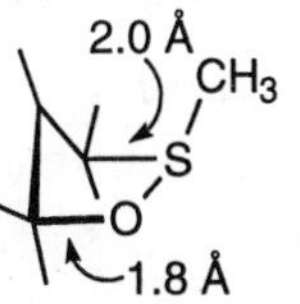

Use the resulting structure to search for the AM1 transition state (remove constraints), and verify that the new model is a true transition state by calculating and displaying its vibrational frequencies. Use the energy of this model to construct an energy surface for racemization via an achiral intermediate. What is the AM1 racemization barrier on this surface? Is the calculated barrier more consistent with the experimental racemization barrier of allyl p-tolyl sulfoxide or phenyl p-tolyl sulfoxide? (As before, a better estimate of the energetics may be obtained by performing a single-point 3-21G$^{(*)}$ calculation on the AM1 transition state.)

Optional

Racemization of benzyl methyl sulfoxide, CH_3SOCH_2Ph, might also proceed *via* a sigmatropic rearrangement pathway in which the benzyl group performs the same function as the allyl group. Use the procedure described above to calculate the AM1 barriers for inversion and sigmatropic rearrangement, and determine which is the preferred pathway. How do these barriers compare to the ones previously calculated for the allylic system? Explain why these differences exist.

NMR and Conformational Analysis of Diastereomers

Semi-empirical AM1 calculations are used to identify the preferred conformers of cis and trans-3,3,5-trimethylcyclohexanol. The energy difference between conformers is then used to interpret each isomer's NMR spectrum.

$^3J_{HH}$, the NMR coupling constant for vicinal hydrogens, depends on the H-C-C-H dihedral angle.[1]

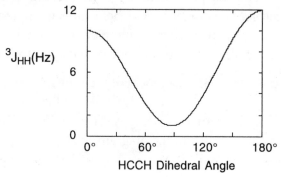

J is largest when the hydrogens are *anti*, and is smallest when the CH bonds are perpendicular. Consequently, coupling constants can be used to determine molecular conformation, providing a molecule exists primarily as a single conformer.[2]

Coupling constants can also be used to identify molecules and assign stereochemistry, as long as the conformational preferences of the molecule are known. In this experiment, you will use AM1 models to determine the conformational preferences of two diastereomeric alcohols, *cis* and *trans*-3,3,5-trimethylcyclohexanol, produced by $NaBH_4$ reduction of the corresponding ketone.

This information, along with the NMR spectra of the alcohols, will then be used to assign their stereochemistry and to determine the preferred course for hydride reduction.

---

1. R.M. Silverstein, G.C. Bassler and T.C. Morill, **Spectroscopic Identification of Organic Compounds**, 5th ed., Wiley, New York, 1991, p.196.
2. Coupling constants in flexible molecules are more difficult to interpret because they are "conformationally averaged", i.e., $J_{obsd} = n_1J_1 + n_2J_2 + ...$, where n_i and J_i are the mole fraction and coupling constant associated with conformer i.

Procedure

Build each of the four alcohol conformers shown on the facing page and optimize its geometry using the semi-empirical AM1 model (note: locating the most stable form of each conformer requires testing different orientations of the OH group). Based on the heat of formation of each conformer, calculate the equilibrium ratio of conformers assuming $K_{eq} \approx e^{-\Delta H/RT}$, where RT at room temperature is approximately 0.6 kcal/mol. Do your results suggest that the NMR spectrum of each compound will be dominated by a single conformer? Which isomer has a stronger conformational preference? Why?

The CHOH NMR signal for each alcohol isomer displays a distinctive pattern. The major isomer (~80% of product mixture) displays a pentet pattern (J = 3.2 Hz), while the minor isomer (~20% of product mixture) displays a triplet of triplets pattern (J = 11.3, 4.3 Hz). Assuming that the CHOH proton is coupled only to vicinal CH protons, and that the relative magnitudes of these coupling constants are determined by HCCH dihedral angles as given in the graph shown on the facing page, determine which alcohol is the major isomer and which is the minor isomer. Explain your reasoning.

What is the preferred pathway for hydride reduction? How do you explain this preference? (Note: you should comment on, and take into account, the conformational preferences of the starting ketone.)

Optional

1. Repeat your calculations using energies from the 3-21G *ab initio* model (use AM1 geometries). Do you see any qualitative differences in results?

2. Build the starting ketone in its most stable conformation, optimize its AM1 geometry and perform a single-point energy calculation using the 3-21G *ab initio* method. Calculate and display a LUMO map. Nucleophilic attack should occur on the face of the carbonyl carbon for which the LUMO is more heavily concentrated (see: **Graphical Models and Graphical Modeling**, and **Stereochemistry of Nucleophilic Addition to Norcamphor**). Is your map consistent with the observed product distributions? If it is, what if anything does it say about the factors (thermodynamic or kinetic) that control product distribution (see: **Thermodynamic vs. Kinetic Control**).

Cyclodecyl Cation. Conformation-Dependent Hydrogen Bridging

Semi-empirical AM1 calculations are used to establish the structure and conformational preferences of the hydrogen-bridged cyclodecyl cation.

Ring strain in cycloalkanes, $(CH_2)_n$, decreases as n increases from 3 to 6 (see: **Ring Strain in Cycloalkanes**). This is because CCC bond angles are able to expand toward an "ideal" tetrahedral value, and because it becomes easier to stagger CC and CH bonds in larger rings. Interestingly, ring strain increases again as n increases from 6 to 12. The usual explanation for this increase in ring strain is that perfect staggering of bonds in the larger (n>6) rings would force some hydrogens to lie "inside" the ring , e.g., in cyclodecane,

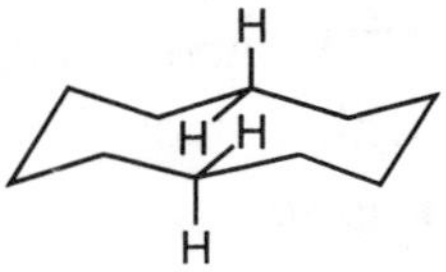

and cause steric repulsion. The ring can reduce these transannular interactions only by adopting a less staggered, slightly strained geometry.

Transannular interactions are responsible for some unusual chemical behavior.[1] For example, the ^1H NMR spectrum of cyclodecyl cation in superacid at -130°C shows a 2H signal at δ 6.8 and a 1H signal at δ -6.85.[2] These unusual chemical shifts, along with the fact that the signals are not temperature sensitive from -117°C to -140°C, and the appearance of a doublet of doublet pattern ($J_{CH} = 32$, 158 Hz) at δ 153 for the "cationic" carbon(s) in the ^{13}C NMR spectrum, have all been cited as evidence for a hydrogen-bridged cation structure, i.e.,

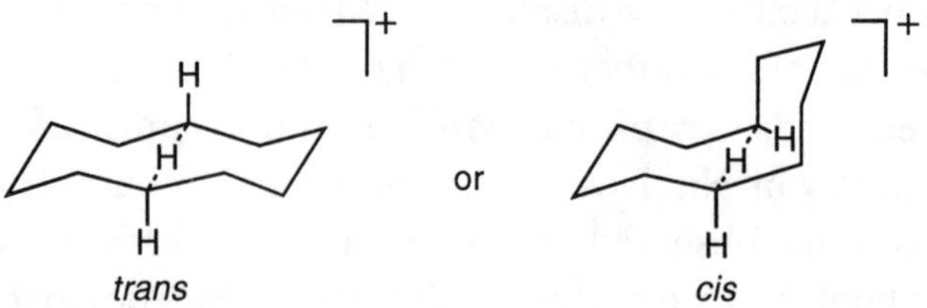

trans   *cis*

Unfortunately, the cation's conformation, and the stereochemistry about the bridging CHC group, cannot be inferred from the spectral data, making molecular models an essential tool for interpreting the NMR data. In this experiment, you will investigate the molecular and electronic structure of hydrogen-bridged cyclodecyl cation using the AM1 semi-empirical model.

---

1. Carey and Sundberg A, p. 317; Lowry and Richardson, p. 484.
2. R.P. Kirchen, T.S. Sorensen, K. Wagstaff, *J. Am. Chem. Soc.*, **100**, 6761 (1978).

Procedure

Build both *cis* and *trans* cyclodecyl cations[3] and optimize each cation's geometry using the AM1 semi-empirical model.[4] Do either of the two cations actually incorporate a bridging hydrogen? What geometrical characteristics of your structures support or reject the idea of hydrogen-bridging? (Note that space-filling models of these cations show that some kind of transannular interaction is unavoidable.) Assuming you obtain at least one bridged cation, what is the geometry about the CHC group? Are your results consistent with the NMR data?

The electrophilic sites in a carbocation can be predicted from the shape of its lowest unoccupied molecular orbital (LUMO). Calculate and display the LUMO of the hydrogen-bridged cation. What is the shape of this orbital? Which atoms (and which face(s) of these atoms) are predicted to be electrophilic? A hydrogen-bridged species must be regarded as a resonance hybrid, i.e., it cannot be described by a single Lewis structure. Calculate and display the electrostatic potential map. Which of the following Lewis structures appear to be major contributors to the resonance hybrid?

---

3. SPARTAN building instructions. Build from the corresponding decalins by breaking the bond common to the two rings, deleting one hydrogen and constraining the remaining hydrogen to lie midway between two carbons i.e., for the *trans* cation.

Constrain the "bridging" hydrogen to be 1.2Å from each of the two carbons, and the CHC bond angle to be 180°.

4. Optimize first with constraints enforced to get good starting geometries, and then reoptimize without the constraints to get the final structures.

Steric Interactions in Enolates

Semi-empirical AM1 calculations are used to identify the preferred conformation of an enolate anion, and electrostatic potential maps are used to predict facial selectivity in the protonation of the enolate.

Conjugate addition of PhMgBr to the α,β-unsaturated ketone shown below yields the *cis* ketone exclusively.[1]

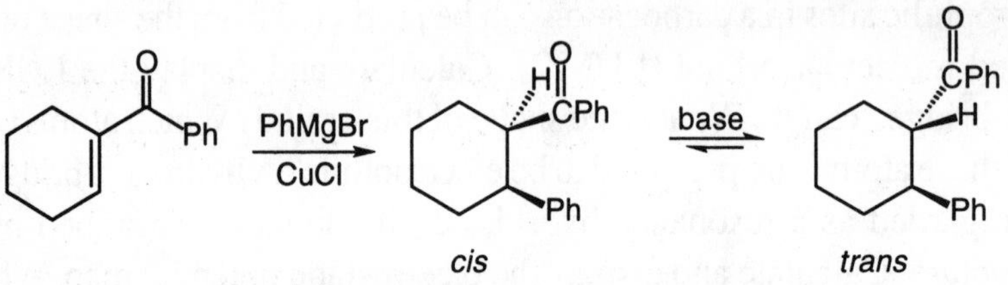

The *cis* ketone is sterically congested, and rearranges to the more stable *trans* isomer when treated with base. Therefore, the question that arises is why does the conjugate addition proceed to the less stable (*cis*) ketone in the first place?

Experimental work by Malhotra and Johnson indicates that the conjugate addition generates an enolate intermediate that can be trapped by acetylation,[2] as shown below. The product is a ~1:1 mixture of *Z* and *E* enol acetates, and NMR analysis shows that both isomers prefer a conformation in which the phenyl group is *axial*.

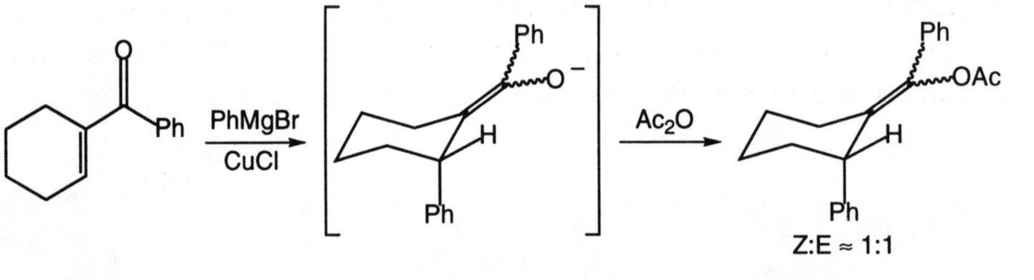

Based on the stereochemistry of the enol acetates, Malhotra and Johnson argued that the enolate itself prefers a conformation with an *axial* phenyl group, and suggested that this preference is due to unfavorable Ph-O⁻ (or Ph-Ph) interactions in the *equatorial* conformer.

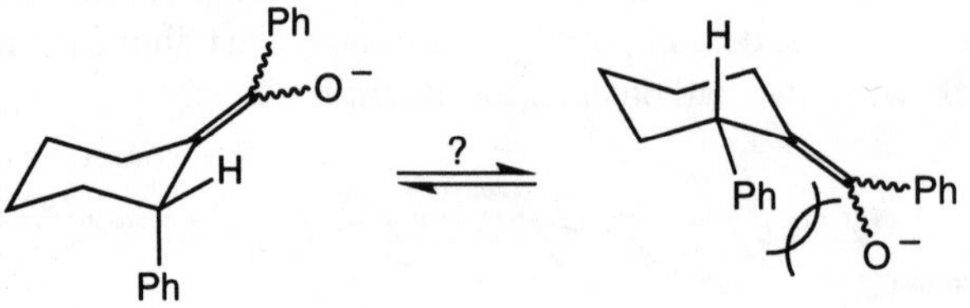

---

1. H.E. Zimmerman, *J. Org. Chem.*, **20**, 549 (1955).
2. S.K. Malhotra and F. Johnson, *J. Am. Chem. Soc.*, **87**, 5493 (1965).

In this experiment, you will test Malhotra and Johnson's hypothesis by using AM1 calculations to examine the conformational preferences of the Z enolate. Note that enolate conformation, by itself, cannot explain selective formation of the *cis* ketone. This is because each enolate has two reactive faces, and, in principle, can react with acid to generate both *cis* and *trans* ketones. Therefore, you will also examine electrostatic potential maps to see if a distinction can be made between the two faces of the preferred enolate (the more reactive face will be more electron-rich and/or more sterically accessible).

Procedure

Build several models of the *axial* and *equatorial Z* enolate conformers, by giving the two phenyl groups different initial orientations (see below).

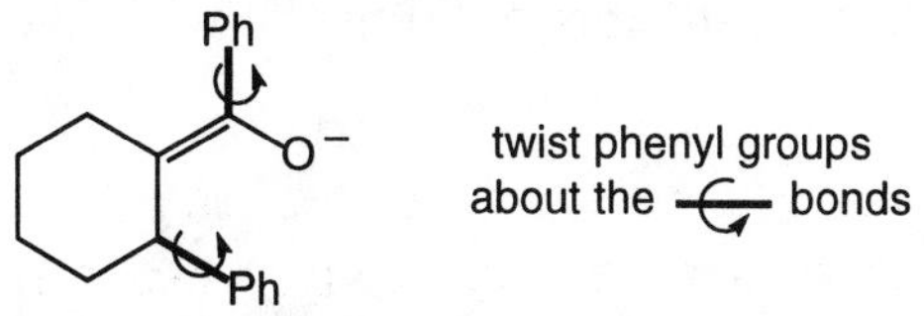

twist phenyl groups
about the ⟵⟶ bonds

Optimize the enolate geometries using AM1 semi-empirical calculations, and discard all but the most stable structure of each conformer. What are the energies of the most stable *axial* and *equatorial* enolates? Which enolate is more stable, and by how much?

Next, calculate and display the electrostatic potential map for each enolate. Focusing on the carbon where protonation needs to occur, which enolate face is more electron-rich? Which face is more sterically accessible? Are your results consistent with exclusive formation of the *cis* ketone? Explain.

Preferred Ring Conformations in Taxol Models

Semi-empirical AM1 calculations are used to determine the conformational preferences of two molecules that model the taxol A-B-C ring system, and to interpret NMR observations.

Chiral natural products that contain medium-sized rings are difficult to synthesize because the rings are flexible and their conformational preferences are not well understood. Taxol, an important anticancer drug that successfully resisted laboratory synthesis for many years, exemplifies this problem in that it incorporates a flexible eight-membered ring (**B**) with five chiral centers.

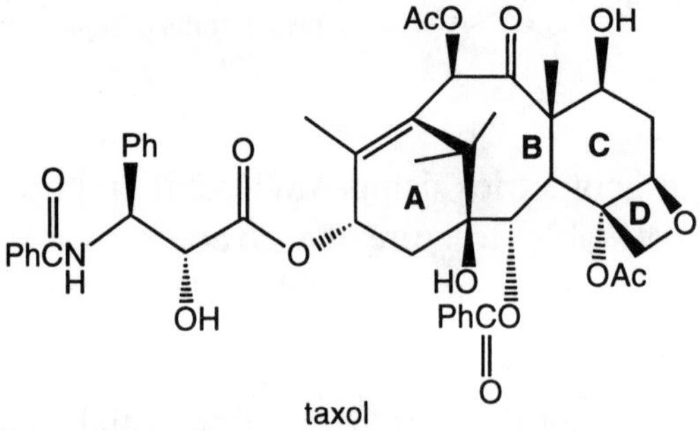

taxol

Some insight into the conformational preferences of taxol-like molecules has been obtained through NMR investigations of the fused-ring ketones, **1H** and **1Me**.[1]

1H   R = H
1Me R = CH$_3$

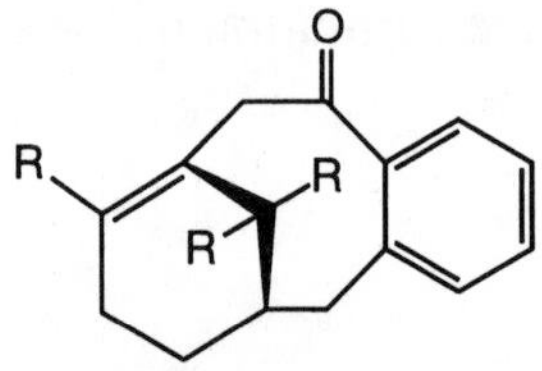

The ^{1}H NMR spectrum of **1H** shows a single vinyl resonance at room temperature, but this signal resolves into two resonances at –50 °C, indicating the presence of two conformers ($\Delta G \approx 0.88$ kcal/mol). The vinyl resonance is strongly shifted upfield in the minor conformer, suggesting that the vinyl hydrogen lies over the face of the benzene ring in this conformer. The spectrum of **1Me** is quite different in that it shows three large and three small methyl resonances at room temperature, indicating the presence of two reasonably stable conformers ($\Delta G \approx 1.24$ kcal/mol). Although one methyl signal is substantially more upfield than the others in each conformer, NMR experiments indicate that a different methyl group is shielded in each conformer (the shielding mechanism, here too, is believed to involve a methyl group lying over the face of the benzene ring).

---

1.   K.J. Shea and J.W. Gilman, *Tetrahedron Lett.*, **25**, 2451 (1984).

Given these observations, it might be anticipated that molecular models would provide evidence of two conformers of similar stability for **1H** and **1Me**, and a clear correlation between structure and energy. In this experiment, you will use semi-empirical AM1 calculations to determine the conformational preferences of **1H** and **1Me**, and look for structural features that can explain the NMR observations.

Procedure

Built *exo* and *endo* conformers of **1H** and **1Me**, and optimize their geometries using semi-empirical AM1 calculations. Record the energy of each conformer, and determine whether its geometry is similar to the drawings shown below (if the optimized geometry is different then sketch the structure of the conformer).

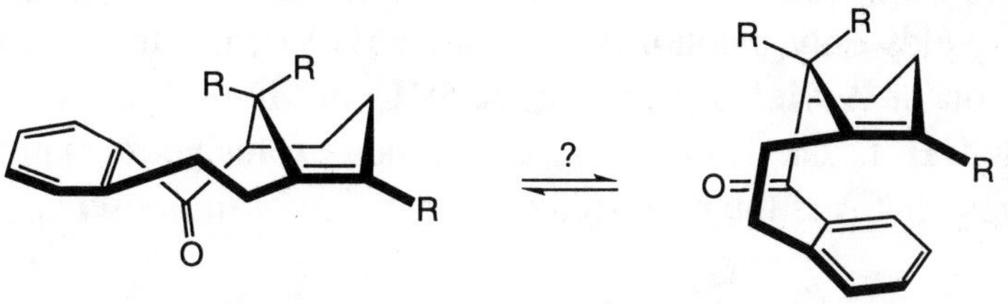

exo   endo

Identify the more stable conformer of **1H** and **1Me**. Are they the same? Calculate the conformer energy differences for each molecule. Are these differences consistent with the experimental energy differences?

Reconcile your results with the NMR observations. Which conformer of **1H** locates the vinyl hydrogen over the face of the benzene ring? Is this result consistent with the NMR data? Which methyl group will be most shielded by the benzene ring in each conformer of **1Me**? Is this result consistent with the NMR data? What effect do methyl groups have on the conformational preferences of the ring system?

Optional

Locate additional conformers of **1H** and **1Me** by performing an automated conformer search using molecular mechanics calculations, followed by geometry optimization of each conformer using AM1 calculations. How, if at all, do these results affect your interpretation of the NMR data?

Thujopsene. Conformational Control of Reactivity

Semi-empirical AM1 calculations are used to establish the conformational preferences of thujopsene and to account for the stereoselective nature of its hydroboration reaction.

Thujopsene is typical of many natural products in that it contains fused rings, multiple substituents, and is conformationally mobile. What makes it novel is the presence of an asymmetric "vinylcyclopropane" unit. Several experimental studies have examined the reactivity of this unit. For example, hydroboration of thujopsene yields only alcohol **A-OH** in >96% yield.[1,2] If it is assumed that hydroboration of **A** yields primarily **A-OH**, and hydroboration of **E** yields primarily **E-OH**, i.e., k_{AA}, $k_{EE} \gg k_{AE}$, k_{EA}, then exclusive formation of **A-OH** might suggest that **A** is the preferred conformation of thujopsene.

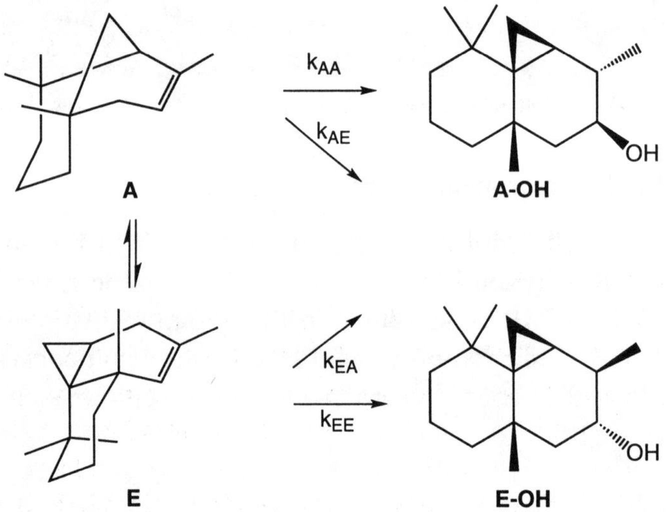

This conclusion is unwarranted, however, because exclusive formation of **A-OH** could also be consistent with **E** being the favored conformer. All that is required is that **A-E** interconversion be faster than hydroboration, and $k_{AA}[A]/[E]$ be much greater than k_{EE} (see: **Chemical Reactivity and Selectivity in Conformationally Mobile Molecules**). It is even possible for **A** to be the preferred conformer, but for product formation to occur via **E** if $k_{EA}[E]/[A]$ is much greater than k_{AA}. Molecular modeling may be useful in clarifying the situation, because it provides a tool for establishing the relative stability and reactivity of **A** and **E**.

---

1.   S.P. Acharya and H.C. Brown, *J. Org. Chem.*, **35**, 3874 (1970).
2.   **A** and **E** refer to the orientation of the bridgehead methyl group relative to the cyclohexane ring, where **A** = *axial* and **E** = *equatorial*.

In this experiment, you will use AM1 semi-empirical calculations to establish the optimal geometries and relative stability of **A** and **E** thujopsene. You will also use electrostatic potential maps of the two conformers to determine which conformer can yield **A-OH** more easily, i.e., you will qualitatively estimate the relative magnitudes of k_{AA} and k_{EA}.

Procedure

Build the **A** and **E** conformers of thujopsene, and optimize each using the semi-empirical AM1 model. Which is the more stable conformer? What is the energy difference between these conformers? Assuming that the free-energy difference between conformers is roughly equal to the difference in their heats of formation, what would be the equilibrium ratio of [**A**]/[**E**] at 298K?

Calculate and display an electrostatic potential map for each conformer (use the same color scale). Orient each molecule so that you can see the alkene face that must react with and electrophile like B_2H_6 to give **A-OH**. Which conformer is more electron-rich? (The more electron-rich conformer is better suited to react with an electrophile.) Which conformer is less sterically hindered? Use your results to explain the preferred route for formation of **A-OH**, and to suggest a reason why **A-OH** forms exclusively.

Optional

Use automated conformation searching and molecular mechanics calculations to generate additional conformers of thujopsene. Optimize each of these structures using semi-empirical AM1 calculations. Are any of these conformers more stable than **A** or **E**? Are any of these conformers likely to be the source of **A-OH** instead of **A** or **E**? (Note: you will need to use space-filling models or electrostatic potential maps to assess the reactivity of these conformers.) How would you describe the shape of the cyclohexane ring in these conformers (see: **The Mechanism of Cyclohexane Ring Inversion**)? Published analyses of thujopsene's conformational behavior have generally ignored these other conformers.[1] Why do you think this is?

Acetylcholine. Conformational Control of Bioactivity

Semi-empirical AM1 calculations are used to establish the conformational preferences of a neurotransmitter, acetylcholine, in the gas phase. AM1-SM2 calculations are then used to estimate conformational preferences in water.

Acetylcholine (ACh), the first neurotransmitter molecule to be identified by scientists, exerts a variety of biological effects by binding to cellular receptors.[1] Experiments suggest that ACh assumes different conformations when bound to different receptors, i.e., each receptor selects for a particular ACh conformer. This is only possible, however, if the energy difference between the "selected" or bound conformer and the most stable conformer in solution is fairly small (<4-6 kcal/mol).

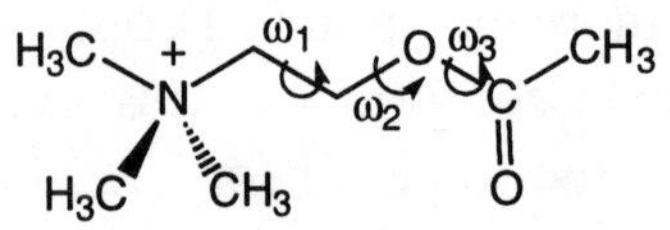

Acetylcholine
($\omega_1 = 180°$, $\omega_2 = 180°$, $\omega_3 = 180°$ shown)

The conformation of ACh is defined by dihedral angles ω_i, where $\omega_1 = \omega(NCCO)$, $\omega_2 = \omega(CCOC)$, and $\omega_3 = \omega(COCC)$ (see diagram). If ω_1 and ω_2 assume staggered values ($\pm 60°$ or $180°$) and ω_3 is eclipsed ($0°$ or $180°$), the maximum number of ACh conformers is 18 ($3 \times 3 \times 2$). The number of unique conformers is lower, however, since some conformers are mirror images or are identical due to symmetry. The ACh conformation shown above is not actually observed experimentally. Crystallographic analysis of solid ACh chloride reveals a coiled molecule and a *gauche* NCCO group: $\omega_1 = 77°$, $\omega_2 = 79°$, $\omega_3 = -167°$. A *gauche* NCCO group is also found in the bromide salt, but the ester group is oriented differently: $\omega_1 = 85°$, $\omega_2 = -167°$, $\omega_3 = -175°$.[2] The variation in these solid-state structures indicates that the medium (in this case, the anion) plays an important role in determining molecular conformation. This conclusion is reinforced by NMR studies of ACh dissolved in water,[3] which indicate that ACh is conformationally mobile, but generally prefers a structure similar to that of crystalline ACh chloride.

1. A.F. Casy, **The Steric Factor in Medicinal Chemistry**, Plenum, New York (1993), chapters 8-10.
2. R.W. Parker, C.H. Chothia, P. Pauling and T.J. Petcher, *Nature*, **230**, 439 (1971).
3. C.C.J. Culvenor and N.S. Ham, *J. Chem. Soc., Chem. Commun.*, 537 (1966).

In this experiment, you will use a combination of molecular mechanics and AM1 semi-empirical calculations to establish the conformational preferences of ACh in the gas phase. You will then recalculate the energies of these conformers using the AM1-SM2 method, and you will use these energies to estimate the conformational preferences of ACh in water.

Procedure

Build ACh, and systematically generate different conformers by adjusting the values of ω_i.[4] Optimize the geometry of each conformer using molecular mechanics, and then delete duplicate and mirror-image conformers so that you have a set of unique conformers. Optimize the geometries of these conformers using the AM1 semi-empirical model, and again delete any new duplicates or mirror-images. Record the energy of each unique conformer, and its dihedral angles, ω. Identify the most stable conformer and calculate the energies of the other conformers relative to this one. Which, if any, of your conformers has the same structure as solid ACh chloride? Is it the most stable one? Try to account for the energy ordering of the conformers in terms of steric, electrostatic, and resonance effects. (It may be useful to examine space-filling models of the conformers, and to calculate and display electrostatic potential maps.)

Use AM1-SM2 calculations to obtain the energy of each conformer in aqueous solution at its AM1-optimized geometry. Identify the most stable conformer, and calculate conformer energies relative to this conformer. What changes in relative conformer energy are caused by solvation? Are the results consistent with the NMR data, i.e., is there a substantial preference (≥ 1 kcal/mol) for the "solid ACh chloride" structure? Should the molecule be conformationally mobile? Do the conformer energies span a wider range in the gas phase or in aqueous solution? How do you explain these results?

---

4. Depending on the program available to you, you can either search the conformational space automatically or manually build individual conformers with dihedral angles of $\omega_1 = \pm 60°$ and $180°$, $\omega_2 = \pm 60°$ and $180°$, and $\omega_3 = 0°$ and $180°$.

Conformational Preferences in Transition States. The Claisen Rearrangement

Semi-empirical AM1 calculations and ab initio 3-21G calculations are used to construct models of chair and boat transition states for the Claisen rearrangement, and to investigate how substituents affect a preference for one transition state conformation over the other.

The Claisen rearrangement is an intramolecular reaction that converts an allyl vinyl ether to the isomeric 4-pentenal. Experimental observations suggest that the reaction is concerted, and involves a cyclic transition state.[1]

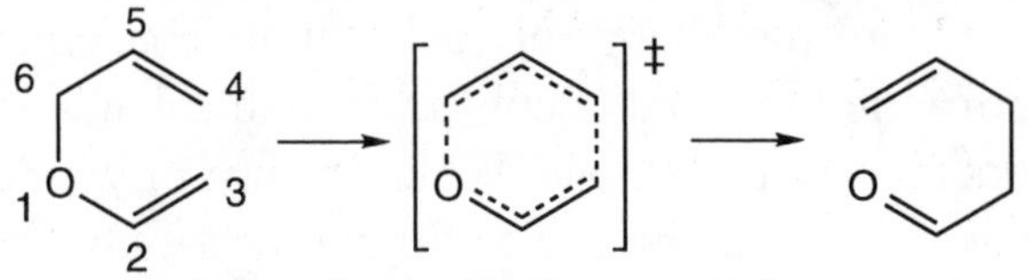

The Claisen rearrangement is synthetically important because it can create chiral centers at C-3 and C-4. The stereochemistry of these atoms will be determined by the geometry of the vinyl groups in the starting material, provided that the conformation of the cyclic transition state is known. Thus, the E,E isomer of the propenyl butenyl ether shown below might give either a *dl* or *meso* product, depending on whether the preferred reaction coordinate passes through a "chair-like" or "boat-like" transition state, respectively. Experiments have shown that the chair transition state is preferred by ~3 kcal/mol over the boat for all isomers of propenyl butenyl ether; the E,E and Z,Z ethers yield the *dl* product, and the E,Z and Z,E ethers yield the *meso* product.[2]

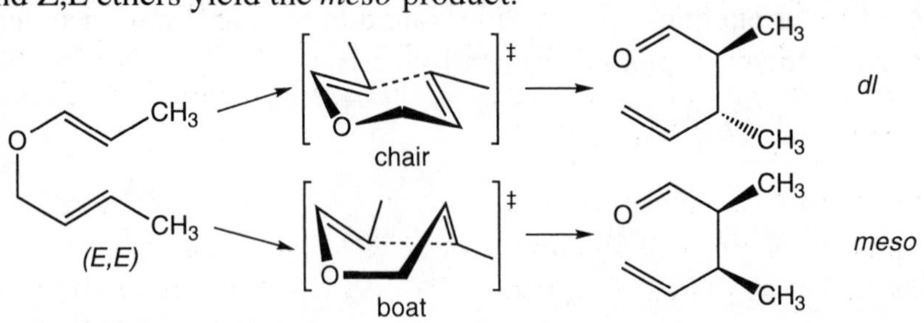

The analogy between Claisen transition state geometry and cyclohexane conformation is conceptually appealing, but is not quantitatively reliable. The interactions that destabilize boat cyclohexane are "1,4-flagpole" interactions, and eclipsing interactions between vicinal hydrogens. Flagpole interactions should not destabilize a boat transition state, however, because C-2 and C-5 each carry only one hydrogen, and eclipsing interactions should also be less severe because the vicinal groups are separated by larger distances in the transition state.

---

1. Carey and Sundberg A., p. 621; Lowry and Richardson, p.968; March, p.1136.
2. P. Vittorelli, T. Winkler, H.-J. Hansen and H. Schmid, *Helv. Chim. Acta,* **51**, 1457 (1968).

In this multi-part experiment, you will build models of chair and boat transition states for different allyl vinyl ethers. These models will help you assess the role of different steric and electronic interactions in the transition state, and will show you how to predict transition state conformations for synthetically interesting rearrangements.

Procedure

Allyl vinyl ether: Build allyl vinyl ether. Optimize its geometry using AM1 semi-empirical calculations, and record its energy. Build models of the chair and boat conformers of the Claisen transition state of allyl vinyl ether. Begin by building a model in which the O-(C-6) and (C-3)-(C-4) distances are constrained to be 1.6 and 1.8 Å, and optimize each model's geometry using AM1 semi-empirical calculations while enforcing these constraints. Then, without enforcing any constraints, use the optimized structures to search for the AM1 transition states. Verify that the new models are true transition states by calculating and displaying their vibrational frequencies (a transition state should have only one imaginary frequency; see: **Finding Transition States**). Also, verify that the imaginary vibrational mode corresponds to a concerted rearrangement, i.e., verify that the O-(C-6) and (C-3)-(C-4) distances change simultaneously and in opposite directions. Record the energy of each transition state. Which one is more stable? Does this agree with the experimental results given above? What is the AM1 barrier for the Claisen rearrangement? How does this compare to the experimental value (30.6 kcal/mol)? What is the distance between the "flagpole" hydrogens in the boat transition state? Are the two hydrogens within the distance (<2.65 Å) where steric repulsion would be anticipated?

To get a better estimate of the Claisen rearrangement barrier, perform single-point 3-21G *ab initio* energy calculations (use AM1 geometries) for both allyl vinyl ether and the two transition states. Are the results qualitatively the same as you obtained using the AM1 model?

The relative stability of chair and boat transition states is better explained, perhaps, in terms of destabilizing orbital interactions.[3] The Claisen transition state can be viewed as two allyl radical fragments, O-(C-2)-(C-3) and (C-4)-(C-5)-(C-6), that are held together by overlap of their singly occupied π orbitals, i.e., each fragment has a π electron configuration $(\psi_1)^2(\psi_2)^1$ (see diagram).[4]

---

3. R. Hoffmann and R.B. Woodward, *J. Am. Chem. Soc.*, **87**, 4389 (1965).
4. This argument is somewhat simplistic, since the oxygen, because of its greater electronegativity, substantially perturbs the shape of the allyl π orbitals.

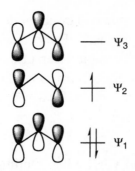

Interaction of the filled ψ_1 orbitals on each fragment yields two new filled orbitals, and is net destabilizing. Therefore, the preferred transition state should be the one that minimizes overlap between the ψ_1 orbitals. You can assess this overlap by examining the shape of the highest-occupied molecular orbital (HOMO) in the transition state (this results from an antibonding combination of the ψ_1 orbitals). Calculate and display the HOMO in each transition state. What combination of ψ_1 (or ψ_2) orbitals do these transition state orbitals most nearly resemble? Which conformation better minimizes the antibonding interactions in this orbital? Cite specific aspects of the orbital shape that support your conclusion. Compare the energies of the highest-occupied orbitals. Do these energies support the conclusion you reached using orbital shape?

The Hammond postulate provides another tool for thinking about transition state geometry (see: **Reactive Intermediates and the Hammond Postulate**). Since the Claisen rearrangement is exothermic, the transition state should resemble the reactant geometry more than the product. Record the CC and OC bond lengths in the two transition states. How do these compare to the bond lengths in the reactant and product? (Get the latter by building this molecule and optimizing its geometry using AM1 semi-empirical calculations.) Are your results consistent with the Hammond postulate?

A cyclic system: Büchi and Powell examined Claisen rearrangements of the *cis* and *trans* propenyldihydropyrans shown below as a general method for preparing cyclohexenes.[5] Their findings are most consistent with: 1) rearrangement via a boat transition state, and 2) facile rearrangement of the *trans* alkene, but slower (or no) rearrangement of the *cis* alkene.

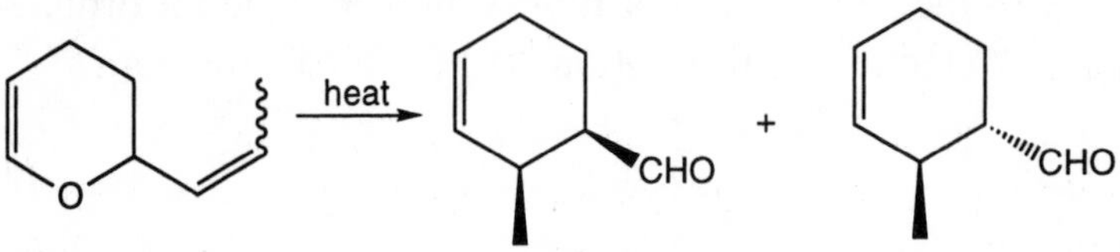

Build the *cis* and *trans* reactants and optimize their geometries using AM1 semi-

---

5. G. Büchi and J.E. Powell, Jr., *J. Am. Chem. Soc.*, **92**, 3126 (1970).

empirical calculations. Record their energies. Next, build models of the chair and boat transition states for the *cis* and *trans* alkenes (make sure that the double bond in the ring is *cis*).[6] Use these models to search for the corresponding AM1 transition states. Verify that the new models are true transition states (one imaginary frequency only), and carefully inspect the imaginary vibrational modes to see if the reaction coordinate corresponds to a concerted rearrangement (simultaneous change in O-(C-6) and (C-3)-(C-4) bond distances), or to a stepwise process (significant change in one bond distance only; see: **Finding Transition States**). Record the energies, and calculate the barriers associated with each pathway. How do these barriers compare to those calculated in the first part of this experiment? Record the O-(C-6) and (C-3)-(C-4) bond distances in your transition states. How do these values compare to those in the first part of this experiment? Is there any correlation between: 1) calculated barriers, 2) transition state geometries, and 3) nature of the transition state (concerted or stepwise)? Are your results in agreement with the conclusions of Büchi and Powell? Cite specific results and their relationship to the experimentalists' conclusions. What factor(s) do you think is responsible for the different behavior seen in these systems?

Optional

1.[7] Ireland and co-workers have developed a technique for generating allyl vinyl ethers of known stereochemistry from allyl esters. The ethers rearrange at relatively low temperatures, hence, the process shown below provides an efficient, stereoselective, and mild technique for constructing a chiral carbon skeleton.[8]

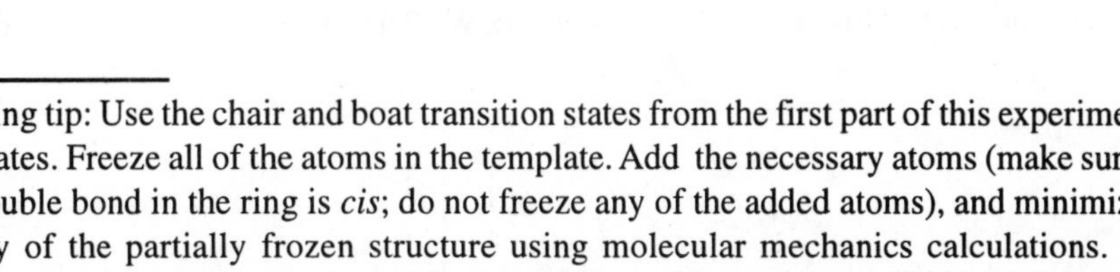

---

6. Building tip: Use the chair and boat transition states from the first part of this experiment as templates. Freeze all of the atoms in the template. Add the necessary atoms (make sure that the double bond in the ring is *cis*; do not freeze any of the added atoms), and minimize the energy of the partially frozen structure using molecular mechanics calculations. Next, optimize the geometry of each model using AM1 semi-empirical calculations subject to the same distance constraints used in the first part, and calculate a Hessian (vibrational frequencies) for the optimized structure. Finally, use this optimized structure and Hessian to search for the AM1 transition state (do not enforce constraints).

7. This experiment is rather time-consuming. Consider doing it as a group project where each student is responsible for two of the eight transition state calculations.

8. S. Pereira and M. Srebnik, *Aldrichimica Acta*, **26**, 17 (1993).

A curious interplay between alkene geometry and heteroatom-substituents was observed in the following system.[9]

alkene	X	$\Delta\Delta G^\ddagger$ (chair-boat, kcal/mol)
E	CH_2	-1.2
Z	CH_2	0.6
E	O	1.0
Z	O	1.6

The E and Z cyclohexenes (X = CH_2) prefer chair and boat transition states respectively, while the E and Z dihydropyrans (X = O) both prefer boat transition states (the dihydropyrans also rearranged more rapidly). In addition, changing X from CH_2 to O increases the preference for a boat transition state, and E alkenes have a weaker boat preference than Z. Tying these observations together requires a good set of transition state models.

Build models of the eight transition states (R=CH_3) resulting from all combinations of E and Z alkenes, rings (X=CH_2, O) substituents, and chair and boat transition states.[10] Use these models to search for AM1 transition states, and verify that these are true transition states for a concerted rearrangement. Record the energies of these transition states.[11] Which transition state is preferred for each reactant? Are these results consistent with the experimental observations? What is the effect of switching $E \rightarrow Z$ and $CH_2 \rightarrow O$ on the chair-boat energy difference? Are these results consistent (qualitatively or quantitatively) with the experimental observations? Chair-boat selectivity is affected by orbital interactions and by steric requirements. Can changes in either type of interaction explain your results? Cite specific observations that support your conclusions.

2. The Cope rearrangement is similar to the Claisen rearrangement in that a chair-like transition state is normally preferred. The rearrangement of E-3-methyl-3-phenyl-1,5-heptadiene gives E-3-methyl-6-phenyl-1,5-heptadiene as the major product, along with a smaller amount of the Z isomer.[12]

9. R.E. Ireland, P. Wipf and J.-N. Xiang, *J. Org. Chem.*, **56**, 3572 (1991).
10. The template approach employed previously is also useful here.
11. *Ab initio* 3-21G calculations may provide more accurate energies than AM1 calculations. If time permits, calculate the 3-21G energy associated with each AM1 transition state, and repeat the analysis described in the text. Which energies are superior, AM1 or 3-21G? Cite specific results that support your conclusion.
12. R.K. Hill and N.W. Gilman, *J. Chem. Soc., Chem. Commun.*, 619 (1967).

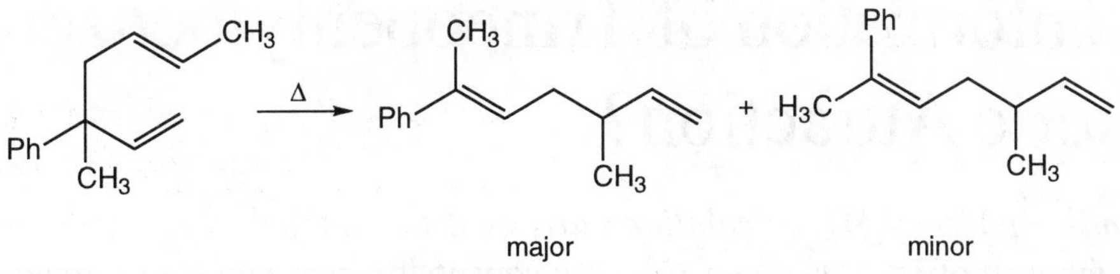

major   minor

Account for this result by building models of the competing transition states, and using them to search for the AM1 transition states. (Note: the desired transition state models are most easily constructed using the "template" strategy described above.) Verify that the resulting models are true transition states, and record their energies. Are chair-like transition states preferred? Which chair-like transition state is lower in energy, *axial* phenyl or *equatorial* phenyl? What role do steric and electronic (orbital) factors play in determining stereoselectivity in this reaction (cite specific observations that support your conclusions). Interestingly, rearrangement of the *R* reactant gives almost exclusively the *S* and *R* enantiomers of the major and minor products respectively. What does this imply about transition state geometry? Is this observation consistent with your results?

116

Conformation of Trineopentylbenzenes. Steric Attraction?

Semi-empirical AM1 calculations are used to establish the conformational preferences of 2,4,6-tribromo-1,3,5-trineopentylbenzene and 1,3-dibromo-2,4,6-trineopentyl-5-nitrobenzene.

Trineopentylbenzenes prefer conformations in which the bulky *tert*-butyl groups lie outside the plane of the benzene ring (**A-D**), and rotation through the plane is hindered by an appreciable barrier. Thus, the –30°C NMR spectrum of 1,3-dibromo-2,4,6-trineopentyl-5-nitrobenzene, **1** (X=Br, Y=NO$_2$), is consistent with two species, the major one being **A/C**, and the minor one being **B** or **D**.[1]

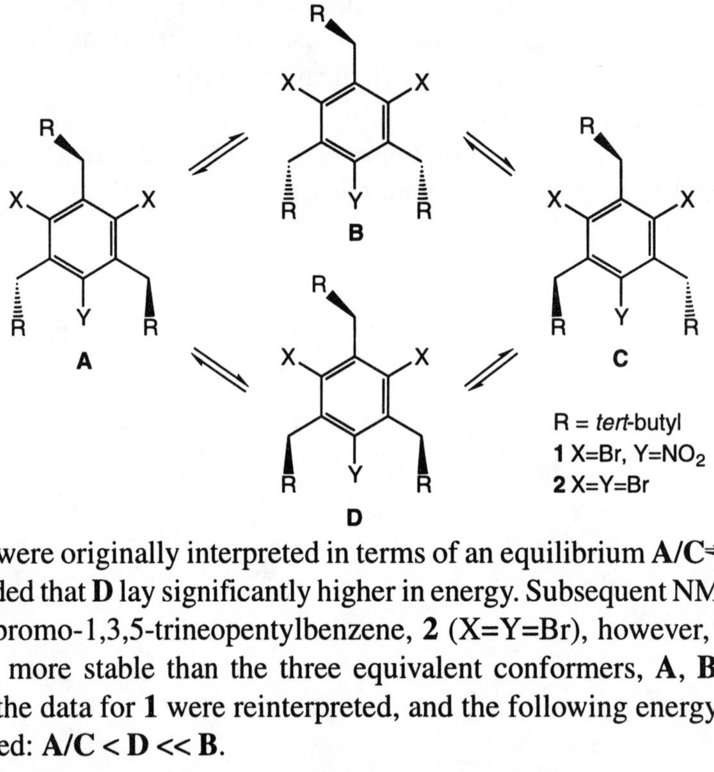

R = *tert*-butyl
1 X=Br, Y=NO$_2$
2 X=Y=Br

These data were originally interpreted in terms of an equilibrium **A/C ⇌ B**, and it was concluded that **D** lay significantly higher in energy. Subsequent NMR studies of 2,4,6-tribromo-1,3,5-trineopentylbenzene, **2** (X=Y=Br), however, indicated that **D** was more stable than the three equivalent conformers, **A**, **B**, and **C**.[2] Therefore, the data for **1** were reinterpreted, and the following energy ordering was proposed: **A/C < D << B**.

The original work on this system was hampered by the lack of adequate molecular modeling tools for estimating conformer stabilities. In this experiment, you will use semi-empirical AM1 calculations to estimate the relative energies of **1A** (= **1C**), **1B**, and **1D**, and also **2A** (= **2B** = **2C**) and **2D**. You will also assess the role of "steric attraction" in **2**.

1. B. Nilsson, P. Martinson, K. Olsson and R.E. Carter, *J. Am. Chem. Soc.*, **95**, 5615 (1973). The free energy difference between the major and minor species was estimated to be 0.3 kcal/mol.
2. R.E. Carter, B. Nilsson and K.J. Olsson, *J. Am. Chem. Soc.*, **97**, 6155 (1975). The free energy difference between the major and minor species was estimated to be 0.5 kcal/mol.

Procedure

Build **1A**, **1B**, **1D**, **2A**, and **2D**, and optimize their geometries using semi-empirical AM1 calculations. Record the energy of each molecule. Which conformer of **1** is the most stable? What are the energies of the other conformers relative to this conformer? Use these energies to calculate the ratio of **1A+1C** : **1B** : **1D** at –30 °C (assume that $\Delta G \approx \Delta H$). Are your results consistent with the NMR observations given above? Repeat the same analysis with **2** (calculate the ratio **2A + 2B + 2C : 2D**).

It has been suggested that **2D** is preferred due to "steric attraction" between the *tert*-butyl groups.[3] The idea is that nonbonding interactions between alkyl hydrogens, while strongly destabilizing if the H-H distance is <2.65 Å, are weakly stabilizing at somewhat longer distances. Assess this argument by examining the H-H distances in **2D**. How many of them are less than 2.65 Å, i.e., likely to produce strong repulsion? What happens to these distances in **2A**?

Another way to assess "steric attraction" is to examine the van der Waals component of the molecular mechanics strain energy. Calculate the strain energies of **2A** and **2D** at their AM1 geometries using molecular mechanics, and compare their van der Waals energies. Do these data support the "steric attraction" hypothesis? What other geometrical and/or electronic distortions might determine the relative stability of **2A** and **2D**? Support your ideas by citing specific features of your AM1 models.

---

3. A brief review of steric attraction can be found in R.R. Sauers, *J. Chem. Ed.*, **73**, 114 (1996).

IX

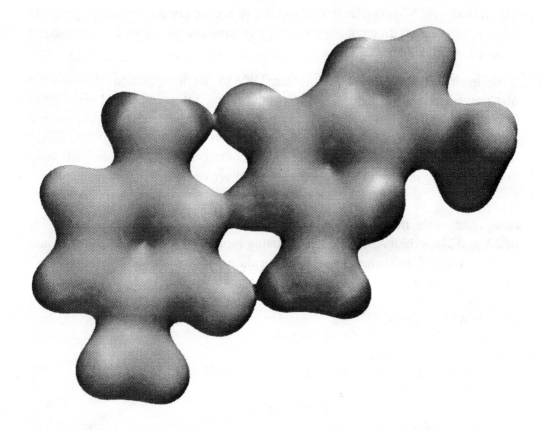

Graphical Models
and
Graphical Modeling

Graphical models play a significant role in the computational experiments contained in this book. Not only do they provide another view of molecular properties, i.e., different from and in addition to geometries, energies, charges, etc., they also provide a convenient format for assessing and interpreting certains types of numerical results.

The graphical models that have found the greatest use include images of molecular orbitals, electron densities, spin densities and electrostatic potentials. All of these quantities are functions in three-dimensions. Therefore, to render them for presentation on a two-dimensional computer display (or on paper) some subset of points must be chosen for display. One convenient method is to select only those points where the function takes on some arbitrarily chosen value. Joining these points together then yields one or more "isovalue surfaces" (or "isosurfaces"), the shape of which conveys information about the function or the molecule (this is particularly true if the value of the function is associated with some molecular property, such as "electronic size"). Another convenient method is to select a specific surface in advance, such as a plane through the molecule, and then map the value of the function on this surface. Such maps have proven especially useful when the selected surface is an isosurface of the total electron density.

Molecular Orbitals

Because of the Woodward-Hoffmann rules,[1] chemists have become interested in the shapes of molecular orbitals and the role that key orbitals play in chemical reactivity and selectivity. Molecular orbitals, $\psi_i(\mathbf{r})$, are written as linear combinations of atomic orbitals, $\phi_\mu(\mathbf{r})$, centered on atomic nuclei,

$$\psi_i(\mathbf{r}) = \sum_\mu c_{\mu i} \phi_\mu(\mathbf{r})$$

where the $c_{\mu i}$ are the molecular orbital coefficients. These follow from *ab initio* and semi-empirical molecular orbital calculations of the type presented in this book. The mathematical definition is reflected in the usual "cartoon" drawings of molecular orbitals which show the orbitals as combinations of atomic orbitals. For example, the π orbital in ethylene is usually drawn as.

Unfortunately, this type of cartoon has two drawbacks. First, it does not convey the true shape of $\psi_i(\mathbf{r})$; most of the isosurfaces of the ethylene π orbital show a single positive surface and a single negative surface, i.e.,

1. R.B. Woodward and R. Hoffmann, **The Conservation of Orbital Symmetry**, Verlag Chemie GmbH, Weinheim, 1970.

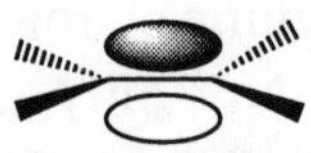

Second, cartoons are only useful for simple molecular orbitals where the number of contributing atomic orbitals is fairly small. Unfortunately, the molecular orbitals in most molecules are delocalized throughout the molecule, and contain contributions from many types of atomic orbitals (even on the same atom). They do not readily correspond to familiar chemical concepts, such as two-center bonds or nonbonded lone pairs.

Computer-generated isosurfaces, on the other hand, are both accurate and easily interpreted, even for relatively complex molecules. The following figures depict isosurfaces of the lowest-unoccupied molecular orbital (LUMO) in planar (left) and perpendicular (right) conformers of benzyl cation.

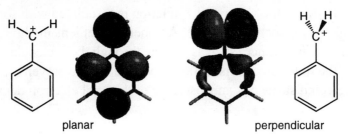

planar   perpendicular

The best way to think about the LUMO is that it identifies electron-deficient sites. In the planar cation, these sites correspond to the benzylic carbon, and also to the *ortho* and *para* ring carbons. In other words, the orbital is delocalized in a fashion similar to the resonance hybrid for this cation.

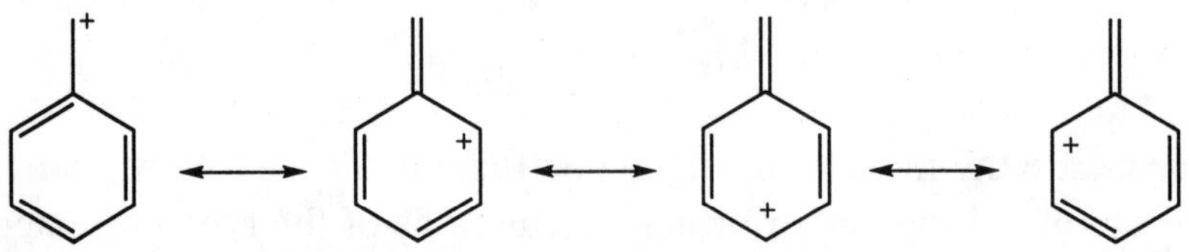

The LUMO in the perpendicular cation, on the other hand, is largely localized on the benzylic carbon, again in accord with resonance arguments.

Total Electron Densities

Where molecular orbitals may indicate how one or two electrons are distributed in a molecule, the total electron density, $\rho(\mathbf{r})$, provides information about all of the electrons. $\rho(\mathbf{r})$ is defined simply as the square of the electronic wavefunction, and in terms of atomic orbitals, ϕ, may be written,

$$\rho(\mathbf{r}) = \sum_\mu \sum_\nu P_{\mu\nu}\phi_\mu(\mathbf{r})\phi_\nu(\mathbf{r})$$

where $P_{\mu\nu}$ are elements of the density matrix. $\rho(\mathbf{r})\mathbf{dr}$, where $\mathbf{dr}$ is a small volume centered on $\mathbf{r}$, gives the probability of finding any electron $\mathbf{r}$ (integrating $\rho(\mathbf{r})\mathbf{dr}$

over all space gives the total number of electrons in the molecule). The total electron density occupies a special place in quantum chemistry because it is an observable quantity that can be measured by X-ray crystallography.[2] Molecular orbitals, by comparison, are useful mathematical fictions, and cannot be observed.

Isodensity surfaces can provide different types of useful information depending on the value of $\rho(\mathbf{r})$ used to define the surface.[3] $\rho(\mathbf{r})$ is largest near atomic nuclei, and isosurfaces corresponding to a high value of $\rho(\mathbf{r})$, e.g., 0.1 electrons/au^3, can be used to distinguish between different types of bonds (ionic vs. covalent, single vs. double vs. triple, etc.). These surfaces are also useful in delineating the bonding in unusual molecules and transition states.

Much lower values of $\rho(\mathbf{r})$ are found near a molecule's outer boundary, and 0.002 electrons/au^3 isodensity surfaces describe molecular volumes and shapes that are very similar to conventional space-filling (CPK) models. The advantage of this isodensity surface over a space-filling model is that it is general and free from unfounded biases, e.g., CPK models assume that every atom is spherical.

Spin Densities

Spin density distributions are used to describe the imbalance between spin-up (α spin) electron distributions and spin-down (β spin) electron distributions in open-shell molecules. The spin density, $\rho^{spin}(\mathbf{r})$, is defined as the difference between the total spin-up electron density, $\rho^{\alpha}(\mathbf{r})$, and the total spin-down electron density, $\rho^{\beta}(\mathbf{r})$.

$$\rho^{spin}(\mathbf{r}) = \rho^{\alpha}(\mathbf{r}) - \rho^{\beta}(\mathbf{r})$$

$\rho^{spin}(\mathbf{r})$ provides useful information about the location of unpaired electrons in free radicals, carbenes, and other open-shell molecules. $\rho^{spin}(\mathbf{r})$ is zero everywhere in closed-shell molecules, however, since their spin-up and spin-down electron distributions are identical.

A spin density isosurface ($\rho^{spin}(\mathbf{r})$ = 0.002 electrons/au^3) for allyl radical is shown below.

This comprises two separate surfaces, each of which is centered on a terminal carbon and is shaped roughly like a 2p atomic orbital. Thus, the unpaired α spin electron is located in the molecule's π system and is delocalized over these two centers, as might have been anticipated from resonance theory.

---

2.   That is to say, X-ray crystallography locates atoms by finding the largest concentrations of electrons.

3.   G.P. Shusterman and A.J. Shusterman, *J. Chem. Ed.*, in press (1997).

The advantage of spin density isosurfaces over resonance structures, of course, is that they are based on a realistic wavefunction, and are easily generated and manipulated, even for complex systems.

Electrostatic Potentials

The electrostatic potential, $\varepsilon(\mathbf{p})$, is defined as the energy of interaction of a point positive charge located at $\mathbf{p}$ with the nuclei and electrons of a molecule, i.e.,

$$\varepsilon(\mathbf{p}) = \sum_{A}^{\text{nuclei}} \frac{Z(\mathbf{A})}{R_{Ap}} - \int^{\text{all space}} \frac{\rho(\mathbf{r})}{R_{rp}} d\mathbf{r}$$

The summation describes the repulsive potential created by point charge-nucleus interactions, and runs over nuclei at $\mathbf{A}$; $Z(\mathbf{A})$ is the nuclear charge (atomic number) and R_{Ap} is the nuclear-point charge distance. The integral describes the attractive potential generated by point charge-electron interactions; $\rho(\mathbf{r})$ is the total electron density and R_{rp} is the electron-point charge distance.

Electrostatic potentials provide information about electron-rich and electron-poor sites inside a molecule, and electrostatic interactions between molecules. Positive isosurfaces tend to be less informative because they encompass all of the nuclei, i.e., the electrostatic potential always becomes positive near each of the nuclei (even in a molecular anion). Positive surfaces that extend beyond a molecule's nonbonded radius, however, identify electron-poor regions where nucleophilic attack is likely. Negative isopotential surfaces are less likely to surround large numbers of atoms (unless the molecule is negatively charged), and can be used to identify electron-rich regions where electrophilic attack is likely. For example, negative (-10 kcal/mol) electrostatic potential surfaces for trimethylamine (left), dimethylether (middle) and fluoromethane (right),

are a consequence of the heteroatom lone pairs, and indicate the preferred directions of protonation.

Multiple Surfaces and Surface Maps

Isosurfaces are somewhat limited in that they provide information about shape only. Thus, a 0.002 isodensity surface may define a molecule's shape and volume, but cannot identify electron-rich or electron-poor regions inside the molecule unless special precautions are taken. For this reason, it may be of use to display multiple surfaces based on several functions. For example, simultaneous display of 0.1, 0.01, and 0.001 isodensity surfaces provides a more complete picture of the total electron density distribution than any one of these isosurfaces alone.

Another useful technique is to simultaneously display the isosurface of a molecular orbital (or spin density or electrostatic potential) and a 0.002 isodensity surface. (Think of this surface much as you would a space-filling model, that is, revealing what an approaching reagent would see.) Most of the orbital's isosurface will be hidden inside the isodensity surface. These regions of the orbital are unavailable for overlap with orbitals on other molecules and can be ignored. Regions of the orbital isosurface that extend outside the isodensity surface, however, identify potentially reactive regions. An example of this type of display is illustrated below for cyclohexanone.

axial   *equatorial*

Here, both a LUMO isosurface and a 0.002 isodensity surface are displayed, and only the "axial" face of the LUMO extends outside the isodensity surface. Therefore, nucleophilic addition should occur primarily from this direction.[4]

An alternative method for assessing intermolecular interactions (and other molecular properties) is to map the value of a molecular orbital, spin density or electrostatic potential on the molecule's 0.002 isodensity surface. Different colors are used to show the value of the property on this map. The use of color coding allows "hot spots" where the property takes on particular high (or low) values to be quickly identified. Thus, a LUMO map of cyclohexanone can be used to identify the more reactive face of the carbonyl group by showing the value of the LUMO over each face.

Electrostatic potential maps are particularly useful.[3] By identifying electron-rich and electron-poor regions within a molecule, and the shapes of these regions, they provide information about bonding, charge distribution, reactivity, and intermolecular interactions. These maps can also be used to make comparisons between molecules. It is essential in this case, however, to make the color scales of the two maps identical, i.e., the colors on two different maps should only be compared if they correspond to the same value of the electrostatic potential.

4.   S. Fielder, PhD Thesis, University of California, Irvine, 1993.

Acidities and Electrostatic Potentials

Electrostatic potential maps obtained from 3-21G ab initio calculations are used to examine changes in gas-phase acidity due to changes in hybridization, ring strain, and substitution by donor and acceptor groups. The ability of electrostatic potential maps to distinguish between weak and strong acids is also examined.

The acidity of an organic acid, HA, should be reflected to some extent in the partial charge on hydrogen. A stronger acid should contain a more electron-poor hydrogen, while a weaker acid should contain a more electron-rich hydrogen. Charge distributions in different acids can be compared by comparing electrostatic potential maps of their nonbonded isodensity surfaces (see: **Graphical Models and Graphical Modeling**). Hydrogens with larger partial positive charges will appear as particularly electron attracting (positive potential).[1]

In this experiment, you will examine electrostatic potential maps for a variety of simple molecules in order to assess qualitatively their relative acidities.

Procedure

Relative acidities of ethane, ethylene, and acetylene: Hydrocarbon acidity increases in the order: alkane < alkene < alkyne. The usual explanation is that the lone pairs resulting from deprotonation increase in energy: sp lone pair < sp^2 lone pair < sp^3 lone pair. Build ethane, ethylene, and acetylene, and optimize their geometries using 3-21G *ab initio* calculations. Calculate electrostatic potential maps and display these side-by-side, and on the same color scale. Note that the potential is most positive near the hydrogens. Record the most positive value of the potential on each surface. Which molecule contains the most electron-poor hydrogen? Is this the most acidic molecule? Which molecule contains the least electron-poor hydrogen? Is this the least acidic?

Effect of ring strain on acidity: Hybridization is commonly invoked to rationalize the higher acidities of strained hydrocarbons relative to unstrained systems. For example, the CH hybrids in cyclopropane have less p character and are lower in energy than the sp^3 hybrids in propane, suggesting that the former is more acidic.[2] Build propane and cyclopropane, optimize their 3-21G geometries, and calculate their electrostatic potential maps. Display the maps side-by-side, and on the same color scale. Note that the potential is most positive near the

---

1. Polarization of the HA bond increases the potential near hydrogen in two ways: it removes electron density from this region of the molecule so that the potential is dominated by the nuclear-probe repulsion term, and it causes the isodensity surface to move closer to the hydrogen nucleus so that nuclear-probe repulsion is enhanced.

2. Isaacs, p.249; Lowry and Richardson, p.293.

hydrogens. Record the most positive value of the potential on each surface. Which molecule contains the most electron-poor hydrogen? Is this the most acidic molecule? Compare the electrostatic potential map of cyclopropane with those of ethylene and acetylene. Do you expect cyclopropane to be more or less acidic than ethylene? Are your predictions consistent with experiment?

Substituent effects on acidity: Donor and acceptor substituents, because they will affect the stability of the lone pair resulting from deprotonation, are expected to affect acidity. Donors destabilize the lone pair and decrease acidity, while acceptors have the opposite effect. Build propyne and 3,3,3-trifluoropropyne, optimize their 3-21G geometries, and calculate their electrostatic potential maps. Display the maps side-by-side and on the same color scale, along with the electrostatic potential map of acetylene. Record the maximum potential found in the vicinity of the alkyne hydrogen. What do you expect to be the relative acidity of these molecules? Based on this result, would you classify a methyl group as a donor or an acceptor? How would you classify a trifluoromethyl group? How do the magnitudes of these substituent effects compare with those caused by changes in hybridization?

Strong and weak acids: Build ethanol (a weak acid), acetic acid (a moderately strong acid), and nitric acid (a strong acid), optimize their 3-21G geometries, and calculate their electrostatic potential maps. Display the maps side-by-side, and on the same color scale. Do the maps correctly anticipate the relative acidities of these molecules?

Optional

Chemists often estimate an acid's strength by examining the properties of its conjugate base. A strong acid has a weak conjugate base, while a weak acid has a strong conjugate base. Relative base strengths can be estimated qualitatively using electrostatic potential maps by comparing the electron-rich regions in different bases. A weak base is less likely to share its electrons (less negative potential), while a stronger base will have a more negative potential. Use this relationship to compare acidity within any of the groups of molecules listed above (build and optimize the anionic conjugate bases of the molecules, and compare electrostatic potential maps side-by-side and on the same color scale).

The Acidity of Quadricyclane

3-21G ab initio calculations and electrostatic potential maps are used to assign the site of deprotonation in quadricyclane, and to rationalize the observed rapid scrambling of hydrogens.

Quadricyclane, **2**, is readily formed by UV irradiation of norbornadiene, **1**.

While much less stable than norbornadiene,[1] quadricyclane persists at room temperature, but reverts to norbornadiene when heated above 140°C. The fact that quadricyclane is highly strained suggests that it will be a stronger acid than "typical" saturated hydrocarbons (see: **Acidities and Electrostatic Potentials**).

The acidity of quadricyclane in the gas phase has been determined,[2] although the site of deprotonation has not been conclusively established. Deuterium labeling studies indicate that six of the eight hydrogens in quadricyclane exchange. This suggests either that the acidities of six of the CH bonds are very similar, or that the anions resulting from deprotonation at different sites rapidly interconvert.

In this experiment, you will use 3-21G *ab initio* calculations together with electrostatic potential maps to predict relative acidities of the CH bonds in quadricyclane. You will then examine the possible anions resulting from deprotonation of quadricyclane, and the pathway interconnecting the anions.

Procedure

Build quadricyclane, **2**. Optimize its geometry using the 3-21G *ab initio* molecular orbital method,[3] and calculate an electrostatic potential map. Display the map and locate the three unique sets of hydrogens (at $C_1 = C_5 = C_6 = C_7$, at $C_2 = C_4$ and at C_3). According to this measure, which set of hydrogens should be the most acidic? Is the acidity of any other set similar? Can you rationalize your results?

---

1. The heat of formation of quadricyclane is estimated at 80 kcal/mol, compared to a heat of formation of 57 kcal/mol for norbornadiene.
2. H.S. Lee, C.H. DePuy and V. Bierbaum, *J. Am. Chem. Soc.*, **118**, 5068 (1996).
3. It might be advantageous for you to first perform a geometry optimization using the AM1 semi-empirical method, and then use the resulting structure as a guess for the 3-21G optimization.

127

Build the different anions resulting from deprotonation of **2**, i.e., **3** from deprotonation at C_1, **4** from deprotonation at C_2 and **5** from deprotonation at C_3. Optimize their 3-21G geometries.[3]

| 3 | 4 | 5 |

Are the valence structures for **3-5** accurate insofar as their depiction of negative charge? Calculate and display the highest-occupied molecular orbital (HOMO) and an electrostatic potential map for each. Which anion is the most stable? Are any of the alternatives close in energy? Are the valence representations for **3-5** consistent with the HOMO for each of these anions? Do the calculated anion stabilities comply with your expectations based on the electrostatic potential map for quadricyclane?

Identify the two most stable anions among **3-5**, and build a transition state connecting them. i.e., involving migration of a hydrogen.[4] Use this structure to search for the 3-21G transition state. Do you find a low or a high barrier to hydrogen migration? Does your result suggest that interconversion among anions will be easy or difficult? Propose an explanation for the observed hydrogen exchange.

---

4. The easiest way to do this is to constrain the "migrating hydrogen" to be equidistant (1.4Å) between the two carbons. It might be advantageous for you to first perform a transition state optimization using the AM1 semi-empirical method, then calculate a Hessian at the AM1 level (frequency calculation), and use both the AM1 structure and Hessian as a guess for the 3-21G transition-state optimization.

pK$_a$'s of Carboxylic Acids

Electrostatic potential maps from semi-empirical AM1 calculations are correlated with pK$_a$'s of carboxylic acids.

One of the most important discoveries in physical organic chemistry was Hammett's observation that the acidity of a substituted benzoic acid (pK$_a$) is often correlated with the reaction rate of an identically substituted aromatic compound.[1] For example, the pK$_a$'s of various acids are correlated with the hydrolysis rates of their methyl esters. Indeed, the construction of quantitative correlations between "structure", as reflected by the pK$_a$ value, and some other property, such as reaction rate, has found application in many areas of chemistry.

Hammett-type correlations have largely been confined to molecules containing substituted benzene rings. However, it should be possible to use molecular models to develop "structural" parameters that could be used to correlate chemical behavior in a wider variety of molecules. In this experiment, you will use semi-empirical AM1 calculations to derive electrostatic potential maps for a series of carboxylic acids (see: **Graphical Models and Graphical Modeling**), and see whether the potential near the carboxylic acid proton can be correlated with pK$_a$.

Procedure

Build the acids shown below and optimize their AM1 geometries.

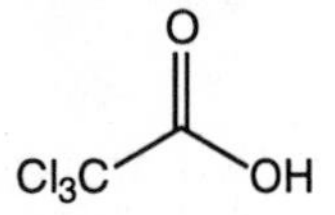

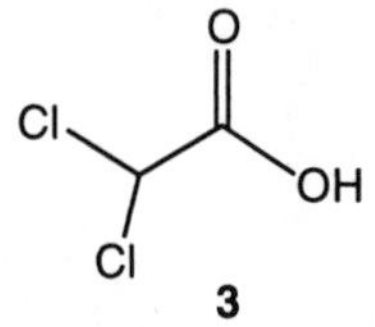

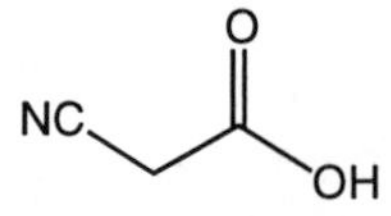

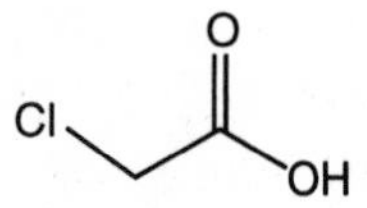

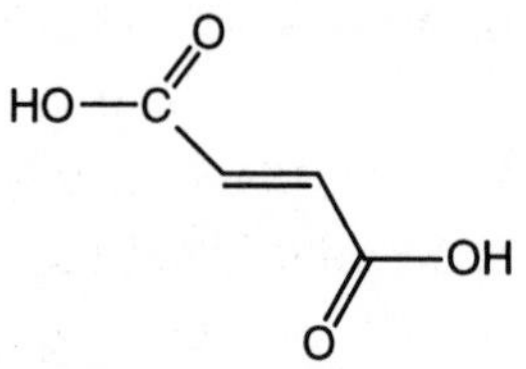

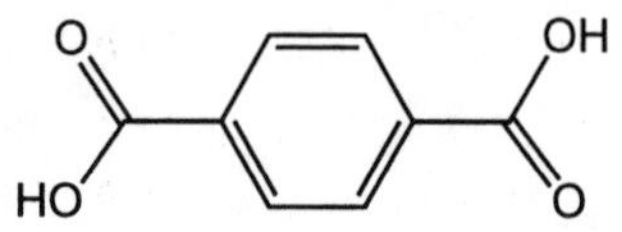

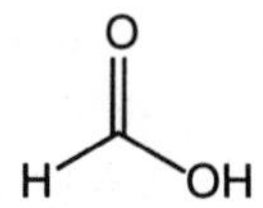

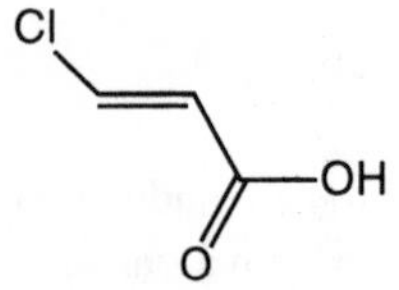

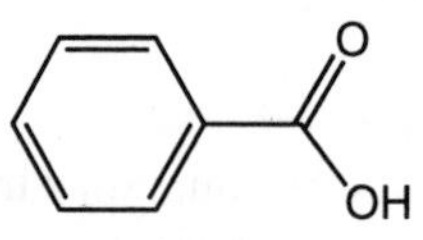

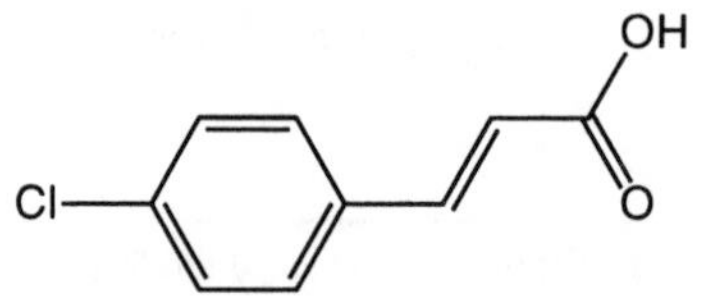

---

1. Carey and Sundberg A, p. 196; Lowry and Richardson, p. 143; March, p. 229.

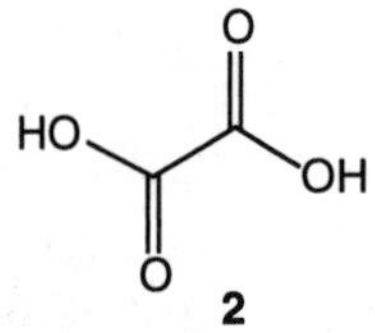

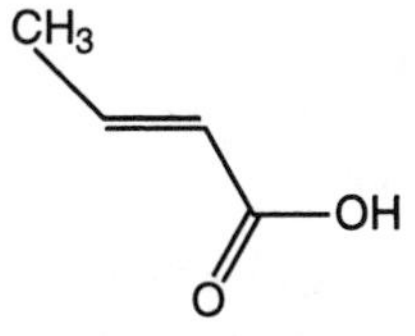

CH₃ ... OH
12

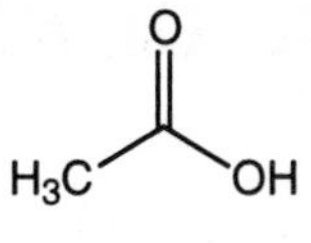

H₃C ... OH
13

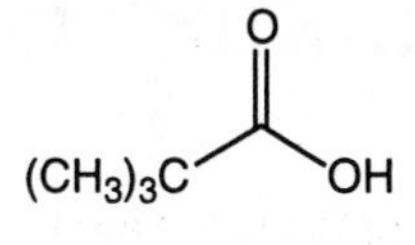

(CH₃)₃C ... OH
14

Calculate the electrostatic potential map for each molecule, and record the most positive potential in the vicinity of the carboxylic acid proton. Plot potential against the acid's aqueous-phase pK_a (given below).

Experimental (aqueous phase) pK_a's or carboxylic acids[2]							
1	0.70	5	2.85	9	3.79	13	4.75
2	1.23	6	3.10	10	4.19	14	5.03
3	1.48	7	3.51	11	4.41		
4	2.45	8	3.75	12	4.70		

Is there a linear correlation? What is the slope of the correlation line? Is this reasonable? Explain. Do you think that these potentials could be used as "structural" parameters in a Hammett-type relationship? Take as a test case, the alkaline hydrolysis rates of the methyl esters of these acids.

Optional

1. Perform single-point-energy and electrostatic potential map calculations using the AM1-SM2 model (AM1 geometries). Does inclusion of solvation alter any of your conclusions based on the gas-phase calculations?

2. Carboxylic acids are moderately strong acids, in part, because their anionic conjugate bases are delocalized. Therefore, pK_a might show a better correlation with the electrostatic potential in these anions. Build the anionic conjugate base of each acid shown above, optimize their AM1 geometries, and calculate their electrostatic potential maps. Record the most negative potential in the vicinity of each carboxylate group, and plot this potential against pK_a. Is there a linear correlation? Is it a better or worse correlation than the one found above? Does the slope of the correlation line have the expected sign? Is the slope of this line larger or smaller in magnitude than the one found above? Is this reasonable? Explain.

---

2. E.P. Sarjeant and B. Dempsey, **Ionization Constants of Organic Acids in Aqueous Solution**, IUPAC no. 23, Pergamon Press, 1979.

130

Ranking Basicities of Amines

Semi-empirical and ab initio molecular orbital calculations along with electrostatic potential maps are used to order basicities in substituted anilines and in cyclohexylamine.

The base strength of an amine is defined by the pK_a of its conjugate acid, an ammonium ion. A weakly acidic ammonium ion corresponds to a strongly basic amine. For example, anilinium ions, $ArNH_3^+$, are generally more acidic than cyclohexylammonium ion. Thus, anilines are weaker bases than cyclohexylamine. Aniline basicity is influenced by substituents on the ring. Electron-withdrawing substituents, such as NO_2, reduce base strength, while electron-donating substituents, such as OCH_3, increase base strength.[1]

There are several ways that molecular orbital calculations can be used to predict the relative basicity of different amines. For example, the most basic site in a molecule should correspond to the atom with the most negative electrostatic potential (see: **Graphical Models and Graphical Modeling**). By comparing electrostatic potentials of different molecules, it should be possible to determine their relative basicity. One could also predict the relative acidity of the corresponding ammonium ions in the same way.

An alternative account of relative basicity follows from the energetics of protonation of one amine, XNH_2, relative to that of another, YNH_2, i.e.,

$$XNH_2 + YNH_3^+ \rightleftharpoons XNH_3^+ + YNH_2$$

$$\Delta G \approx \Delta H = \Delta H_f (XNH_3^+) + \Delta H_f (YNH_2) - \Delta H_f (XNH_2) - \Delta H_f (YNH_3^+)$$

It is necessary to assume that the experimental relative free energies of protonation (ΔG) in solution closely approximate relative enthalpies of protonation (ΔH) in the gas phase, i.e., that entropy differences are very small, and that solvation effects will mostly cancel for different amines. Both assumptions are likely to be valid where comparisons are made between systems with similar size and shape, e.g., substituted anilines and cyclohexylamines, but are questionable for comparisons involving greatly different molecules.

In this experiment, you will use semi-empirical and *ab initio* molecular orbital calculations, together with electrostatic potential maps to examine the relative basicities of a number of substituted anilines and of cyclohexylamine. Your objectives will be: 1) to compare the effects of the two different ring systems on amine basicity, 2) to relate substituent effects on the basicity of aniline to both the nature and position of the substituents, and 3) to determine which modeling methods properly anticipate the observed relative basicities of amines.

---

1.   March, p. 263.

amine (conjugate base of ammonium ion)	pK$_a$ (ammonium ion)
4-nitroaniline	1.00
3-nitroaniline	2.47
aniline	4.63
4-methoxyaniline	5.34
cyclohexylamine	10.66

Procedure

Electrostatic potential maps for amines: Build aniline, 3- and 4-nitroaniline, 4-methoxyaniline and cyclohexylamine, and optimize their geometries using the AM1 semi-empirical method. Calculate and display electrostatic potential maps on the same color scale. Do the maps correctly anticipate the known ordering of basicities? Do they suggest that cyclohexylamine is a stronger base than aniline? Are they consistent with the observed substituent effects?

Electrostatic potential maps for ammonium ions: Build the conjugate acids of aniline, 3- and 4-nitroaniline, 4-methoxyaniline and cyclohexylamine and optimize their AM1 geometries. Calculate electrostatic potential maps and display them on the same color scale. Do the maps anticipate the observed ordering of ammonium acidities?

Thermochemistry of protonation: Using the AM1 heats of formation from your calculations, evaluate the enthalpies of reactions,

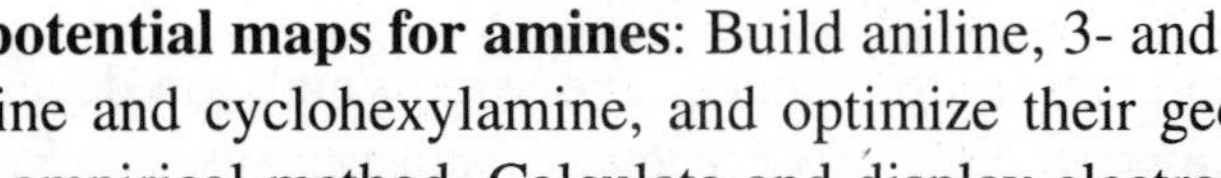

$$XNH_2 + \text{(benzene ring)}-NH_3^+ \longrightarrow XNH_3^+ + \text{(benzene ring)}-NH_2$$

where XNH$_2$ is 3- and 4-nitroaniline, 4-methoxyaniline and cyclohexylamine. These provide a measure of base strength relative to aniline as a standard. Do your data properly reproduce the experimental ordering of base strengths in these compounds? Focus, in particular, on the difference in basicity between aniline and cyclohexylamine and on the effect of substituents on the basicity of aniline.

Optional

Repeat all of your calculations using the AM1-SM2 method (which accounts for aqueous solvent) in place of AM1; use the AM1 geometries and perform single-point energy calculations. Does the solvent change the ordering of amine basicities? Does it amplify or lessen basicity differences noted in the gas phase? (See: **The Role of Solvent**.)

Electrostatic Potentials and the Rate of Radical Addition to Carbon-Carbon Double Bonds

Electrostatic potential maps obtained from semi-empirical AM1 calculations are used to examine if the rate of radical addition to alkenes depends on the electron deficiency of the double bond.

Free radicals add readily to alkenes. The rate of addition depends both on the electron donating/withdrawing character of substituents attached to the double bond, e.g.,

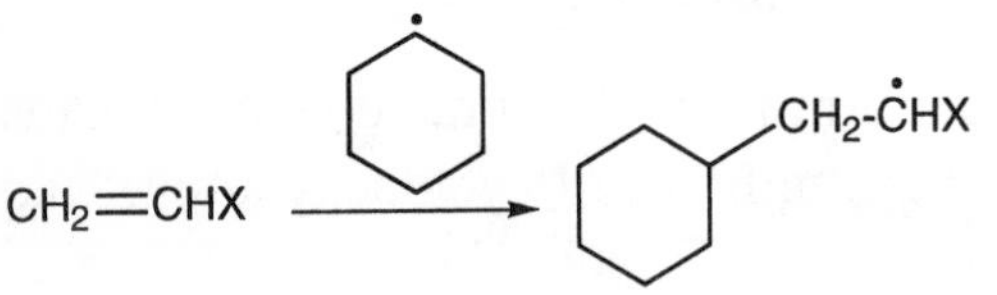

CH$_2$=CHX

		relative rate
X=	n-Bu	1
	Ph	84
	CO$_2$Me	3000
	CN	6000
	CHO	8500

and on the steric environment of the alkene, e.g.,

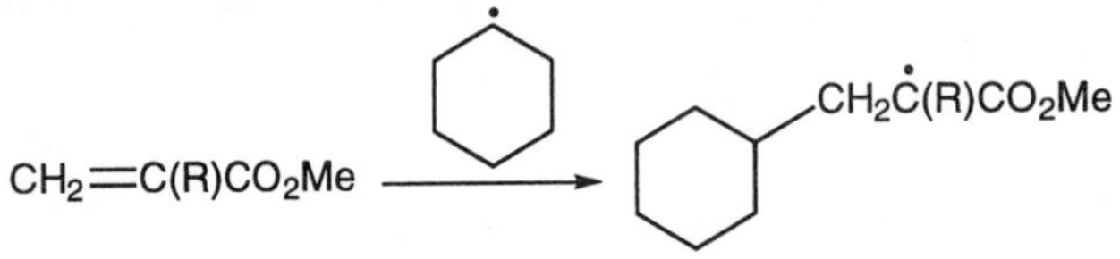

CH$_2$=C(R)CO$_2$Me

		relative rate
R=	H	1
	Me	0.75
	tert-Bu	0.26

In this experiment, you will use electrostatic potential maps obtained from semi-empirical AM1 calculations to ascertain whether the observed rate effects are due to electron deficiency of the carbon-carbon double bond or to steric crowding (see: **Graphical Models and Graphical Modeling**).

Procedure

Electrostatic potentials for unencumbered alkenes: Optimize the geometries of alkenes H$_2$C=CHX (X= n-Bu, Ph, CO$_2$Me, CN and CHO) using the semi-empirical AM1 method, and for each calculate an electrostatic potential map. Measure and record the value of the potential over the π face of the β carbon, and plot potential versus the log of the relative reaction rate. Do you observe an overall correlation between electron deficiency of the alkene (as measured by the electrostatic potential) and reaction rate? Point out any significant deviations and try to rationalize them.

Electrostatic potentials for sterically encumbered alkenes: Optimize the AM1 geometries of alkenes $H_2C=C(R)CO_2Me$ (R= Me, *tert*-Bu), and for each calculate an electrostatic potential map. Measure and record the value of the potential over the π face of the β carbon. Do your results agree with the previous correlation? How do you explain this?

Optional

Are electrostatic potential maps able to predict relative acid-catalyzed hydration rates of alkenes?

$$H_2C=CH_2 \; < \; CH_3CH=CH_2 \; < \; CH_3CH_2CH=CH_2 \; < \; (CH_3)_2C=CH_2 \; < CH_3)_2C=C(CH_3)_2$$
$$\quad\; 1 \qquad\qquad\quad 2 \qquad\qquad\qquad 3 \qquad\qquad\qquad 4 \qquad\qquad\qquad 5$$

The overall mechanism is presumed to involve a carbocation intermediate, e.g., formation of *tert*-butyl cation in the hydration of isobutene,

$$(CH_3)_2C=CH_2 \xrightarrow{\;H_3O^+\;} (CH_3)_2\overset{+}{C}\text{-}CH_3 \xrightarrow{\;H_2O\;} (CH_3)_2\overset{\overset{\textstyle OH}{|}}{C}\text{—}CH_3$$

the stability of which might be expected to be related to the electrostatic potential in the vicinity of the double bond.

Build **1-5**, optimize using the AM1 method, and calculate electrostatic potential maps. Display all on the same color scale. Is the ordering of the potentials you observe consistent with the relative rates of hydration? For the asymmetrically substituted systems **2-4**, do you see any asymmetry in the electrostatic potential map that could be used to predict reaction regioselectivity? Are the potentials consistent with Markovnikov's rule, i.e., that protonation occurs at the less-substituted carbon?

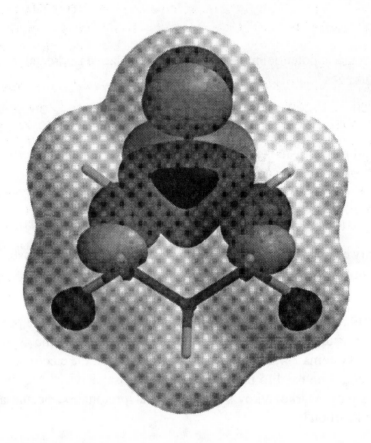

Visualizing
Molecular Orbitals

C hemists are adept at using qualitative models. They use steric models to help them identify molecules that are crowded and may assume unusual (and unfavorable) geometries, and chemical reactions that are likely to be very slow simply because the reactants can't get close. At the other extreme, chemists employ qualitative resonance arguments to spot molecules, that because they are able to delocalize their charge, are likely to be exceptionally stable. The pioneering work of Woodward and Hoffmann[1] prompted chemists to think in terms of qualitative molecular orbital theory. They showed clearly how the shapes and particularly symmetries of key molecular orbitals can anticipate whether a particular chemical reaction will be favorable, and what the likely products will be. Actually, the use of molecular orbitals as qualitative models predates Woodward and Hoffmann by many years. For example, very early work by Walsh[2] showed how the shapes of the valence molecular orbitals of AH_2 and AH_3 molecules allowed rationalization of their geometries.

To some extent, the advent of practical quantitative molecular orbital models should reduce chemists' reliance on qualitative reasoning. Another perspective is that the availability of quantitative molecular orbital calculations will encourage the development of better qualitative models. This is the point of view taken in this book.

Interpretation of the molecular orbitals that emerge from quantitative calculations is a key step. Here, the stepwise construction of the valence molecular orbitals of a methyl group both in a planar geometry (as in methyl cation) and in a pyramidal geometry (as in methyl anion) is illustrated.[3]

Planar CH₃

The lowest-energy of the three molecular orbitals describing "CH bonding" results from a combination of the 2s atomic orbital on carbon and 1s atomic orbitals on each hydrogen:

$$\psi_1 = .83\ 2s^C + .32\ 1s^{H_1} + .32\ 1s^{H_2} + .32\ 1s^{H_3}$$

---

1. R.B. Woodward and R. Hoffmann, **The Conservation of Orbital Symmetry**, Verlag Cheme GmbH, Weinheim, 1970.
2. A.D. Walsh, *Prog. Stereochem.*, **1**, 1 (1954).
3. Molecular orbital coefficients are from AM1 semi-empirical calculations. There are seven atomic orbitals (2s, $2p_x$, $2p_y$, $2p_z$ on carbon and 1s on each hydrogen).

This molecular orbital reflects the three-fold symmetry of the methyl group. No other bonding combinations of atomic orbitals that do this are possible. The remaining CH bonding orbitals may only reflect part of the symmetry and, because of this, must be of equal energy (they are said to be "degenerate").

$$\psi_2 = .72 \ 2p_x^C - .47 \ 1s^{H_2} + .47 \ 1s^{H_3}$$

$$\psi_3 = .74 \ 2p_z^C + .54 \ 1s^{H_1} - .27 \ 1s^{H_2} - .27 \ 1s^{H_3}$$

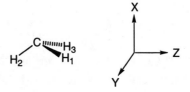

The next highest energy orbital is simply the $2p_y$ atomic orbital on carbon.

This orbital is empty in methyl cation, half filled in methyl radical and doubly occupied in methyl anion.

Pyramidal CH$_3$

The valence molecular orbitals of pyramidal methyl are formed in a similar way. The lowest-energy CH bonding combination,

$$\psi_1 = .77 \ 2s^C - .11 \ 2p_x^C + .36 \ 1s^{H_1} + .36 \ 1s^{H_2} + .36 \ 1s^{H_3}$$

results first from hybridization of 2s and 2p atomic orbitals on carbon, followed by mixing of this hybrid with a fully symmetric combination of hydrogen 1s functions.

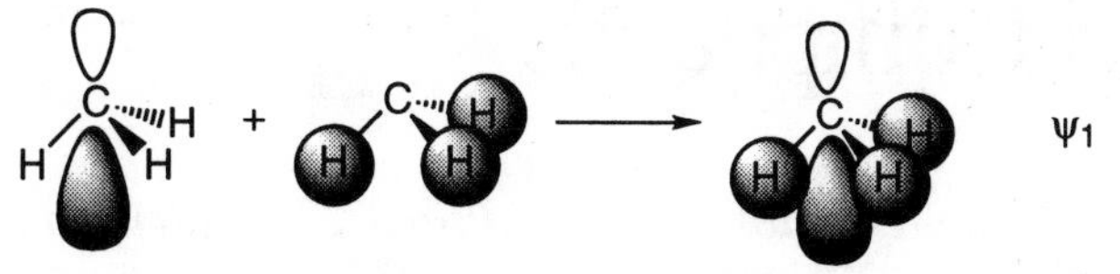

ψ_1

As with the planar methyl group, no further combinations of CH bonding orbitals which reflect the threefold symmetry of the molecule are possible, and a degenerate pair of orbitals, each of which partially reflects the molecule's symmetry, form instead.

$$\psi_2 = .65 \, 2p_z^C + .31 \, 1s^{H_1} - .62 \, 1s^{H_2} + .31 \, 1s^{H_3}$$

$$\psi_3 = .65 \, 2p_y^C + .53 \, 1s^{H_1} - .53 \, 1s^{H_3}$$

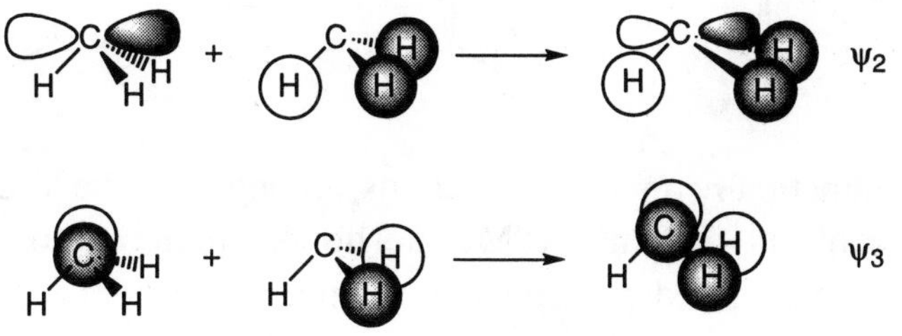

ψ_2

ψ_3

The remaining valence orbital in pyramidal CH_3 is primarily an sp hybrid on carbon pointing away from the hydrogens.

$$\psi_n = -.29 \, 2s^C - .93 \, 2p_x^C + .12 \, 1s^{H_1} + .12 \, 1s^{H_2} + .12 \, 1s^{H_3}$$

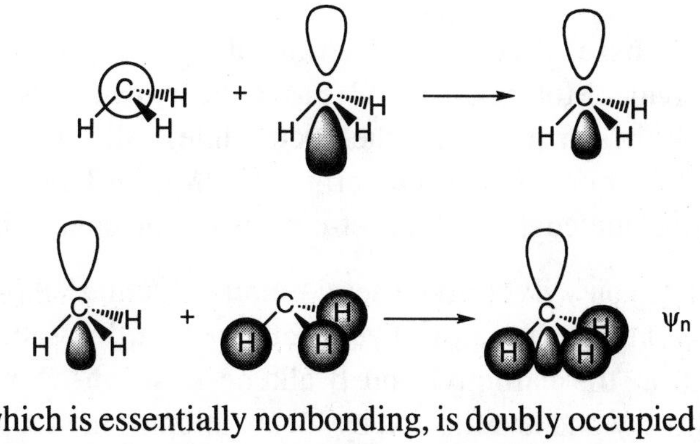

ψ_n

This orbital, which is essentially nonbonding, is doubly occupied in methyl anion, half occupied in methyl radical and empty in methyl cation.

Nucleophilic Addition to α,β-Unsaturated Carbonyl Compounds

Semi-empirical AM1 calculations and ab initio 3-21G calculations, and Frontier Molecular Orbital theory are used to rationalize why nucleophilic attack on an α,β-unsaturated ketone may occur either at the carbonyl carbon or at the β carbon.

α,β-Unsaturated ketones, such as cyclohexenone, may undergo nucleophilic addition at either the carbonyl or β carbons, i.e.,

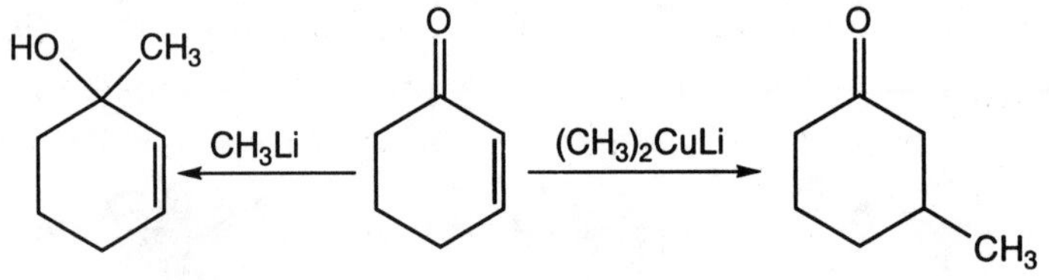

A number of synthetically important reactions, among them the Michael reaction,[1] involve addition of a nucleophile, a "Michael donor", to the carbon-carbon double bond of an α,β-unsaturated carbonyl compound, a "Michael acceptor", e.g.,

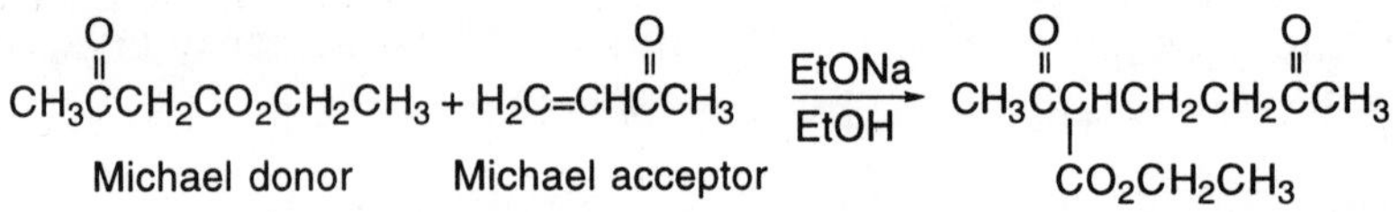

The selectivity of these reactions is determined by both thermodynamic and kinetic factors. Frontier Molecular Orbital (FMO) theory can be used to investigate the latter. Reactive sites in the ketone (the electrophile) will correspond to regions where its lowest-unoccupied molecular orbital (LUMO) is large and can overlap effectively with the nucleophile's highest-occupied molecular orbital (HOMO).

In this experiment, you will use both semi-empirical and *ab initio* calculations together with FMO theory to rationalize reactivity of α,β-unsaturated carbonyl compounds both at the carbonyl and β alkene positions, both in terms of equilibrium geometry and as reflected by localization of the LUMO.

---

1. Carey and Sundberg B, p. 39; March, p. 741.

Procedure

Build cyclohexenone and optimize its geometry using AM1 calculations. Examine the resulting geometry for evidence of resonance interaction between the unsaturated groups, i.e.,

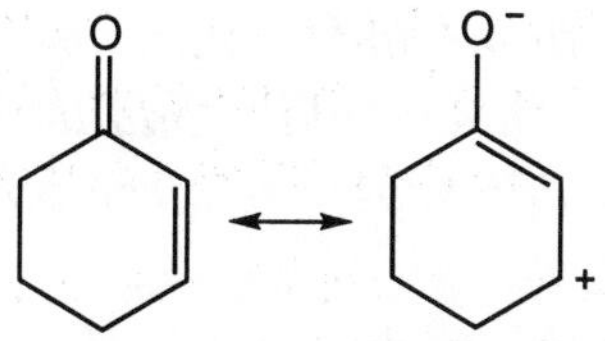

In particular, a resonance interaction might cause changes in the bond distances in the two multiple bonds and in the CC single bond separating them (the bond distances in cyclohexanone and in cyclohexene may be regarded as "normal"). Do the distances indicate a resonance interaction? Explain.

Perform a single-point 3-21G *ab initio* calculation on cyclohexenone using the AM1 optimized geometry. Calculate and display the LUMO and the LUMO map. Examine the LUMO first. Note that this orbital is delocalized over several atoms, and this makes it difficult to identify reactive sites. The LUMO map provides a much clearer picture. Regions where the LUMO value is large indicate sites where HOMO-LUMO overlap will be effective with a minimal degree of steric interference. Do you see at least two sites for nucleophilic attack? Which of these sites appears best suited for HOMO-LUMO overlap? (Note that the outcome of a nucleophilic addition also requires favorable thermodynamics, and this factor is not considered in the FMO approach.)

Optional

Repeat your calculations on 3-methylcyclohexenone.

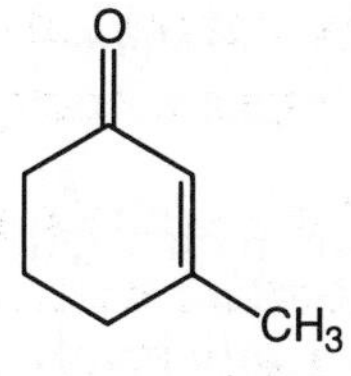

Compared with cyclohexenone, would you expect this molecule to be more or less susceptible to Michael addition?

140

Stereochemistry of Nucleophilic Addition to Norcamphor

Semi-empirical AM1 calculations and ab initio 3-21G calculations are used to predict the stereoselectivity of nucleophilic addition to norcamphor. The results are analyzed in terms of a balance between frontier orbital and steric effects.

Borohydride reduction of norcamphor, **1** (R=H), yields a mixture of bicyclic alcohols, **2** and **3**.[1]

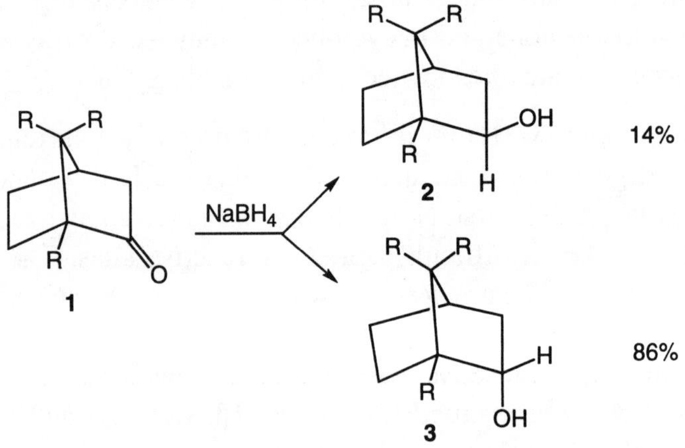

Here, product stereochemistry is kinetically controlled, i.e., it is determined primarily by the relative rates of attack of the nucleophile (BH_4^-) on the two faces of norcamphor. The question is, "how can this stereoselectivity be predicted?"

A useful model for predicting stereoselectivity in kinetically controlled nucleophilic additions considers three factors:

1) steric interactions due to electron-electron repulsion between electrons in filled orbitals,

2) electrostatic interactions between charged sites, and

3) frontier orbital interactions typically involving a bonding interaction between the highest-occupied molecular orbital (HOMO) on the nucleophile and the lowest-unoccupied molecular orbital (LUMO) on the substrate.

Each of these affects stereoselectivity by raising or lowering the barriers for formation of different stereoisomers. The favored pathway will be the one which has the best combination of low steric interactions, favorable electrostatic interactions, and favorable frontier orbital interactions.

---

1.   H.C. Brown and J. Muzzio, *J. Am. Chem. Soc.*, **88**, 2811 (1966).

Each of the interactions can be assessed using molecular modeling. Steric hindrance at the reaction site can be determined by examining a space-filling model, or by examining the shape of total electron density isosurfaces. Electrostatic interactions can be determined by examining electrostatic potential maps. Frontier orbital interactions involve both the energies of the key frontier molecular orbitals, and the overlap between these orbitals in the transition state. Maximum stabilization occurs when the donor and acceptor orbital energies are similar, and there is good overlap between them. Since formation of **2** and **3** involves the same nucleophile-substrate combination, product selectivity does not depend on the orbital energy difference, but only on differences in orbital overlap as a function of "approach geometry". There may be better orbital overlap on one face of the substrate than the other, or more to the point, there may be a better balance between steric hindrance and orbital overlap on one side of the substrate than the other.

Procedure

Build norcamphor, optimize its geometry using AM1 semi-empirical calculations, then use this geometry for a single-point 3-21G *ab initio* calculation. Calculate the LUMO, and the LUMO map.

The reactive sites in **1** are revealed by displaying the LUMO. Where is this largest? The importance of sterics can be assessed by simultaneously displaying the LUMO and the "nonbonded" isodensity surface (display the latter as a mesh or transparent solid). Most of the LUMO lies "inside" the molecule's "electron cloud" and is unavailable for effective overlap. The best direction for nucleophile attack is the face where the LUMO extends furthest beyond the isodensity surface (see: **Graphical Models and Graphical Modeling**). Which face is this? Display the LUMO map. Identify regions where the LUMO has its largest (absolute) value; this is where good overlap with the nucleophile is possible. Which face should be more reactive? Which alcohol should be the major product? Is this consistent with experiment?

Optional

Reduction of camphor, **1** (R=CH₃), with sodium borohydride gives the opposite stereoselectivity as the corresponding reduction of norcamphor. Following the same procedure employed above, determine the preferred face for nucleophilic addition in camphor. Do your models correctly reproduce the observed reversal? What do you think is the cause?

Reduction of camphor, **1** (R=CH_3)

Stereochemistry of Base-Induced Elimination

Semi-empirical AM1 calculations and ab initio 3-21G$^{(*)}$ calculations and Frontier Molecular Orbital theory are used to rationalize the stereochemistry of base-induced elimination in ethyl chloride and norbornyl chloride.

Base-induced elimination in alkyl halides proceeds by a concerted E2 mechanism involving simultaneous cleavage of the carbon-halogen and vicinal carbon-hydrogen bonds. The stereochemistry of elimination normally favors removal of the hydrogen *antiperiplanar* to the halide.[1]

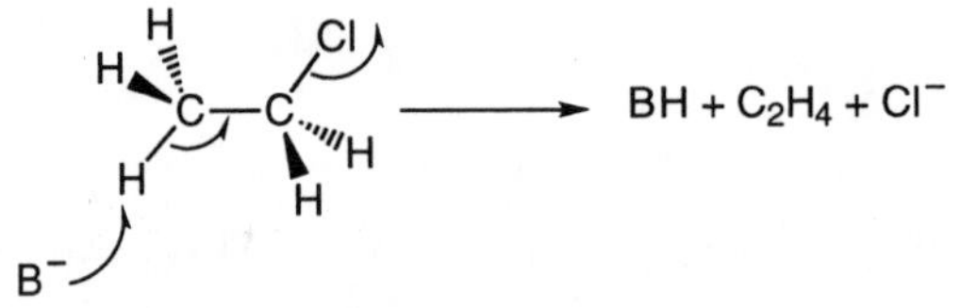

$$\text{BH} + C_2H_4 + Cl^-$$

The norbornyl ring skeleton in norbornyl chloride prevents an *antiperiplanar* arrangement of Cl and H. Labeling studies show that the elimination proceeds with *syn* stereochemistry, i.e.,[2]

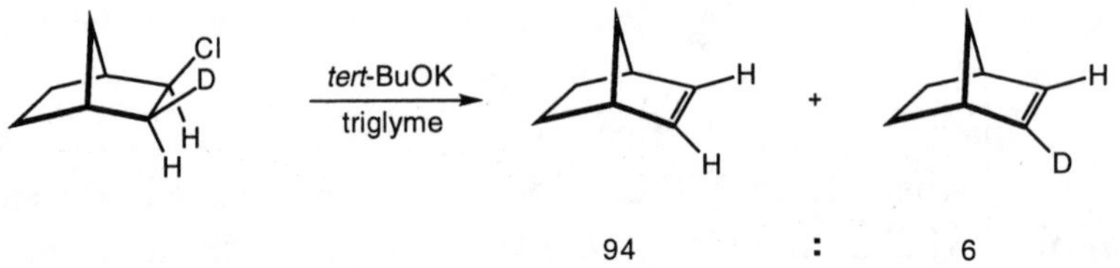

94 : 6

The changeover from *anti* to *syn* elimination can be rationalized using Frontier Molecular Orbital (FMO) theory. The alkyl halide is the electrophile in this reaction, and its reactive sites can be identified by examining the shape of its lowest-unoccupied molecular orbital (LUMO); reactive sites correspond to regions where the alkyl halide LUMO can overlap most effectively with the highest-occupied molecular orbital (HOMO) of the incoming base (of course, a reaction must also have favorable thermodynamics in order to proceed).

In this experiment, you will examine the shape of the LUMO in ethyl chloride in an attempt to rationalize the *anti* elimination pathway. You will then examine the shape of the LUMO in norbornyl chloride to see if the changeover in stereoselectivity can be predicted.

---

1.   Carey and Sundberg A, p.377; Isaacs, p.574; Lowry and Richardson, p. 595; March, p.983.
2.   R.A. Bartsch and J.G. Lee, *J. Org. Chem.*, **56**, 212 (1991).

Procedure

Build ethyl chloride and *exo* norbornyl chloride. Optimize their geometries using semi-empirical AM1 calculations, and then use these geometries for single-point 3-21G$^{(*)}$ *ab initio* calculations. Calculate a LUMO map for each molecule. Examine the LUMO map of ethyl chloride. Which of the vicinal hydrogen sites is most prominent, i.e., appears better suited for HOMO-LUMO overlap? Is this consistent with *anti* stereoselectivity? What other sites in the molecule are suited for HOMO-LUMO overlap? What products might be obtained if the base attacks these sites? (Note that the "base" may act as a nucleophile, and a chemical reaction also requires favorable thermodynamics.) Repeat this analysis using the LUMO map of norbornyl chloride. Does the map allow you to predict a changeover in stereoselectivity? What other reactive sites can you identify (and what products might be obtained)?[3]

Optional

1. Base-induced elimination from **1** leads predominantly to **2** and not **3** or **4**.[4]

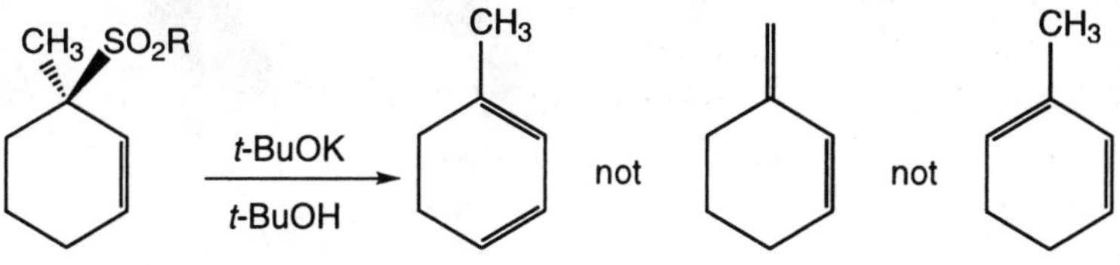

Can you rationalize this result using FMO analysis? Optimize the geometry of **1**, R=CH$_3$, using AM1, and then use 3-21G$^{(*)}$ calculations (AM1 geometry) and generate a LUMO map.

2. Experimentally, base-induced *syn* elimination from *trans*-**5** leads to **6** and not **7**.

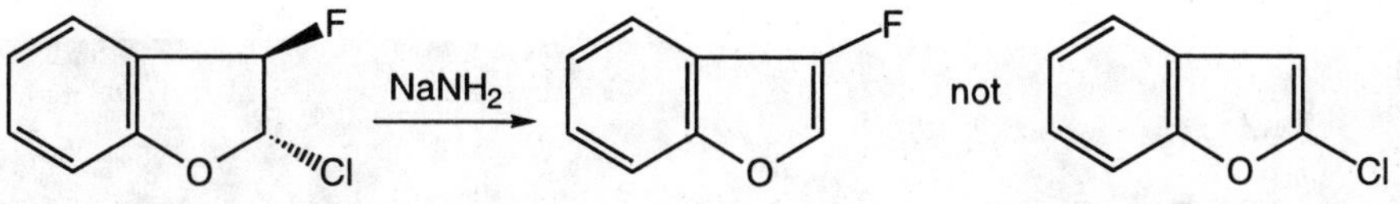

Use FMO analyses and electrostatic potential maps to rationalize this result. (Optimize the AM1 geometry of **5**, and use *ab initio* 3-21G$^{(*)}$ calculations to generate the LUMO and electrostatic potential maps.) What other factors might affect the selectivity of the reaction?

---

3. Ref. 2 reports that a minor by-product of elimination is a saturated tricyclic hydrocarbon with formula C_7H_{10}.

4. M. Sellen, J.E. Bäckvall and P. Helquist, *J. Org. Chem.*, **56**, 835 (1991).

XI

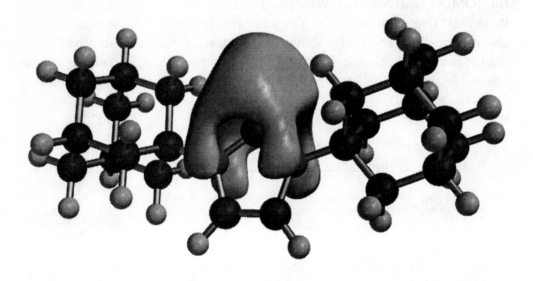

Reactive Intermediates and the Hammond Postulate

The transformation of one molecule into another during a chemical reaction can be a complex process. Intervening between the reactant and the product are many starts and stops, places where progress seems rapid, and occasions where a newly formed molecule suddenly reverts to its prior structure. In order to predict the outcome of a chemical reaction, it is necessary to take into account the route or "mechanism" that is followed, and to identify any "intermediates", that may be involved in the overall transformation. Intermediates, among them carbocations, carbanions, free radicals and carbenes, are molecules that are not sufficiently stable to be isolated and characterized using normal methods. A great deal of effort has been directed at developing experimental methods for generating and detecting reactive intermediates. The unusual structures and properties of short-lived intermediates have also made interpretation of experimental data difficult, and the chemical literature abounds with incorrect conclusions based on "hard" experimental facts.

On the other hand, reactive intermediates are no more difficult to examine on a computer than are "normal" stable molecules. Computer models can be used both to explore the properties of entirely new, unusual reactive intermediates and to predict the properties of intermediates for which experimental data is limited. This latter use has made modeling an invaluable tool for experimental chemists faced with the problem of interpreting exotic laboratory observations.

Why should reactive intermediates be of any concern to chemists? Why should there be any connection between the stability of some molecule on a reaction path and the rate at which the eventual product is formed? After all, chemists are taught early on that thermodynamics and kinetics have nothing to do with each other, and the fact that a process is exothermic does not necessarily mean that it will occur more rapidly, i.e., with a lower energy barrier, than a process that is endothermic. Nevertheless, overall thermochemistry and overall rate are sometimes linked, at least for closely-related processes. This observation has been codified in the "Hammond postulate". This states that the transition state for a highly endothermic single-step reaction "resembles" the product, while the transition state for a highly exothermic single-step reaction "resembles" the reactant.

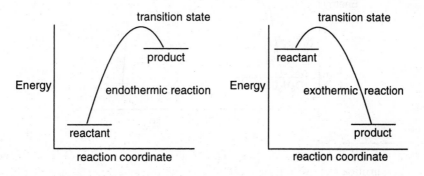

146

The word resembles has several meanings. Clearly the energy of the transition state for an endothermic reaction more closely resembles the product than the reactant, and that for an exothermic reaction more closely resembles the reactant than the product. Structural similarities between the reactant and transition state for an exothermic reaction, and between the transition state and the product for an endothermic process are also to be expected.

Reactions involving high-energy reactive intermediates present an important and extreme case. Here the step: reactant $\rightarrow$ intermediate is necessarily highly endothermic meaning that the transition state will closely resemble the reactive intermediate both in energy and in overall geometry.[1] To illustrate this situation, and the utility of the Hammond postulate, consider the possible products resulting from electrophilic bromination of toluene.

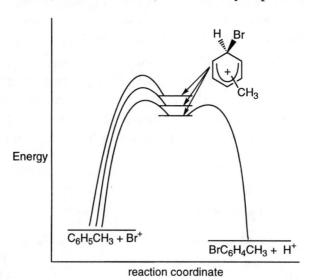

The relative amounts of the *ortho*, *meta* and *para* isomers will be determined by their respective relative rates of formation, which in turn are given by the relative energies of the three transition states for the rate-determining step. Bromination appears to involve two steps: electrophilic attack by "Br^+" to give an high energy cationic intermediate (a benzenium ion), followed by deprotonation.

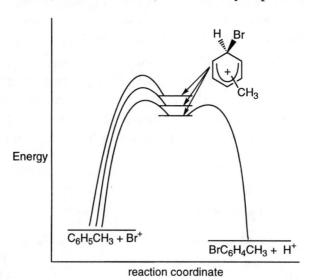

---

1. Correspondingly, the step: intermediate $\rightarrow$ product is necessarily highly exothermic, meaning that the transition state will again closely resemble the intermediate. In either event, the reactive intermediate takes on special significance.

The initial step is highly endothermic, and if it is rate-determining, the Hammond postulate suggests that the energy of the transition state will resemble the energy of the corresponding cationic intermediate (the "product" of the initial step). The second step is highly exothermic, and if it is rate determining, the Hammond postulate again suggests that the energy of the transition state will resemble the energy of the intermediate. In either case, the relative stabilities of the *ortho-*, *meta-* and *para*-substituted intermediates should anticipate the relative energies of the respective transition states.

The Hammond postulate, as it relates to processes involving unstable intermediates, is a valuable tool for the study of reactivity and selectivity.

Substituent Effects on Reactive Intermediates[1]

Ab initio 3-21G calculations are used to evaluate the substituent effects on carbocations, carbanions, free radicals, and singlet and triplet carbenes.

Bond making and breaking at carbon often involves intermediates in which the carbon atom contains an incomplete valence shell, nonbonding electrons, or both. Substituents can affect the stability of these intermediates by interacting with the "nonbonding" orbital on carbon (see: **A Molecular Orbital Description of Substituent Effects**). For example, resonance arguments suggest that π donors will stabilize carbocations, and π acceptors will stabilize carbanions.

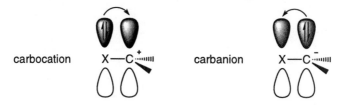

The effect of an acceptor substituent on a carbocation, or a donor substituent on a carbanion, is more difficult to anticipate. Both electron donors and acceptors might be expected to stabilize free radicals which have a singly-occupied orbital.

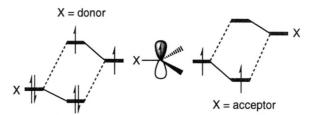

Substituent effects on carbenes present a special problem because carbenes can exist in either of two electronic states, singlet or triplet, and substituents will affect the relative energies of these states. The singlet state, with its empty p orbital, might be expected to behave like a carbocation, while the triplet state, with its two singly-occupied orbitals, might behave more like a diradical.

In this experiment, you will use *ab initio* 3-21G calculations to obtain energies for the following *isodesmic* reactions for X = -CH$_3$, -NH$_2$, -CN (see: **Isodesmic Reactions**).

1.   W.J. Hehre, L. Radom, P.v.R. Schleyer and J.A. Pople, **Ab Initio Molecular Orbital Theory**, Wiley, New York, 1986, p.346.

$$X\text{-}CH_2^+ \;+\; CH_4 \;\rightarrow\; X\text{-}CH_3 \;+\; CH_3^+$$

$$X\text{-}CH_2^- \;+\; CH_4 \;\rightarrow\; X\text{-}CH_3 \;+\; CH_3^-$$

$$X\text{-}CH_2^{\cdot} \;+\; CH_4 \;\rightarrow\; X\text{-}CH_3 \;+\; CH_3^{\cdot}$$

$$X\text{-}\ddot{C}H \;+\; CH_4 \;\rightarrow\; X\text{-}CH_3 \;+\; \ddot{C}H_2$$

$$X\text{-}\dot{C}H \;+\; CH_4 \;\rightarrow\; X\text{-}CH_3 \;+\; \dot{C}H_2$$

You will also use orbital energies of HX molecules to interpret calculated substituent effects in terms of orbital interactions.

Procedure

Reaction energies: Build all the reactive intermediates involved in reactions above ($X = CH_3$, NH_2, CN), and optimize their geometries using *ab initio* 3-21G calculations. Record all energies, and calculate each reaction energy.[2] What is the (de)stabilizing effect of each substituent on each type of intermediate? (An endothermic reaction indicates that the substituent has a stabilizing effect relative to H). What analogies exist between different intermediates? Do substituents have the same effect on carbocations and singlet carbenes? On free radicals and triplet carbenes? On free radicals and carbocations? On free radicals and carbanions?

Orbital interactions: Build CH_4, NH_3 (planar and pyramidal), and HCN, and optimize their geometries using *ab initio* 3-21G calculations. Record all energies, and record the energies of the highest-energy donor and lowest-energy acceptor orbitals (Note: an acceptable donor/acceptor orbital must be able to engage in π overlap with a nonbonding orbital on the intermediate. Therefore, examine the various molecular orbitals, beginning with the HOMO and LUMO, in order to identify the correct orbitals for each molecule). Use the orbital energies to sequence the substituents in terms of their relative donor ability and their relative acceptor ability. Are the two sequences the same? How do you account for the ordering in each sequence? Are your orbital-based descriptions of each substituent (π donor vs. π acceptor) consistent with the substituent effects calculated above? Explain.

The strength of a π interaction depends on the energies of the interacting orbitals and on their overlap. The latter is affected by the intermediate's geometry, i.e., CX bond distance, bond angles at C, and conformation. Examine the geometry of each intermediate. Is it consistent with the calculated substituent effect, and with a π interaction mechanism for the substituent effect? Explain. If it isn't, then what other mechanisms might be operating?

---

2. 3-21G total energies (hartrees): methane, -39.9769; methyl cation, -39.0091; methyl anion, -39.2394; singlet methylene, -38.6519; triplet methylene, -38.7091; ethane, -78.7940; methylamine, -94.6817; acetonitrile, -131.1918.

Hyperconjungation and the Structures and Stabilities of Carbocations

Semi-empirical AM1 calculations and ab initio 3-21G calculations are used to determine the structure of isopropyl and tert-amyl carbocations, and to look for evidence of hyperconjugation.

Carbocation stability increases in the order: methyl << primary << secondary < tertiary. One rationale for this ordering is that alkyl groups stabilize carbocations by hyperconjugation or "no bond" resonance.[1] The idea here is that contributions from resonance structures in which a CH (or CC) σ bond is replaced by a CC π bond stabilize the carbocation by delocalizing the positive charge onto H (or C).

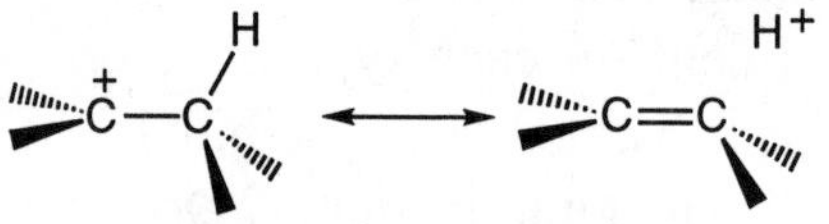

"no bond"
resonance contributor

Hyperconjugation can also be thought of in terms of orbital interactions between the empty p orbital on C^+ and an occupied π-type alkyl orbital (see: π **Interactions Involving σ Electrons**). A methyl group has two suitable π orbitals (see: **Visualizing Molecular Orbitals**) and, depending on conformation, effective p-π overlap can occur with either or both.

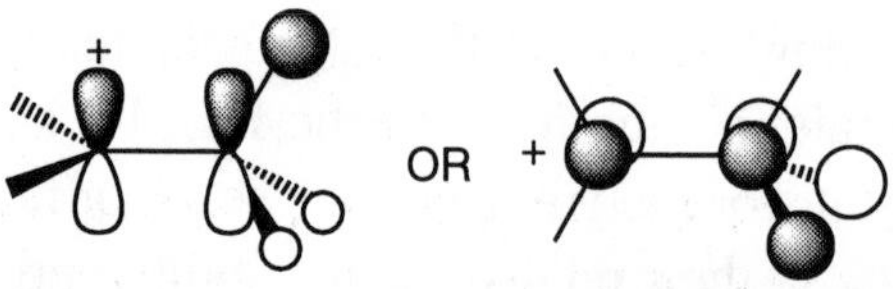

OR

Because p-π overlap implies transfer of electron density (from the methyl group to the carbocation center), it is to be expected that the methyl CH bonds involved will weaken (lengthen) while the CC bonds will strengthen (shorten).

In this experiment, you will use *ab initio* 3-21G calculations to obtain the structure of isopropyl cation and look for evidence of hyperconjugation. You will also use semi-empirical AM1 calculations to determine the conformational preference of *tert*-amyl cation, $[CH_3CH_2C(CH_3)_2]^+$, and so assess the relative effectiveness of CH and CC hyperconjugation. In particular, you will investigate whether a CH bond or a CC bond is more effective in transferring its electron density to the carbocation center.

---

1. Carey and Sundberg A, p.54; Lowry and Richardson, p.431; March, p.143.

Procedure

Isopropyl cation: Build isopropyl cation in the two different conformations shown below, and optimize their geometries using *ab initio* 3-21G calculations.

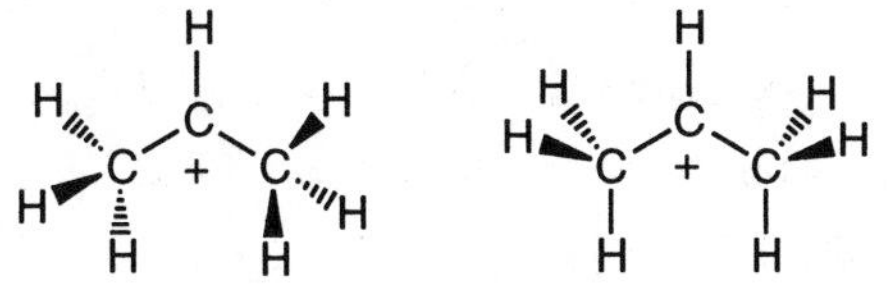

isopropyl cation

Record the energies of the two conformers. Which one is more stable? For each conformer, calculate and display an electrostatic potential map. Which hydrogen(s) are most electron-poor in each cation? In each methyl group? Are any of these results consistent with hyperconjugation? What geometric evidence is there for (or against) hyperconjugation? Cite specific bond distances and make appropriate comparisons to show evidence of bond lengthening or shortening. (Note: the 3-21G CC bond distance in propane is 1.541 Å.)

tert-Amyl cation: Build conformers of *tert*-amyl cation in which the CCCC dihedral angle is constrained to be 0°, 30°, 60°, and 90°, and optimize the geometry of each constrained conformer using semi-empirical AM1 calculations.

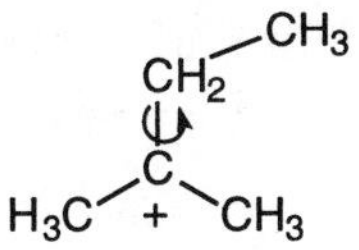

tert-amyl cation

Record the energy of each conformer. Which conformer is the most stable? What are the energies of the other conformers relative to this one? What do these results suggest about the relative effectiveness of CH and CC hyperconjugation? Examine the variation in the $C_{methylene}$-C_{cation} and C_{methyl}-C_{cation} bond distances as a function of dihedral angle. Account for your results.

Optional

Overman and Burk have developed a method for forming nitrogen-containing rings that involves electrophilic addition of an iminium cation to a vinyl silane.[2] For example, refluxing Z-pyrroline, **1**, with CF_3CO_2H gives the *cis* hexahydroindole, **2**, in 81% yield. The mechanism of this reaction presumably involves protonation of nitrogen, followed by cyclization to give a β-silyl cation, and finally nucleophilic attack on silicon. Interestingly, identical treatment of the *E*-pyrroline did not give any **2**, just protodesilylation.

---

2. L.E. Overman and R.M. Burk, *Tetrahedron Lett.*, **25**, 5739 (1984).

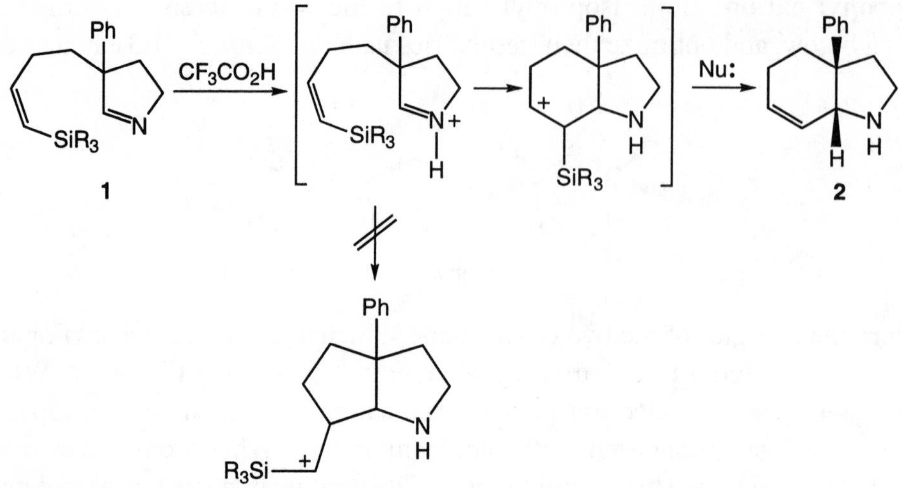

Two questions that naturally arise are: 1) why do the Z and E reactants behave differently? and 2) what factors control regioselectivity in the cyclization process, i.e., why is formation of a β-silyl cation, $R_3Si-C-C^+$, preferred to formation of an α-silyl cation, R_3Si-C^+? The answers offered to both questions by Overman and Burk assume an important role for CSi hyperconjugation. Thus, they presented models that suggest that the Z isomer can more readily align its CSi bond with the empty p orbital on the incipient carbocation, and they suggested that CSi hyperconjugation in a β-silyl cation is more stabilizing than any α-silicon substituent effect.

You can evaluate the relative importance of CSi hyperconjugation and an α-silicon effect by examining some of the carbocations formed by protonation of vinyltrimethylsilane. Build the three cations (**0β**, **90β** and **α**) shown below and optimize their geometries using the AM1 semi-empirical method (note: build **0β** with C_s symmetry, and use this symmetry constraint during the optimization).

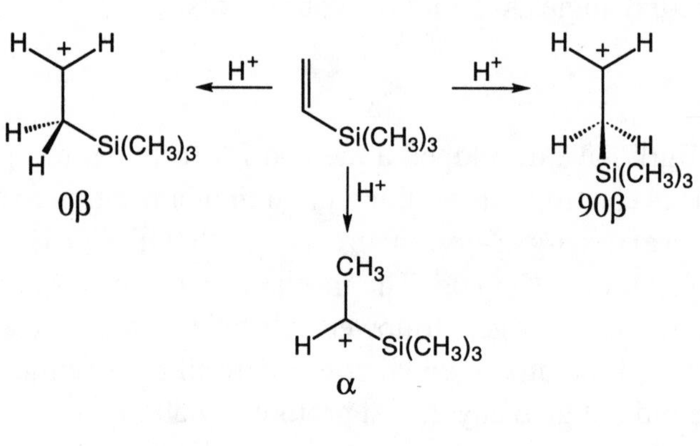

153

Record the energy of each cation. Which one is the most stable? Record the CC and CSi bond lengths in each cation. Can each cation be described adequately by a single resonance structure, or must additional "no-bond" resonance structures be included? Explain. How do these additional structures account for the differences in cation stability? Which is more effective, CSi, CH, or CC hyperconjugation? Which of your results support this conclusion? How do you account for the ordering of the three cation energies?

Allylic Cations. The Interplay of Stability and Charge Delocalization

Semi-empirical AM1 calculations and ab initio 3-21G calculations are used to investigate the relationship between stability and charge delocalization in allyl cation.

Resonance stabilizes allyl cation by delocalizing the positive charge and by lowering the energy of the π-bonding electrons. The resulting ion is more stable than other primary carbocations, such as 1-propyl cation and may even be more stable than secondary cations, such as 2-propyl cation.

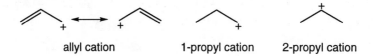

| allyl cation | 1-propyl cation | 2-propyl cation |

The molecular orbital description of allylic resonance is based on a stabilizing two-electron interaction involving the empty p orbital of the $-CH_2^+$ group and the high-energy π donor orbital of the vinyl group. The energy gap between these orbitals is relatively small, and the orbitals overlap strongly in a planar geometry, but not in a perpendicular geometry. A strong stabilizing interaction is expected, and this interaction should be more effective than hyperconjugation (see: **A Molecular Orbital Description of Substituent Effects**).

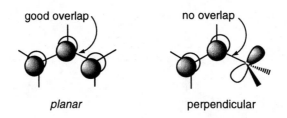

planar   perpendicular

In this experiment, you will use semi-empirical AM1 calculations to determine the geometries of allyl, 1-propyl, and 2-propyl cations, and *ab initio* 3-21G calculations to evaluate their relative stabilities, as defined by the energy of the *isodesmic* reactions describing hydride transfer to methyl cation, i.e.,

$$RH + CH_3^+ \longrightarrow R^+ + CH_4$$

Hydride transfer will be more exothermic for more stable cations. You will also compare planar and perpendicular allyl cations for evidence of resonance stabilization.

Procedure

Build each of the molecules that participate in the hydride transfer reaction shown on the facing page, where RH = propane or propene, and R⁺ = allyl (planar and perpendicular), 1-propyl, and 2-propyl cations. Optimize their geometries using semi-empirical AM1 calculations (build perpendicular allyl cation with C_s symmetry and optimize its geometry subject to this symmetry constraint), record their energies, and calculate ΔH for each hydride transfer reaction. What is the relative stability of the cations according to this definition? Next, calculate the *ab initio* 3-21G energy of each molecule at its AM1 geometry. Record these energies, and recalculate ΔH. Now what are the relative cation stabilities? Assuming that the 3-21G ΔH values are more realistic than the AM1 values, which AM1 values appear to be unrealistic?

What is the 3-21G barrier for CH_2 rotation in allyl cation? How does this barrier compare to the rotation barriers in ethane (~3 kcal/mol) and ethylene (~65 kcal/mol)? How do the CC bond lengths in the two allyl cations compare with those in propene? What do these results imply about resonance stabilization in each cation? Cite specific results that support your conclusions.

Calculate electrostatic potential maps for planar and perpendicular allyl cations, and compare them on the same color scale. In which ion is the positive charge more localized? Compare the highest-occupied molecular orbitals (HOMO's) for the two ions. Which ion better delocalizes its π electrons? Compare the lowest-occupied molecular orbitals (LUMO's). Which ion better disperses the positive charge?

Optional

Tropene is a cyclic conjugated hydrocarbon that might undergo protonation in strong acid at any of five different carbons.

tropene

Four of the resulting cations should be resonance stabilized, and it is not obvious which, if any, might form preferentially. Build each of the five conjugate acids of tropene, and optimize their geometries using AM1 calculations. Calculate 3-21G energies using the AM1 geometries. Identify the preferred ion. Why is this ion preferred? Use structural information, such as CC bond distances and electrostatic potential maps, to support your explanation.

156

Dimethylcyclopropylcarbinyl Cation

Semi-empirical AM1 calculations are used to investigate the conformational preferences of dimethylcyclopropylcarbinyl cation, and to interpret NMR observations of this cation.

Dimethylcyclopropylcarbinyl cation, **1**, is a stable carbocation that can be prepared in superacid (SO_2-SbF_5-HSO_3F) and studied by NMR spectroscopy. The ^{13}C NMR spectrum of this ion at –65 °C shows two methyl signals of equal height at δ 30.1 and 38.9. Warming the solution causes the two methyl groups to exchange their chemical environments with an exchange barrier (E_a) of 13.7 kcal/mol.[1] These results suggest that the cation prefers a bisected conformation in which the methyl groups are nonequivalent and the ion is stabilized by CC hyperconjugation.

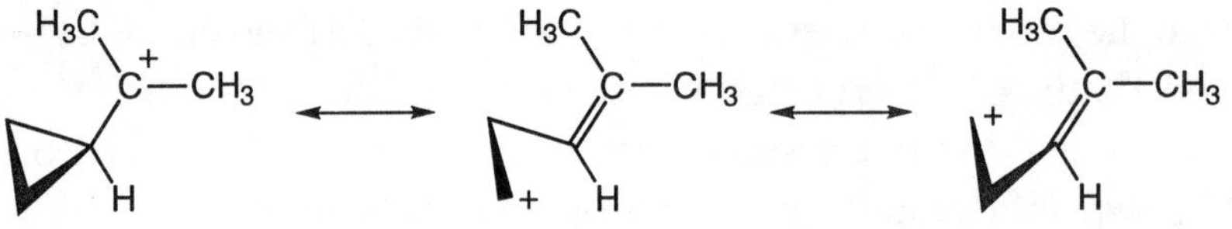

Two methyl exchange mechanisms have been suggested. One involves rotation about the CC bond and a perpendicular cation (note that this ion should be stabilized by CH hyperconjugation; see: **Hyperconjugation and the Structures and Stabilities of Carbocations**). The other involves ring expansion-contraction via a cyclobutyl cation.

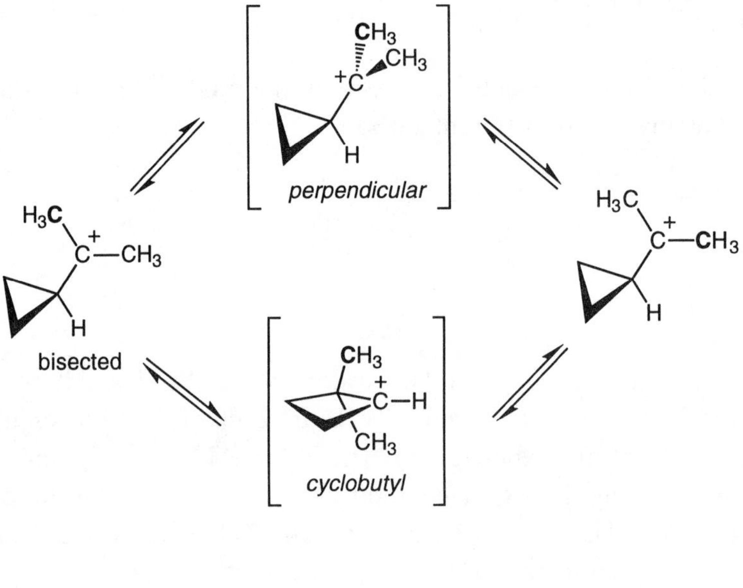

---

1.   D.S. Kabakoff and E. Namanworth, *J. Am. Chem. Soc.*, **92**, 3234 (1970).

In this experiment, you will use semi-empirical AM1 calculations to establish the conformational preferences of dimethylcyclopropylcarbinyl cation, and determine whether either a perpendicular cation or cyclobutyl cation can lie on the energy surface for methyl exchange.

Procedure

Build bisected and perpendicular dimethylcyclopropylcarbinyl cations, and dimethylcyclobutyl cation. Optimize the cation geometries using semi-empirical AM1 calculations (build the perpendicular cation with C_s symmetry and optimize its geometry subject to this symmetry constraint), and record their energies. Which ion is the most stable? Is this consistent with the NMR observations? Which of the higher-energy ions are sufficiently low enough in energy that they can participate in methyl exchange?

The relative importance of CC and CH hyperconjugation in the bisected and perpendicular cations can be assessed by examining the geometry and electron distribution in these cations. Build methylcyclopropane and optimize its AM1 geometry. Compare the CC and CH bond lengths in this neutral molecule with the analogous bond lengths in the two cations. What differences are there? Are these consistent with a hyperconjugation model, or must other types of electronic interactions be invoked? Explain.

Calculate electrostatic potential maps for all three cations. In which cation is the charge most localized? Most delocalized? Which dimethylcyclopropylcarbinyl cation delocalizes more charge onto the methyl groups? Onto the cyclopropane ring? Are these results consistent with the hyperconjugation model, or must other types of interactions be invoked? Explain.

CC hyperconjugation in the bisected cation involves mixing the empty p orbital with the filled donor orbital shown below (see: **A Molecular Orbital Description of Substituent Effects**).[2]

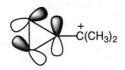

Are your results, geometric and electronic, consistent with this orbital mixing? Cite specific results that support your conclusion. (Note: you can also look for evidence of this orbital mixing by calculating and displaying the highest-energy occupied orbital of the bisected cation.)

---

2. This orbital is one of the two degenerate highest-energy occupied orbitals of cyclopropane, and is known as a Walsh orbital. See: Lowry and Richardson, p. 31. A.D. Walsh, *Trans. Faraday Soc.*, **45**, 179 (1949).

Ortho Benzyne

Ab initio 3-21G calculations are used to describe ortho benzyne and to characterize its reactivity.

The commercially important synthesis of phenol from chlorobenzene is believed to involve an elimination-addition mechanism in which *ortho* benzyne is an electrophilic intermediate.[1]

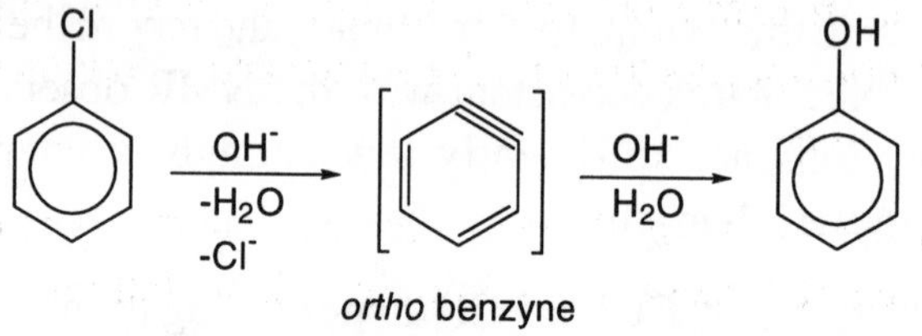

ortho benzyne

This mechanism is supported by the observation that the reaction of 1-^{14}C-chlorobenzene and KNH_2 gives equal amounts of two labeled anilines.

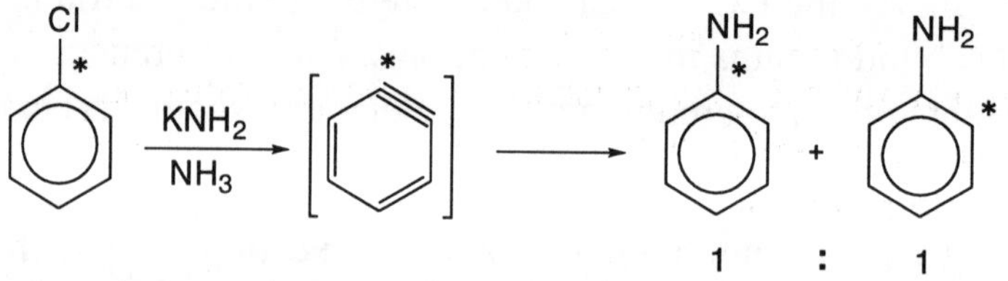

Ortho benzyne can be produced in other ways too, and can be trapped as a Diels-Alder adduct with various dienes.

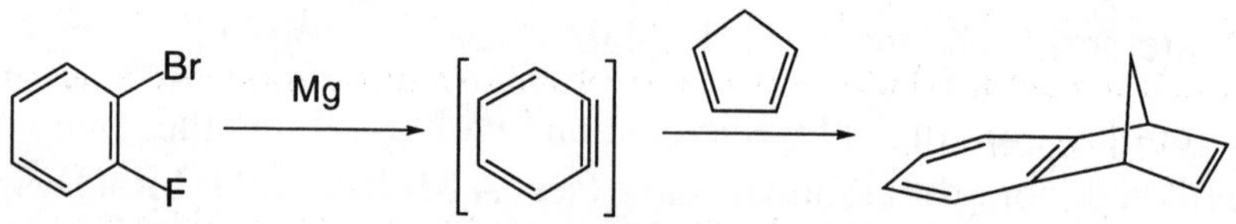

Ortho benzyne is sufficiently stable that it can even be isolated and studied in a low temperature inert matrix. The IR spectrum of *ortho* benzyne contains a peak at 2085 cm^{-1}, which has been assigned to a stretching vibration of the CC "triple bond" (alkyne stretching frequencies are typically 2200 cm^{-1}).

In this experiment, you will use *ab initio* 3-21G calculations to determine the geometry of *ortho* benzyne. You will also use the valence molecular orbitals of *ortho* benzyne to describe its electronic structure and characterize its chemical reactivity.

---

1. Cary and Sundberg A, p. 583; Lowry and Richardson, p. 643; March, p. 580; the classic account of benzynes is: R.W. Hoffmann, **Dehydrobenzene and Cycloalkynes**, Academic Press, New York, 1967.

Procedure

Build *ortho* benzyne and benzene, and optimize their geometries using *ab initio* 3-21G calculations. Record the CC bond distances in each molecule, and compare them to the 3-21G CC distances in hexa-1,5-dien-3-yne and *trans*-1,3-butadiene.

$$1.322 \quad / \quad \overset{1.429}{\underset{1.193}{\equiv}} \quad \backslash \quad 1.467 \quad \backslash \quad 1.321$$

Assuming that the latter CC distances are "normal", how would you describe each of the bonds in *ortho* benzyne (single, double, aromatic, etc.)? What resonance structure(s) appears to make the greatest contribution to *ortho* benzyne? Explain your reasoning.

Ortho benzyne and benzene both have three occupied π-type orbitals (benzyne also has an occupied π-type orbital in the ring plane, but this will be referred to as a σ orbital). Examine the occupied orbitals of each molecule and identify the π-type orbitals. What are their energies? Comparing orbital shapes in *ortho* benzyne and benzene, would you describe the π system in benzyne being made up of isolated π bonds or as an aromatic system similar to benzene? Explain the difference in orbital energies in *ortho* benzyne and benzene. (Hint: the short "triple" bond in benzyne (de)stabilizes molecular orbitals that have (anti)bonding interactions at this bond.)

Identify the σ and σ* orbitals which make up the "triple" bond in *ortho* benzyne. In order to act as a "good" dienophile in Diels-Alder cycloadditions, *ortho* benzyne must have a low-energy σ* orbital. Identify it from your plots, and compare its energy with those found in typical "good" dienophiles, acrylonitrile (0.104 a.u.), acrolein (0.096 a.u.), and order the three molecules according to their expected reactivity.

Based on the shape of the *ortho* benzyne σ* orbital, what would be the orientation of benzyne and cyclopentadiene in a Diels-Alder transition state? Use a sketch to show the expected transition state geometry. What would be the transition state geometry for the corresponding cycloaddition of cyclohexene and cyclopentadiene? Which transition state will probably experience less steric hindrance?

Optional

Use AM1 semi-empirical calculations to obtain a transition state for Diels-Alder cycloaddition of cyclopentadiene and *ortho*-benzene. Is its geometry in line with your expectations based on the analysis above?

Nucleophilic Substitution of Aryl Chlorides. Benzyne Intermediate vs. Addition-Elimination

Semi-empirical AM1 calculations are used to investigate the mechanism of nucleophilic substitution in an aryl chloride by comparing the energetics of benzyne formation and addition-elimination.

5-Chloroacenaphthylene, **1A**, reacts with sodium ethoxide to give the corresponding 5-ethoxy compound as the major product.[1] 2-Chloronaphthalene, **1N**, on the other hand, is inert under these conditions.

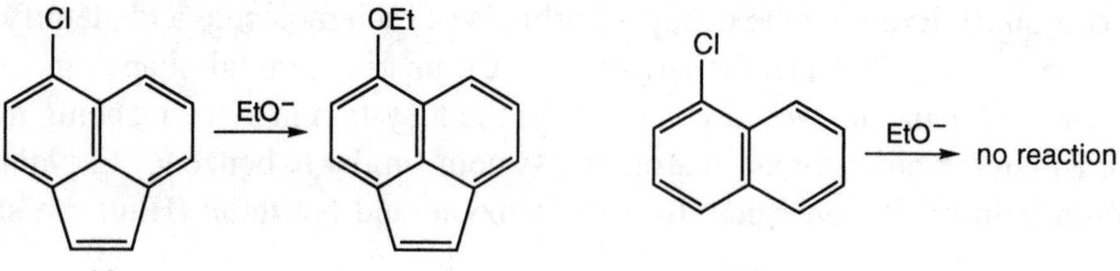

Nucleophilic substitution of aryl halides[2] often proceeds via an *ortho*-benzyne intermediate (top pathway), but this mechanism does not provide an obvious explanation for why **1A** is so much more reactive than **1N**. Substitution can also occur via a nucleophilic addition-elimination mechanism (bottom pathway), but this type of reaction is more characteristic of aryl halides containing electron-withdrawing substituents, such as NO_2.

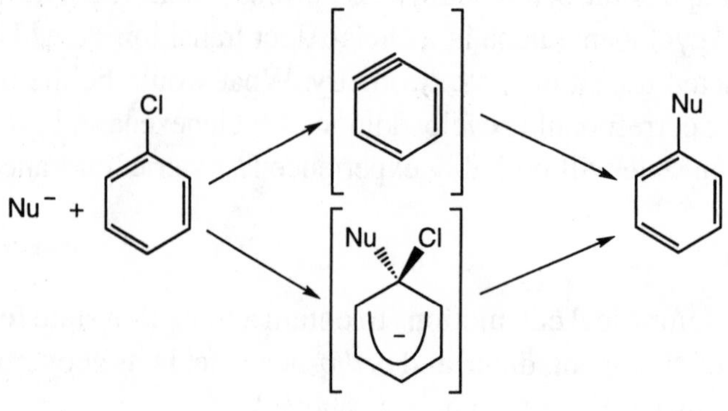

1. M.J. Perkins, *J. Chem. Soc., Chem. Commun.*, 231 (1971).
2. Carey and Sundberg A, p. 579 and 583; Lowry and Richardson, p. 640.

In this experiment, you will use semi-empirical AM1 calculations to determine the energetics for benzyne formation from **1A** and **1N**, and to determine the energetics for nucleophilic addition (Nu$^-$=H$^-$) to **1A** and **1N**. A direct comparison of reaction energetics for each pathway is not necessary in this case. Instead, you will combine experimental observations and computational results to suggest a likely reaction path.

Procedure

Benzyne pathway: Build **1A**, **1N**, and the *ortho* benzynes that would be produced by elimination of HCl from each (the "triple bond" in each benzyne should have a bond length between 1.2 and 1.3 Å). Optimize each molecule's geometry using semi-empirical AM1 calculations, and record its energy. Then calculate ΔH_{rxn} for the *isodesmic* process: **1A** + benzyne derived from **1N** → benzyne derived from **1A** + **1N**. What does ΔH_{rxn} tell you about the relative ease of benzyne formation from **1A** and **1N**? Given that **1A** undergoes substitution and **1N** is inert, can this mechanism be responsible for the observed substitution reaction?

Addition-elimination pathway: Build the anionic adducts that would be produced by addition of hydride (Nu$^-$=H$^-$) to **1A** and **1N**, and optimize their AM1 geometries. Calculate ΔH for the *isodesmic* process: **1A** + anion derived from **1N** → anion derived from **1A** + **1N**. What does ΔH tell you about the relative ease of nucleophilic addition to **1A** and **1N**? Can this mechanism be responsible for the observed substitution reaction?

You should find that one pathway does not appear to favor either **1A** or **1N**, while the other pathway clearly favors **1A**. Examine the structures of the intermediates for the favored pathway for some indication of why **1A** is more reactive than **1N**. (Hint: if you are comparing benzynes, then compare the lengths of the CC "triple" bonds, and compare the shapes and energies of the benzyne "π" orbital. However, if you are comparing anions, then compare the lengths of the CCl bonds, and compare electrostatic potential maps.) Suggest an explanation for why this pathway seems to favor **1A**.

1,3-Di-1-adamantylimidazol-2-ylidene. A Stable Singlet Carbene

Semi-empirical AM1 calculations are used to characterize the title molecule and to compare it to the analogous 1,3-dimethyl derivative. The energetics of CH bond insertion is calculated, and used to rationalize the title molecule's stability.

Carbenes are highly-reactive species that rapidly undergo several types of reactions, including cycloaddition with alkenes and insertion into alkane CH bonds. The title molecule, **1A**, is exceptional, however, in that it forms a stable solid, the structure of which has been determined by X-ray crystallography.[1]

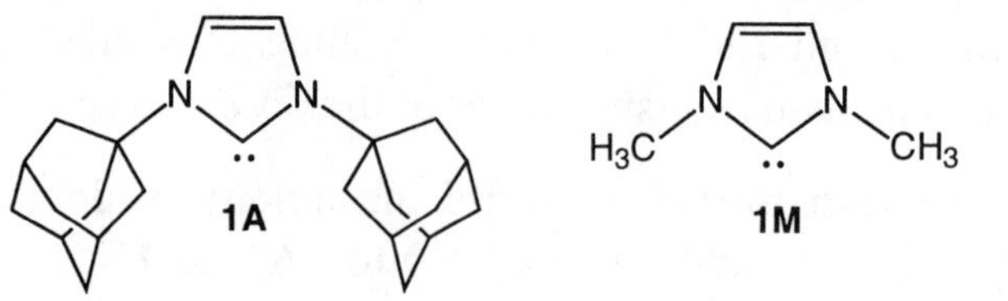

The apparent stability of **1A** raises two questions: 1) is its stability kinetic or thermodynamic? and 2) is it stabilized by steric factors (hindrance by the adamantyl groups), by electronic factors (**1A** contains a six electron, "aromatic", π system), or by some combination of the two? The first question can be answered by calculating the energetics of a typical carbene reaction, such as insertion into the CH bond of CH_4, i.e.,

$$X_2C: + CH_4 \longrightarrow H\text{-}CX_2\text{-}CH_3$$

"Kinetic stability" requires a larger barrier for **1A** insertion than for a "normal" singlet carbene, such as CCl_2, while "thermodynamic stability" requires **1A** insertion to be endothermic. The second question can be answered by comparing **1A** with another carbene that shares either its steric or electronic characteristics. Since the methyl groups in **1M** are substantially smaller than the adamantyl groups in **1A**, these two carbenes should have similar electronic properties, but very different steric properties.

In this experiment, you will use semi-empirical AM1 calculations to compare the structure and electronic properties of **1A** and **1M**, and to establish the energetics of the CH bond insertion reaction shown above.

---

1.   A.J. Arduengo, III, R.L. Harlow and M. Kline, *J. Am. Chem. Soc.*, **113**, 361 (1991).

Procedure

Build **1A**, **1M**, and CCl_2, optimize their geometries using semi-empirical AM1 calculations, and record their energies. (Assume that all three carbenes are singlets.) Next, build CH_4 and the products obtained from CH insertion of each carbene. Optimize their AM1 geometries (note: stagger the ring substituents on alternate sides of the ring in the products derived from **1A** and **1M**), record their energies, and calculate ΔH_{rxn} for each insertion. Use these results to assess whether **1A** is thermodynamically stable with respect to CH insertion. The AM1 barriers for CH insertion are approximately 20, 50, and 50 kcal/mol for CCl_2, **1M**, and **1A**, respectively. Use these results to assess whether **1A** is kinetically stable with respect to CH insertion. What relationship exists between the AM1 barriers and reaction enthalpies? What do these results suggest about the relative role of steric and electronic factors in stabilizing **1A**? (Assume, for the sake of argument, that **1A** and **1M** are electronically similar.)

Molecular models provide a better test of electronic/steric similarity than simple molecular formulas. What are the ring bond lengths in **1A** and **1M**? Do these imply similar π interactions in each? Compare space-filling models of the two carbenes. Do the adamantyl and methyl groups offer significantly different steric environments?

Optional

Another way to compare the electronic properties of different carbenes is to examine the shapes and energies of their reactive orbitals. A singlet carbene has two reactive orbitals, a filled σ-type orbital and an empty p-type orbital (see: **Substituent Effects on Reactive Intermediates**). Identify the reactive orbitals in each carbene by calculating and displaying various high-energy occupied orbitals and low-energy unoccupied orbitals. Record the energies of these orbitals and decide whether **1A** and **1M** are electronically similar in an "orbital" sense. How do these carbenes differ electronically from CCl_2? Which carbene is a better donor? A better acceptor? Given the relatively low barrier for CCl_2 insertion into CH bonds, what do these orbital energies suggest about the direction of electron transfer in the insertion transition state?

Structure and Reactivity of "Cyclic" Bromonium Ions

Semi-empirical AM1 calculations are used to characterize the "cyclic" bromonium ions that might form from reaction of Br₂ and various alkenes.

Bromine, Br_2, adds to alkenes in a stepwise fashion. The first step involves formation of a *cyclic* bromonium ion that then undergoes backside attack by Br⁻, or another nucleophile, to give *trans* products.

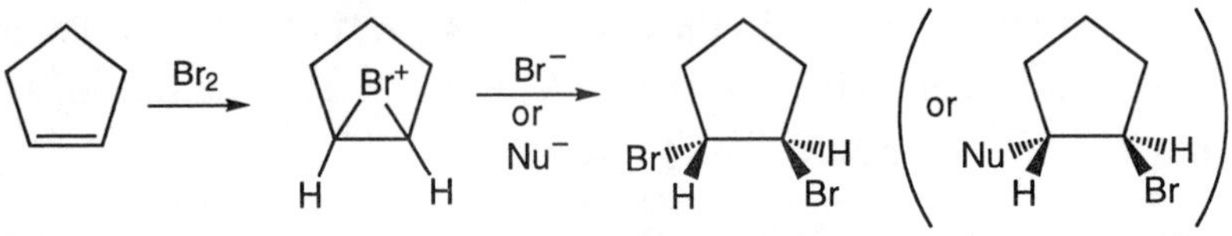

The existence of cyclic bromonium ions is supported by NMR observations in strong acid,[1] and by the determination of a cyclic ion's crystal structure.[2] However, substantial evidence[3] (selective formation of one regioisomer, formation of *cis* products) suggests that bromine can also bridge the two alkene carbons unsymmetrically (**A**), or not bridge at all (**B,C**), when one of the carbons is substituted by an electron-donor substituent, X.

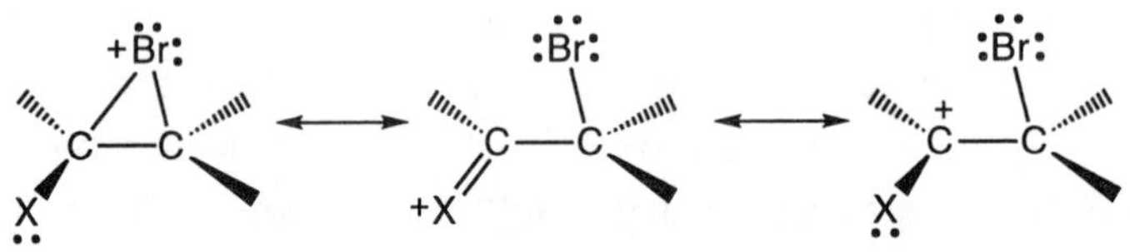

In these cases X and Br "compete" as electron-donor groups, and the structure of the ion depends on their relative donor abilities.

In this experiment, you will use semi-empirical AM1 calculations to characterize the structures and electronic properties of bromonium ions derived from symmetric and unsymmetric alkenes.

1. G.A. Olah, P.W. Westerman, E.G. Melby and Y.K. Moe, *J. Am. Chem. Soc.*, **96**, 3565 (1974); G.A. Olah, **Halonium Ions**, Wiley, New York, 1975.
2. H. Slebocka-Tilk, R.G. Ball and R.S. Brown, *J. Am. Chem. Soc.*, **107**, 4504 (1985).
3. G. Bellucci, R. Bianchini, C. Chiappe, R.S. Brown and H. Slebocka-Tilk, *J. Am. Chem. Soc.*, **113**, 8012 (1991).

Procedure

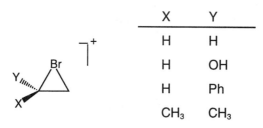

X	Y
H	H
H	OH
H	Ph
CH$_3$	CH$_3$

Build the four bromonium ions listed above, and optimize their geometries using semi-empirical AM1 calculations. Record the three "ring" bond distances and the larger of the two CCBr bond angles in each ion. Which ions are symmetrically bridged, and which are not? Do any of the ions appear to be unbridged? (Hint: bond distances and angles in an unbridged ion should be very similar to those in other "normal" molecules, e.g., the AM1 distance for the "normal" CBr bond in CH$_3$Br is 1.965 Å.)

Calculate an electrostatic potential map for each ion, and display the four maps on the same color scale. Use these maps to compare the charge distribution in each ion (in particular, focus on the change in electrostatic potential at Br and the substituted carbon). Given your geometry and charge distribution results, which resonance structures, **A**, **B**, or **C**, appear to make the greatest (least) contribution to each ion? Cite specific geometrical and electronic features that support your conclusions.

The preferred site of nucleophilic attack in each ion can be inferred from the shape of its lowest-energy unoccupied orbital (LUMO). Calculate and display a LUMO map for each ion, and use it to identify the preferred site(s) for nucleophilic attack (this site will be the region where the LUMO magnitude is largest). Which ions would be attacked in the most (least) regioselective fashion, i.e., attack at one carbon is favored over the other? Which ions would be attacked in the most (least) stereoselective fashion, i.e., *trans* attack is favored over *cis* attack? What connections exist between regioselectivity, stereoselectivity, and bromonium ion structure?

XII

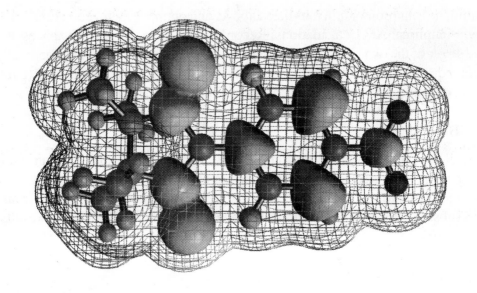

A Molecular Orbital
Description of
Substituent Effects

M any chemical reactions do not go in a single step, but rather involve several steps, and one or more reactive intermediates. In this case, the most favorable reaction pathway will be that which leads through the lowest-energy intermediates. It is useful, therefore, to understand how substituents stabilize (and destabilize) intermediates.

The most familiar approach is resonance theory. Simply by "counting" resonance structures a chemist can judge the stability of a reactive molecule. For example, benzyl cation should be more stable than methyl cation simply because it offers a greater number of resonance structures, i.e.,

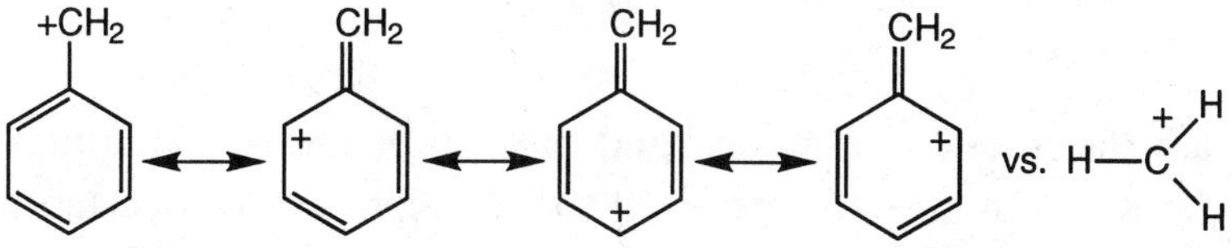

p-Dimethylaminobenzyl cation will be more stable yet because of a further increase in number of resonance structures, i.e.,

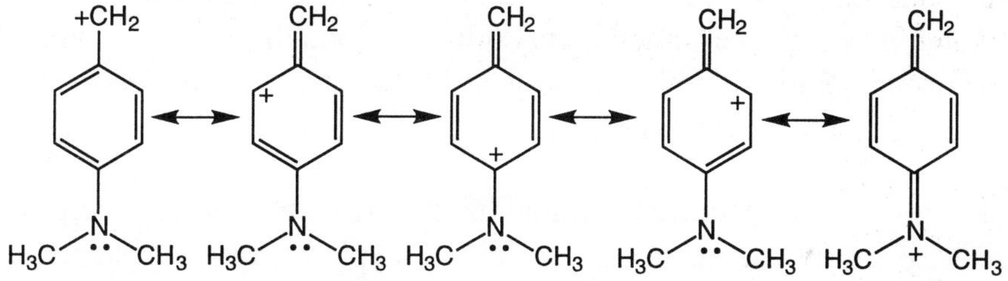

While resonance arguments are satisfactory in many cases, they have serious limitations. For one, they are difficult to apply to any but the simplest molecules and are usually restricted to planar molecules. Also, resonance arguments are inherently qualitative. For example, even though a resonance structure may assign a formal charge to a particular atom, it cannot say how much charge actually accumulates.

Less familiar to chemists is the description of substituent effects provided by molecular orbital theory. A molecular orbital analysis begins by dividing a molecule into two fragments, the substituent and the rest of the molecule.[1] The key orbitals of the two fragments are described, and then mixed together to generate the orbitals of the original molecule. Since mixing changes the orbital energies, the substituent's effect can be inferred from the change in energy of the occupied orbitals. If mixing stabilizes the occupied orbitals, the substituent has a stabilizing effect. On the other hand, if mixing destabilizes these orbitals, the substituent has a destabilizing effect.

---

1. This division is somewhat arbitrary. For example, one might choose to view *p*-dimethylaminobenzyl cation as a combination of a *p*-dimethylaminophenyl group and a methyl cation, or as a combination of a dimethylamino group and a benzyl cation.

The key fragment orbitals can be identified, mixed, and the energy changes calculated using perturbation theory. This theory says that, to a first approximation, mixing a substituent orbital ϕ_s° with energy ε_s°, with an intermediate orbital ϕ_i° with energy ε_i°, will give two new orbitals with the following energies:

$$\varepsilon_s = \varepsilon_s^\circ + \frac{(\varepsilon_{si} - \varepsilon_s^\circ S)^2}{\varepsilon_s^\circ - \varepsilon_i^\circ}$$

$$\varepsilon_i = \varepsilon_i^\circ + \frac{(\varepsilon_{si} - \varepsilon_i^\circ S)^2}{\varepsilon_i^\circ - \varepsilon_s^\circ}$$

ε_{si} and S are the interaction energy and overlap of the two fragment orbitals, respectively. ε_{si} is roughly proportional to $-S$, i.e., positive overlap makes the interaction energy negative. Application of these equations to commonly encountered situations yields the following conclusions:[2]

1. Mixing increases the difference in orbital energies. The lower-energy orbital is stabilized, and the higher-energy orbital is destabilized. Therefore, mixing an occupied orbital with an empty orbital stabilizes the molecule since the occupied orbital is stabilized.

2. The higher-energy orbital is destabilized more than the lower-energy orbital is stabilized. Therefore, mixing two occupied orbitals destabilizes the molecule.

3. The largest change in orbital energy (and the largest substituent effect) occurs when there is: i) a good match in the energies of the interacting orbitals, and ii) there is good overlap between the orbitals. Good overlap is sufficiently important that fragments will often change their geometry in order to increase stabilizing orbital overlap.

4. Combining conclusions #1 and #3 leads to the observation that the largest stabilizing substituent effect occurs when one fragment has a relatively low-energy empty orbital and the other fragment has a relatively high-energy occupied orbital. Therefore, these orbitals are probably the most important ones to consider in evaluating substituent effects.

---

2. The reader can "prove" each conclusion by working out the relationship between ε_s and ε_s° (and also between ε_i and ε_i°) for a sample situation in which $\varepsilon_s^\circ < \varepsilon_i^\circ$. The first conclusion is equivalent to the statement: $\varepsilon_s < \varepsilon_s^\circ$ and $\varepsilon_i > \varepsilon_i^\circ$. The second conclusion follows from the fact that $|\varepsilon_{si}|$ is normally greater than $|\varepsilon^\circ S|$, and $|\varepsilon^\circ S|$ is normally larger for the lower-energy orbital. And the third conclusion is based, in part, on the fact that ε_{si} is proportional to $-S$.

Similar conclusions obtain when degenerate (of the same energy) orbitals mix. Mixing produces two new orbitals, one that is lower in energy and one that is higher in energy than the original fragment orbitals. As before, the higher-energy orbital is destabilized more than the lower-energy orbital is stabilized, and the magnitude of the orbital energy change is proportional to the orbital overlap.

The following diagram illustrates the orbital energy changes associated with orbital mixing between two fragments, A and B, to give a molecule, AB.

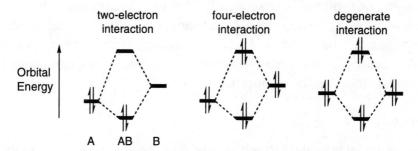

"Two-electron interactions," such as the mixing of an empty orbital with an occupied orbital, stabilize the molecule (conclusion #1), but "four-electron interactions" have the opposite effect (conclusion #2). Furthermore, these conclusions hold whether or not the fragment orbitals are degenerate.

Carbocations

The key fragment orbital for a carbocation, e.g., CH_3^+, is its low-energy empty p orbital. Mixing this orbital with an occupied substituent orbital (also known as the donor orbital) stabilizes the latter,[3] and a stabilizing substituent effect is expected. This situation is identical to the "two-electron" interaction depicted above, where A is the substituent and B is the carbocation.

Different substituents bring about different degrees of stabilization because of differences in the energies of their donor orbitals. The most stabilizing substituents are those that have a relatively high-energy donor orbital (conclusion #3). These substituents include groups with loosely held nonbonding electrons, such as $-NR_2$, -OR, and -SR, and groups with loosely held π electrons, such as vinyl and phenyl.

Alkyl groups, such as $-CH_3$, have filled π-type orbitals (see: **π Interactions Involving σ Electrons**), and can stabilize carbocations. This effect is smaller, however, because the alkyl donor orbitals correspond to σ bonds and are lower in energy. One exception to this rule is provided by cyclopropyl substituents. The donor orbitals in a cyclopropyl group are higher in energy because the bond angles in the cyclopropane ring prevent good σ overlap. Hence, these orbitals are a better match for the low-energy carbocation p orbital (see: **Dimethylcyclopropylcarbinyl**

---

3. This result follows from the fact that a filled orbital is always lower in energy than an empty orbital.

Cation). Another exception is provided by β-substituted cations, X-C-C⁺, where X involves an electropositive atom, such as silicon. In this case, the electrons in the CX bond are held more loosely, and the CX bonding orbital is polarized toward carbon, making this a better donor orbital (see: **Hyperconjugation and the Structures and Stabilities of Carbocations**).

One important consequence of orbital mixing in stabilized carbocations is that the new occupied orbital is a bonding combination of the substituent donor orbital and the empty p orbital. Consequently, the substituent-carbon bond will be stronger and shorter than normal. At the same time, the chemical bonds inside the substituent will be weaker and longer (assuming the donor orbital is a bonding orbital) because electron density that was associated with these bonds is now partially transferred to the carbocation center.

Carbanions

The key fragment orbital for a carbanion, e.g., CH_3^-, is the occupied nonbonding orbital. This orbital mixes with both occupied (donor) and empty (acceptor) substituent orbitals, and each type of mixing has a different effect on the orbital energy. Therefore, both types of mixing must be considered, i.e.,

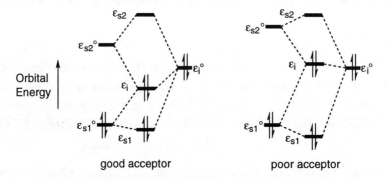

Perturbation theory suggests that, to a first-approximation, orbital energy changes will be additive. Therefore, the substituents that produce the largest stabilizing effect are those that maximize stabilizing two-electron interactions, while minimizing destabilizing four-electron interactions. Since the carbanion contributes two electrons, these substituents must have a low-energy acceptor orbital, $\phi_{s2}°$, that can overlap well with $\phi_i°$, and very low energy donor orbitals, $\phi_{s1}°$, that cannot overlap. Examples of good acceptor substituents include $-NO_2$, $-C(=O)X$, and $-C\equiv N$. Each of these substituents has a low-energy π^* acceptor orbital that is concentrated on the atom that is bonded to the carbanion (N in NO_2, and C in $-C(=O)X$ and $-C\equiv N$), i.e., these orbitals are ideally positioned for overlap with the carbanion. The donor orbitals in these substituents are also very low in energy, and are concentrated on the terminal substituent atoms (O in $-NO_2$ and $-C(=O)X$, N in $-C\equiv N$). Therefore, these orbitals do not destabilize the carbanion.

Vinyl groups and phenyl groups also have π^* acceptor orbitals that can stabilize a carbanion. However, the donor orbitals in these substituents are relatively high in energy and interact strongly with the carbanion. Therefore, these substituents are poorer acceptors and produce a combination of stabilizing and destabilizing interactions that is usually net stabilizing, but much smaller than the stabilization provided by the good acceptors.

The orbital interactions produced by good acceptors lead to a new occupied orbital, ϕ_i, that is, a bonding combination of ϕ_i° and ϕ_{s2}°. Therefore, a stronger, shorter substituent-carbon bond is to be expected in these carbanions. Since ϕ_i also incorporates some of the π^* characteristics of ϕ_{s2}°, weaker and longer substituent bonds are also expected. Both geometrical effects are seen in enolates; interaction of the carbonyl group and the carbanion makes the CC bond stronger and shorter, while the CO bond becomes longer and weaker.

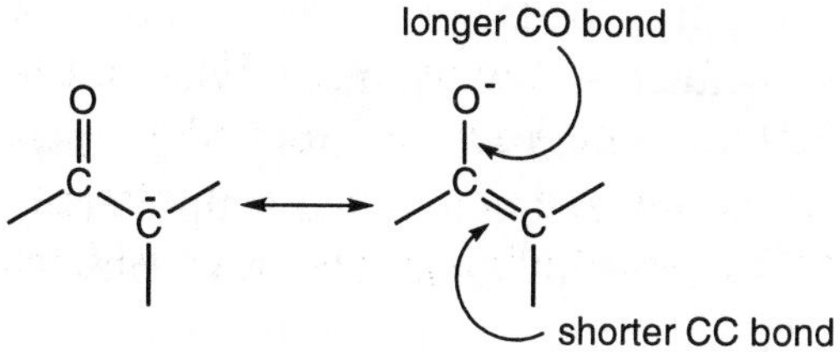

Free Radicals

The key fragment orbital for an unsubstituted free radical, e.g., $^{\cdot}CH_3$, is the singly occupied orbital (SOMO). Since this orbital is only singly occupied, net stabilizing interactions can arise in two different ways: from interaction with an electron donor substituent, or from interaction with an electron acceptor substituent.

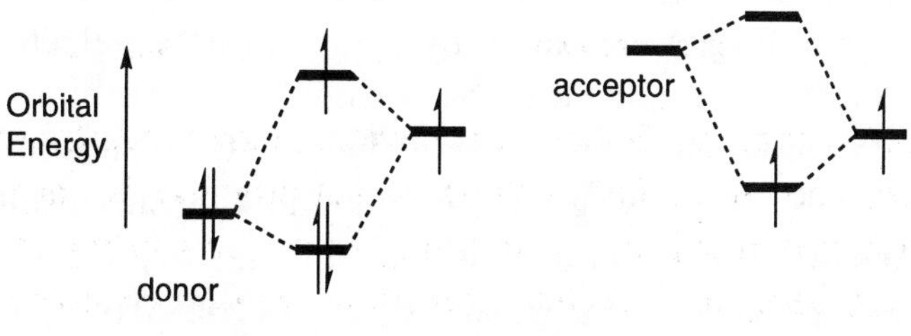

Electron donors stabilize free radicals by means of a three-electron interaction involving the SOMO and a relatively high-energy donor orbital on the substituent (see above). This interaction is reminiscent of the two-electron interaction that donors use to stabilize carbocations, and suggests that substituents that stabilize carbocations will also stabilize radicals. Substituent effects on radicals should not be as large, however, partly because donors destabilize the SOMO (this does not happen in carbocations), and partly because the energy gap between the fragment orbitals, $|\varepsilon_s^\circ - \varepsilon_i^\circ|$, is inherently larger in radicals (the positive charge on CH_3^+ lowers all of its orbital energies relative to those in $^{\cdot}CH_3$).

Electron acceptors stabilize free radicals by a different mechanism involving a one-electron interaction between the SOMO and an acceptor orbital on the substituent (see diagram on previous page). This interaction is similar to the two-electron interaction that acceptors use to stabilize carbanions, but the effect on a free radical should be smaller since only one electron is involved, and the energy gap between the fragment orbitals, $|\varepsilon_s^\circ - \varepsilon_i^\circ|$, is larger in radicals (the negative charge on CH_3^- raises all of its orbital energies relative to those in $\cdot CH_3$).

The preceding analysis suggests that radicals have a dual nature, and can interact favorably with either an electron-donor or an electron-acceptor substituent. On the other hand, smaller substituent effects are expected relative to those seen in carbocations and carbanions (see: **Substituent Effects on Reactive Intermediates**).

Multiple Substituents

The molecular orbital approach also provides key insights into the cumulative effect of multiple substituents. For example, what would be the stabilizing effect of two carbonyl groups on a carbanion? Certainly, the resulting anion would be expected to be more stable than a simple enolate, but exactly how much more stable? Experimental pK_a's are available. They suggest that the stabilization afforded to the carbanion center by two carbonyl groups is greater than that provided by one, and furthermore that the second group provides less stabilization than the first.

$$CH_4 \qquad CH_3\overset{\overset{\displaystyle O}{\|}}{C}CH_3 \qquad CH_3\overset{\overset{\displaystyle O}{\|}}{C}CH_2\overset{\overset{\displaystyle O}{\|}}{C}CH_3$$

$$pK_a \sim 50 \qquad pK_a \sim 20 \qquad \qquad pK_a \sim 9$$

An orbital interaction model provides a good rationale for this behavior. A carbonyl group stabilizes a carbanion by means of a two-electron interaction involving the CO π^* orbital and the carbanion nonbonding orbital, ϕ_i°. The effect of two carbonyl groups can be predicted by dividing the disubstituted carbanion into an enolate and a carbonyl group, and then mixing the resulting fragment orbitals. The key orbital interaction is still a stabilizing two-electron interaction, but it now involves a CO π^* orbital and the highest-occupied orbital of the enolate, ϕ_i, where the properties of this orbital are obtained by mixing together ϕ_i° and π^*. Since the energy of ϕ_i is less than that of ϕ_i°, the energy gap between ϕ_i and π^* is larger than the energy gap between ϕ_i° and π^* (see diagram on the facing page). Also, ϕ_i is a delocalized orbital (ϕ_i° is not), and this reduces the overlap between ϕ_i and π^*. Thus, on both energy and overlap grounds, it is to be expected the second carbonyl group will produce less stabilization.

Susceptibility to Substituent Effects

Substituent effects on reactive intermediates are often much larger than analogous effects on "normal molecules". An example of this can be seen by comparing rotation in allyl cation (carbocation + vinyl group) with rotation in acrolein (carbonyl + vinyl group). The cation rotation barrier is many times larger than the aldehyde rotation barrier, strongly suggesting a much larger role for the vinyl group in the cation.

This general kind of behavior can be easily rationalized using a molecular orbital model. In both allyl cation and acrolein, the main orbital interaction involves a stabilizing two-electron interaction between the occupied π orbital of the vinyl group and a low-energy empty orbital on the other fragment. In the case of allyl cation, the other fragment is $-CH_2^+$, and its empty orbital is a low-energy p orbital (note that the positive charge lowers this orbital's energy considerably relative to a p orbital on a neutral carbon atom). Therefore, the π-p energy gap is small and a strong interaction is expected. In the case of acrolein, however, the other fragment is a neutral formyl group, -CH=O. The empty orbital is now a higher-energy CO π^* orbital (the π^* energy is higher partly because the formyl group is neutral and partly because this orbital has antibonding character). Consequently, the π-π^* energy gap is large, and the vinyl-formyl interaction is weak.

This behavior is quite general. The key orbitals in a reactive intermediate are usually nonbonding, and of moderate energy, i.e., lower in energy than antibonding orbitals and higher in energy than bonding orbitals. This behavior is exaggerated in a charged intermediate, where an empty orbital might be shifted downward in energy (cation) or an occupied orbital might be shifted upward (anion). Therefore, the energy gap between an intermediate's orbital and a stabilizing substituent orbital is likely to be much smaller than the energy gap between a neutral fragment orbital and a stabilizing substituent orbital.

Directing Effects in Electrophilic Aromatic Substitution

Semi-empirical AM1 calculations are used to compare the stabilities of σ complexes resulting from nitration of substituted benzenes, and to compare the directing and activating effects of different substituents.

Electrophilic aromatic substitution is the most important chemical reaction of aromatic compounds.[1] The reaction occurs in two steps: initial electrophilic addition to give a σ complex (benzenium ion), followed by deprotonation and formation of a substituted benzene.

σ complex
(benzenium ion)

The first step is product-determining, and is usually (but not always) rate-determining too. Thus, substituents can influence both the product distribution and the reaction rate by favoring the formation of one σ complex over another.[2] Since the positive charge in the σ complex is delocalized over the *ortho* and *para* positions, donor substituents at these positions stabilize the σ complex, and act as *ortho/para* directors and activators. Acceptor substituents, on the other hand, destabilize the σ complex, and tend to be *meta* directors and deactivators.

In this experiment, you will use semi-empirical AM1 calculations to determine the preferred site of nitration ($E^+ = NO_2^+$) on toluene ($X = CH_3$), aniline ($X = NH_2$), and nitrobenzene ($X = NO_2$). To keep the number of calculations manageable, you will only compare *meta* vs. *para* attack. You will also use two types of graphical models to explain the substituent's effect on σ complex stability.

---

1. Carey and Sundberg A, p. 554; Lowry and Richardson, p. 623; March, p. 447.
2. The relationship between substituents and directing effects is a consequence of the Hammond postulate (see: **Reactive Intermediates and the Hammond Postulate**). σ complex formation is very endothermic, so transition state energies and reaction barriers for complex formation should parallel σ complex energies. Alternatively, reaction of σ complexes to products is highly exothermic, so transition state energies and reaction barriers to products should parallel σ complex energies. In either case, substituent effects on σ complex energies should mimic reaction rate differences.

Procedure

Reactants: Build benzene, toluene, aniline, and nitrobenzene, and optimize their geometries using semi-empirical AM1 calculations. Record their energies. Build NO_2^+ and optimize its AM1 geometry. Record its energy.

σ Complexes: Build the σ complex resulting from nitration of benzene ($E^+=NO_2^+$, X=H). Pick a reasonable conformation for the nitro group and optimize the cation's AM1 geometry. Record the energy of this ion.

Next build the σ complexes resulting from *meta* and *para* attack of NO_2^+ on toluene, aniline, and nitrobenzene.[3] Optimize the AM1 geometry of each cation, and record its energy. Compare the energies of the *meta* and *para* cations for each system; which cation is the more stable? Are your results consistent with the known directing effects of methyl, amino, and nitro substituents? Calculate ΔH_{rxn} for the process: $NO_2^+ + PhX \rightarrow \sigma$ complex (do this for all four aromatic molecules, but consider only the more stable σ complex for each substituted cation). Are these ΔH_{rxn} values positive or negative? Why? (Hint: the AM1 method calculates the energy of an isolated, i.e., gas-phase, molecule.) Assuming the relative values of ΔH_{rxn} reflect the relative barriers for σ complex formation, order the aromatic molecules according to their reactivity toward NO_2^+. Are your results consistent with the known (de)activating effects of these substituents?

Calculate and display an electrostatic potential map for each σ complex. Compare maps of the *meta* and *para* σ complexes for each substituent on the same color scale. In which σ complex is the positive charge more delocalized? Is there a correlation between charge delocalization and σ complex stability?

Calculate and display the lowest-unoccupied molecular orbital (LUMO) of the parent σ complex (X=H). Substituents stabilize the σ complex due to a stabilizing two-electron interaction between the LUMO and a substituent donor orbital (see: **A Molecular Orbital Description of Substituent Effects**). What are the best sites for a donor substituent? Is this analysis consistent with the directing effects calculated above? Is it consistent with the (de)activating effects calculated above? (Hint: you also need to consider the shape and energy of the substituent's donor orbital.)

Optional

Repeat all your calculations using the AM1-SM2 model to account for aqueous solvation (use AM1 geometries). Describe and rationalize any changes which you find (see: **The Role of Solvent**).

---

3. Use the σ complex corresponding to nitration of benzene as a "template" for building complexes corresponding to nitration of toluene, aniline and nitrobenzene. Make copies and add substituents (CH_3, NH_2 or NO_2) without altering the geometry.

Sulfonation of Naphthalene

Semi-empirical AM1 calculations are used to interpret the temperature-dependence of naphthalene sulfonation.

Sulfonation of naphthalene, **1**, gives two different products, depending on reaction conditions. Lower temperatures favor formation of α-naphthalenesulfonic acid, **2α**, while higher temperatures favor formation of the β isomer, **2β**.[1]

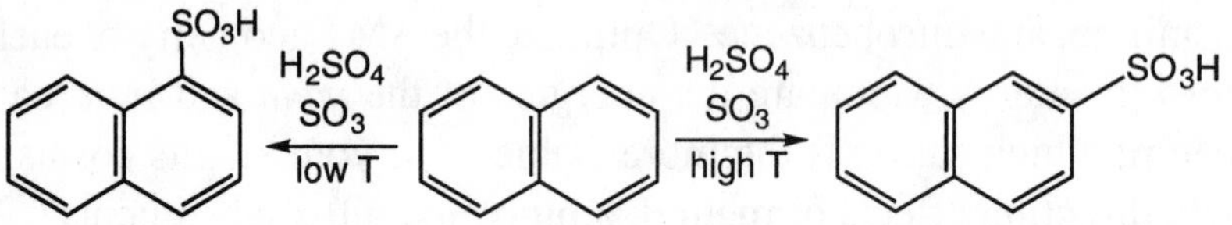

This behavior is consistent with a "normal" two-step electrophilic substitution mechanism (see: **Directing Effects in Electrophilic Aromatic Substitution**) if: 1) the barrier for formation of the α isomer is the lower (this would explain selective formation of **2α** at low temperature), 2) the β isomer is the more stable, and 3) formation of the α isomer is reversible under the conditions used to form **2β** (this would allow initial formation of **2α**, followed by reversion to **1**, and slower, but irreversible, formation of **2β**). Thus, the low-temperature reaction would be under kinetic control, while the higher-temperature reaction would be under thermodynamic control (see: **Thermodynamic vs. Kinetic Control**).

In this experiment, you will use semi-empirical AM1 calculations to compare both the stability of the products, **2α** and **2β**, as well as the stability of the intermediate σ complexes, **3α** and **3β**.

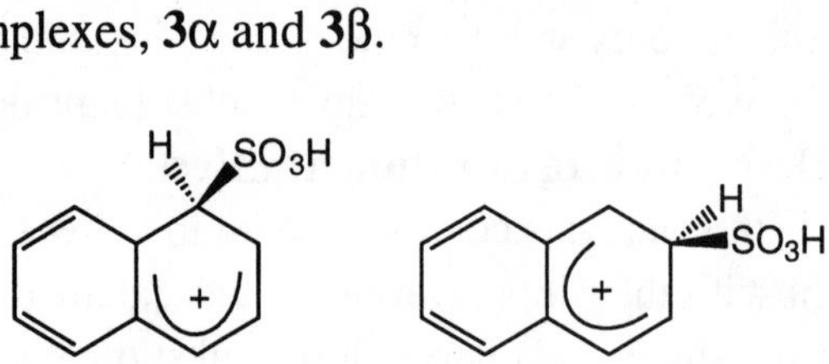

Note, you will assume that the actual electrophile (SO_3) is completely protonated in strong sulfuric acid. This being the case, the relative stabilities of cationic σ complexes (**3α** and **3β**) should anticipate the relative transition state energies for their formation, i.e., the lower-energy σ complex will form preferentially under conditions of kinetic control (see: **Reactive Intermediates and the Hammond Postulate**).

---

1. H. Beyer and W. Walter, **Handbook of Organic Chemistry**, Prentice Hall, New York, 1996, p. 628.

Procedure

Products: Build the two naphthalenesulfonic acids, **2α** and **2β**, and optimize their geometries using semi-empirical AM1 calculations. Record their energies. Which isomer is more stable? What would be the equilibrium ratio of these two molecules at room temperature? Can thermodynamic control account for selective formation of **2β** at high temperature?

σ Complexes: Build σ complexes, **3α** and **3β**, and optimize their AM1 geometries. Which complex is more stable? Is this result consistent with kinetic control of the reaction leading to **2α**?

Optional

Phenanthrene, **4**, undergoes nitration at C-9. Sulfonation of **4** (low temperature), however, chiefly occurs at C-2 and C-3.

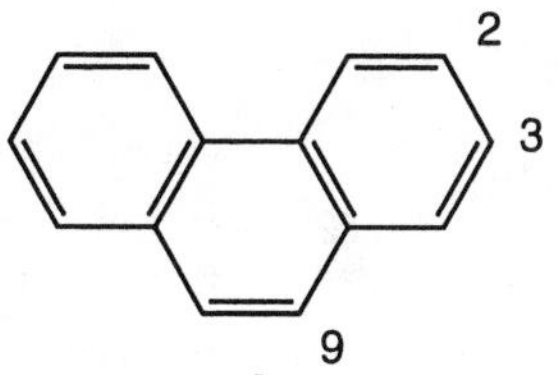

4

Build 2-, 3-, and 9-phenanthrenesulfonic acids, and optimize their AM1 geometries. Record their energies. Do these results support thermodynamic control of sulfonation? Build the corresponding σ complexes and optimize their AM1 geometries. Record their energies. Do these results support kinetic control of sulfonation? Use an analogous procedure to determine whether nitration is under kinetic or thermodynamic control.

Bromination of 2-Naphthol. Directing Effects in Polycyclic Aromatics

Semi-empirical AM1 calculations are used to investigate the mechanism of bromination of 2-naphthol by comparing the energies of various intermediate σ complexes.

2-Naphthol undergoes bromination to give selectively 1,6-dibromo-2-naphthol. The reaction is remarkable for several reasons. First, the reaction does not appear to follow the "normal rules" of benzene substitution. Bromination occurs once in each ring, rather than twice in the "activated" ring. Second, the site of bromination is specific within each ring, and although bromination at C-1 is consistent with the *ortho/para* directing properties of the hydroxyl substituent, it is not apparent why substitution does not occur at C-3.

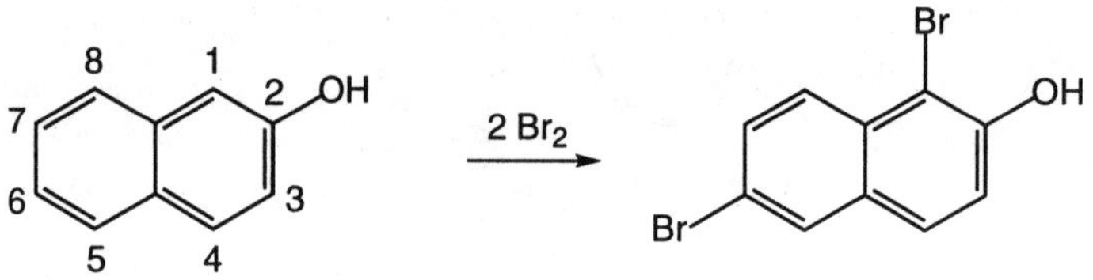

In this experiment, you will use semi-empirical AM1 calculations to determine the bromination sequence, and to determine the factors responsible for selectivity at each step. Some assumptions are necessary: 1) that the reaction is under kinetic control, 2) that formation of a σ complex is rate-determining, and 3) that the transition states for bromine attack are similar in energy and geometry to σ complexes, **1** and **3-8**. (Note: the label identifies the carbon attacked by bromine as shown below for molecules **1**, **3**, and **6**.)

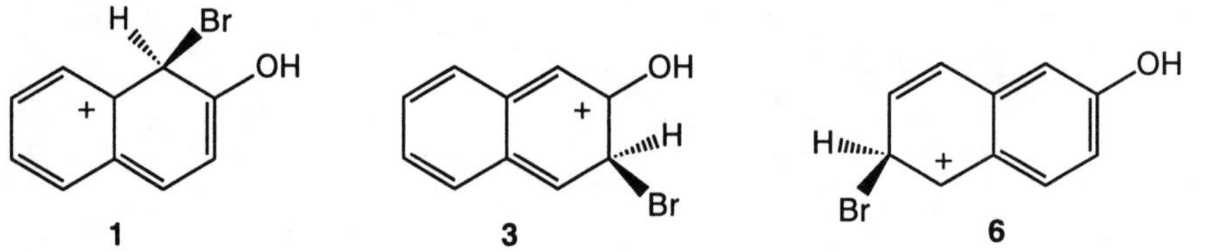

That is to say, you will assign the preferred (kinetic) site of bromination by identifying the most stable σ complex.

Procedure

First site of attack: Build the cationic σ complexes, **1** and **3-8**, and optimize their geometries using semi-empirical AM1 calculations. Record their energies. Which ion should form most rapidly? Is this result consistent with the ultimate formation of 1,6-dibromo-2-naphthol?

There are several possible ways to rationalize the selectivity of bromine attack. The simplest (but most tedious) might be to draw all of the important resonance contributors to each σ complex. Another would be to assume that the main effect controlling σ complex energy is the stabilizing two-electron interaction between an occupied hydroxyl orbital and an empty σ complex orbital ("σ complex", is now taken to mean the fragment obtained by mentally breaking the CO bond in any of the σ complexes). This interaction creates an additional CO bonding interaction (see: **A Molecular Orbital Description of Substituent Effects**), and the CO bond distance should be inversely related to the stabilizing effect of the hydroxy substituent. Record the CO bond distance in **1** and **3-8**. Is the expected relationship between distance and energy observed? What other factors might intrude on this simple picture? Support your arguments using resonance theory or molecular orbital theory.

Second site of attack: Assume that the favored first site of attack identified above gives the corresponding bromo-2-naphthol. Now build the six possible dibromo σ complexes derived from this molecule, optimize their AM1 geometries, and record their energies. Which site of attack is kinetically preferred? Is this result consistent with formation of 1,6-dibromo-2-naphthol? What other factors might control selectivity here? (Hint: consider steric effects.) Clearly identify any changes in mechanism that are required by your explanation.

Free Radicals from 1,3-Dioxolan-4-ones

Semi-empirical AM1 calculations are used to determine the relative stabilities of free radicals obtained by hydrogen abstraction from 1,3-dioxolan-4-ones.

The selectivity of hydrogen abstraction reactions is guided primarily by the stability of the incipient radical. This can be difficult to predict in polyfunctional molecules, such as 1,3-dioxolan-4-ones, **1**, where each radical is stabilized by a different combination of functional groups. For example, radical **2** is a captodative radical,[1] i.e., it is stabilized by an acceptor group ($>$C=O) and a donor group (OR), while radical **3** is stabilized by two donor groups (OR).

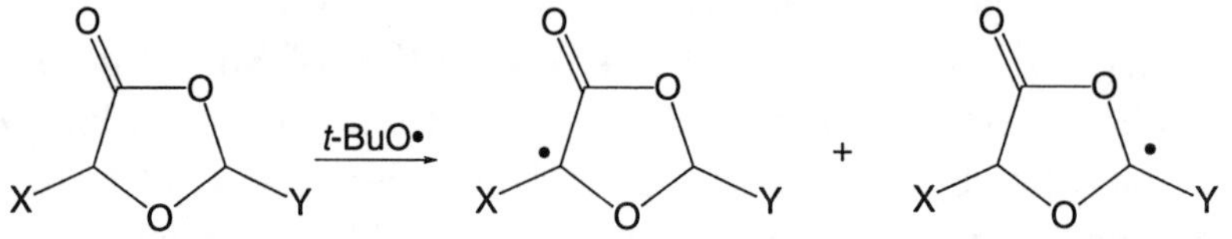

| **1** | **2** | **3** |

ESR measurements on radicals derived from substituted 1,3-dioxolan-4-ones indicate that the ratio of products also depends on the nature of these substituents.[2]

In this experiment, you will use semi-empirical AM1 calculations to determine the relative stability of these substituted radicals, and you will try to correlate these energy differences with the experimentally observed product ratios.

Procedure

Build radicals **2** and **3** for the substituent patterns specified in the following table.

Relative radical abundancies for H abstraction from 1,3-dioxolan-4-ones[2]		
X	**Y**	**[2]/[3]**
H	H	1.42
H	CH_3	0.87
H	$C(CH_3)_3$	1.10
CH_3	H	14.00
CH_3	CH_3	6.57
CH_3	$C(CH_3)_3$	11.20

1. Carey and Sundberg A., p. 681; March, p. 190.
2. A.L.J. Beckwith and A.A. Zavitsas, *J. Am. Chem. Soc.*, **117**, 607 (1995).

Optimize their geometries using semi-empirical AM1 calculations. Is the radical with the lower AM1 energy also the radical that forms preferentially? AM1 energies may not accurately reflect the relative stabilities of **2** and **3**, but they may describe the changes in radical stability that accompany changes in X and Y. Plot $\Delta(\Delta H_f)$ against log ([**2**]/[**3**]) for each pair of radicals. Is there a linear relationship? Which aspect of $\Delta(\Delta H_f)$ is more useful, its absolute value or its relative value?

Build the radical precursors, **1**, optimize their AM1 geometries, and record their energies. Use these energies to determine what effect the X group has on the two CH bond energies, i.e., how much do the bond energies change when X is changed from H to CH_3? Is this consistent with experimental observations on CH bond energies in simple alkanes? In a similar manner, determine the effect the Y group has on the CH bond energies. Is this effect consistent with experiment?

Intermediates in the Ozonolysis of Ethylene

Semi-empirical AM1 calculations are used to explore the mechanism of ozone addition to ethylene, and in particular, to examine the cyclic ozonide intermediates.

Ozonolysis, the addition of ozone across a carbon-carbon double bond and subsequent cleavage of that bond, is a convenient and synthetically important way of introducing carbonyl functionality, e.g.,

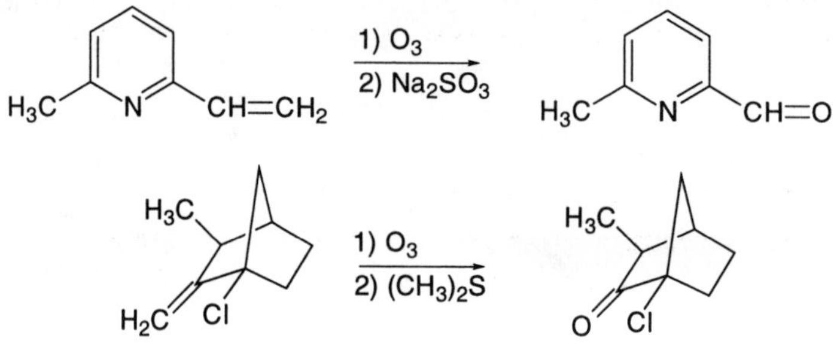

While the overall mechanism for ozonolysis remains to be established, it is now known that two distinct "cyclic ozonides" are involved,[1] e.g., **1** and **2** for ozonolysis of ethylene.

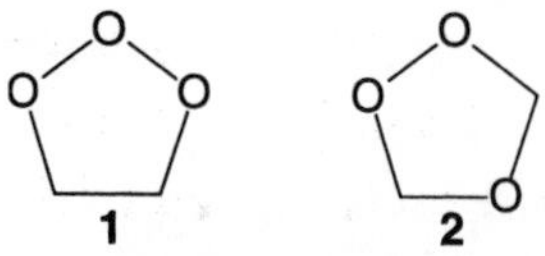

The first ozonide, formed by direct addition of ozone across the carbon-carbon double bond, rearranges to the (presumably) more stable second ozonide, which then goes on to form products, e.g., for ozonolysis of ethylene.

Such a mechanism would seem to be particularly attractive if: 1) every step was exothermic, 2) the activation barriers for all steps were relatively small.

---

1. M.B. Smith, **Organic Synthesis**, McGraw Hill, New York, 1994, p.306; Carey and Sundberg B, p.645; Lowry and Richardson, p. 917; March, p.1177.

Procedure

Build ethylene, ozone and ozonides **1** and **2**. Optimize the geometry of each using AM1 semi-empirical calculations. Is **1** more stable than separated reactants? Is **2** more stable than **1**? Try to account for your results regarding the relative stabilities of the two ozonides. (Cite specific geometrical features to support your rationalization.) Are your thermochemical results in line with the mechanism for ozonolysis proposed above?

Locate the transition state for initial addition of ozone to ethylene. Start with **1** and elongate the two CO bonds to 2.0 Å. Use this structure to search for the AM1 transition state, and verify that this is a transition state by calculating its vibrational frequencies (a transition state has one and only one imaginary frequency; see: **Finding Transition States**) What is the barrier for this reaction? Is it consistent with the mild reaction conditions used in ozonolysis?

Optional

Interconversion of **1** and **2** may occur by fragmentation of **1** to formaldehyde and **3** (a so-called carbonyl oxide), followed by recombination, i.e.,

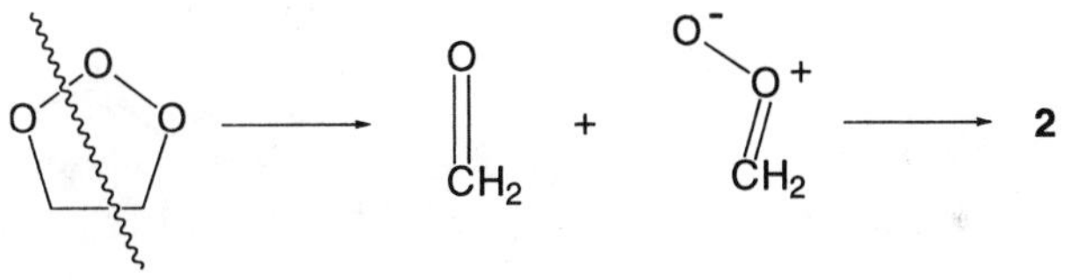

(see: **1,3-Dipolar Cycloadditions of Azomethine Ylides** for a discussion of this class of reactions.) Build formaldehyde and **3**, and optimize their AM1 geometries. Are these steps exothermic? Using AM1 calculations, locate transition states both for fragmentation of **1** and for recombination to **2**. Do you find low or high activation barriers? Based on your results, would you say that the fragmentation/recombination mechanism is reasonable?

Selectivity in Cyclopropanone Cycloadditions

Semi-empirical AM1 calculations and Frontier Molecular Orbital Theory are used to characterize the intermediate formed by the ring-opening of 2,2-dimethylcyclopropanone, and to evaluate its periselectivity.

Cyclopropanones are strained, highly-reactive molecules, and readily undergo cycloaddition reactions with other unsaturated molecules. For example, 2,2-dimethylcyclopropanone adds to furan to give a seven-membered ring product, **1**, and to trichloroacetaldehyde to give a five-membered ring product, **2**.[1]

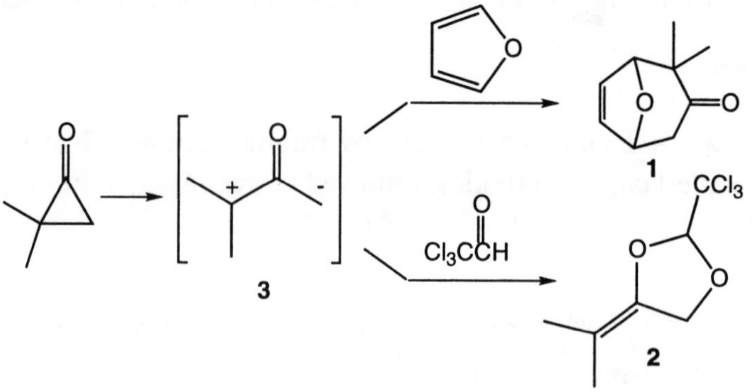

It is believed that each of these reactions proceeds in a stepwise fashion; the cyclopropanone opens to give a zwitterion **3**, which is then trapped in a concerted cycloaddition. What is particularly curious about **3** is its apparent ability to: 1) employ different groups of atoms in cycloadditions (CCC to form **1** and CCO to form **2**), and 2) its ability to react efficiently with groups whose frontier molecular orbitals have different symmetries. Frontier Molecular Orbital (FMO) theory asserts that the favored mode of cycloaddition will be the one that creates the best overlap between the interacting frontier orbitals in the two reactants. Dienes and carbonyls have frontier orbitals of different symmetries, and interactions that are favored by a diene should not be favored by a carbonyl and *vice versa*.

One problem here is that the electronic structure of **3** is far from obvious. Several resonance structures can be written, and although the π system contains only four atoms, it is somewhat difficult to infer the shapes and energies of its orbitals. In this experiment, you will use semi-empirical AM1 calculations to examine the frontier orbitals of **3**, furan, and trichloroacetaldehyde, and you will determine whether FMO theory is able to reconcile the orbital properties with the experimental observations.

1. T.L. Gilchrist and R.C. Storr, **Organic Reactions and Orbital Symmetry**, 2nd. ed., Cambridge University Press, 1979, p. 166.

Procedure

Frontier molecular orbitals: Build **3** and allyl cation, $CH_2CHCH_2^+$, and optimize their geometries using AM1 semi-empirical calculations. Examine the highest-occupied molecular orbital (HOMO) and the lowest-unocupied molecular orbital (LUMO) of each molecule. What similarities do you see? Compare their orbital energies. How do you account for the shapes and energies of the frontier orbitals of **3** (see: **A Molecular Orbital Description of Substituent Effects**; hint: treat **3** as an allyl cation containing an O^- substituent).

Build furan and trichloroacetaldehyde, and optimize their AM1 geometries. Record the shapes and energies of their frontier orbitals. Focus on the two atoms in each molecule that eventually become bonded to **3**. Do the orbitals in these molecules have the same or opposite symmetries? Which molecule is the better electron-donor (higher-energy HOMO)? The better electron-acceptor (lower-energy LUMO)?

Orbital interactions: FMO predictions are based on the orbital interaction characterized by the smaller orbital energy gap. Which gap is smaller for **3** and furan, i.e., which orbital interaction is more important, the HOMO of **3** and the LUMO of furan, or the HOMO of furan and the LUMO of **3**? What geometrical arrangement of **3** and furan will create the best overlap between these orbitals? Is this consistent with experiment?

Repeat this analysis for the cycloaddition of **3** and trichloroacetaldehyde. Identify the more important orbital interaction by calculating the two energy gaps, and then determine the transition state geometry that will give the best overlap. Is this result consistent with experiment?

Intermediates in the Benzylic Acid Rearrangement

Semi-empirical AM1 calculations are employed to investigate the mechanism of the benzilic acid rearrangement, and to explain the "migratory ability" of phenyl as a function of substitution.

The benzilic acid rearrangement,

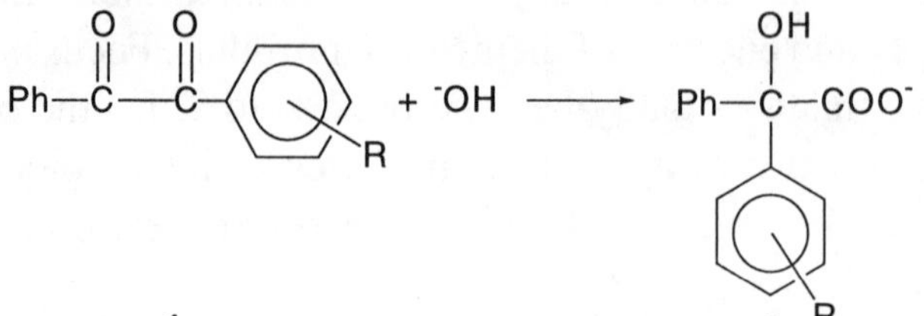

discovered by Liebig in 1838, was the first example of an intramolecular rearrangement. Even today it enjoys widespread use in organic synthesis.[1] The currently accepted mechanism involves addition of OH⁻ to **1** to produce a mixture of two distinct intermediates, **3a** and **3b**. Migration of the substituted phenyl ring in **3a** or the unsubstituted ring in **3b** leads to the same second intermediate, **4**, which then rapidly tautomerizes to **2**.

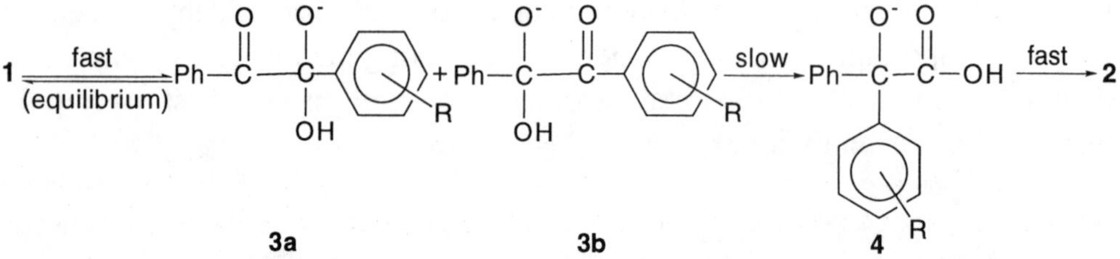

[14]C labeling experiments have shown that ring substituents influence which phenyl group migrates. When R = 3-Cl, 81% of the product results from migration of the substituted ring, while with R = 4-OMe, 69% of product results from migration of the unsubstituted ring. There are two possible (limiting) interpretations. One is that ring substitution affects the relative stabilities of **3a** and **3b**, but not the relative migratory abilities of the rings. In this case, product will be formed from **3a** and **3b** in proportion to their equilibrium abundances. The other interpretation is that substitution does not alter the relative stabilities of **3a** and **3b**, but does affect their relative migratory abilities. In this case, the ratio of products will follow the heights of the energy barriers connecting **3a** and **3b** to **4**.

In this experiment, you will use semi-empirical AM1 calculations to examine the first of these interpretations.

---

1. Lowry and Richardson, p. 486.

Procedure

Build both intermediates **3a** and **3b** for both R = 3-Cl and R = 4-OMe (four structures in total), and optimize their geometries using AM1 semi-empirical calculations.[2] Which of the two intermediates, **3a** or **3b**, is more stable for each ring substituent? For each of the two substituents, calculate the equilibrium ratio of **3a** to **3b** at room temperature, i.e.,

$$[\mathbf{3a}]/[\mathbf{3b}] = \exp\left[(\Delta H_f(\mathbf{3b}) - \Delta H_f(\mathbf{3a}))/RT\right]$$

Compare your results with the experimental ratios. Does this hypothesis satisfactorily account for the experimental observations? By what mechanism do these substituents affect ΔH_f?

Optional

The second hypothesis regarding substituent effects can be evaluated using transition state models. Construct a transition state for the phenyl migration **3 → 4** for R=H (**3a = 3b**), using the distances shown below.

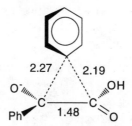

Optimize the AM1 geometry of this structure (constrain distances), then use the resulting structure to search for the AM1 transition state (no constraints). Next substitute the two rings with 3-Cl and 4-OMe and perform single-point energy calculations using the AM1 method (four calculations in total). Calculate barrier heights (relative to **3a** and **3b**). Do these account for the observed product ratios?

Calculate electrostatic potential maps for **3** (R=H) and for the corresponding transition state, and compare these on the same color scale. Is phenyl migration accompanied by electron transfer? Use these changes in electronic structure to account for substituent effects on phenyl migration barriers.

---

2. First optimize the parent system (R = H; **3a = 3b**) and use its geometry for a starting point for calculations on the substituted systems. This will ensure that similar conformations are used throughout.

XIII

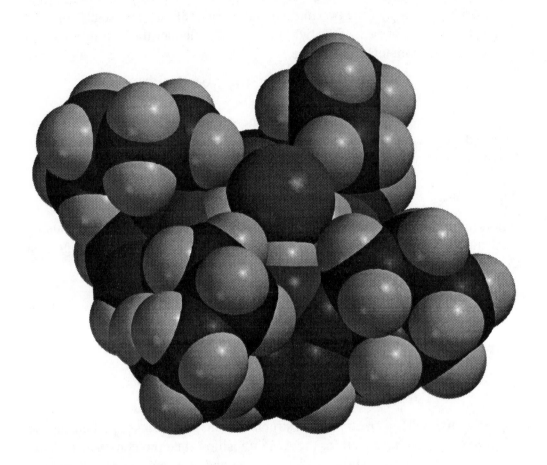

Weakly Bonded and Not-So-Weakly Bonded Molecules

O rganic chemists are trained to think about matter in terms of its "molecular" structure, that is, to assume that the most interesting way to think about matter is to pay attention to those groups of atoms that are connected by covalent bonds. In recent years, however, it has become obvious that Nature does not always share this same "molecular" bias. Many important structures in Nature are held together by other types of bonds, i.e., noncovalent bonds, and organic chemists are now beginning to explore the structural properties of these "weakly bonded molecules".[1]

The most obvious place to look for "weakly bonded molecules" is in any liquid. Although chemists do not normally think about liquid "structure", i.e., the geometric arrangement of molecules inside a liquid, the mere fact that liquids exist tells them that chemically significantly noncovalent bonding can occur, and that liquids have "a structure".

Much more dramatic examples of "weakly bonded molecules" are also found throughout biology. The famous DNA double helix consists of two molecules held together by noncovalent bonds. Hemoglobin, the body's oxygen carrier, consists of four proteins held together by noncovalent bonds. Most remarkable perhaps is the tobacco mosaic virus (TMV) which consists of a single RNA molecule surrounded by a helical array of 2130 separate, but identical, protein molecules. TMV, which is a cylinder 3000 Å long and 180 Å in diameter, dissociates into its component molecules in concentrated acetic acid. However, under the proper conditions, the protein and RNA can spontaneously reassemble into an active virus. This highly specific arrangement of molecules is not held together by any covalent bonds!

Although chemists cannot, as yet, compete with Nature when it comes to the synthesis of "weakly bonded molecules", they have now constructed a wide variety of such molecules, and in so doing have identified and characterized many different types of noncovalent bonds. Of these, the most common type appear to involve electrostatic interactions: ion-ion, ion-dipole, dipole-dipole, ion-induced dipole, and so on. A number of important practical applications have come out of these efforts. For example, crown ethers greatly enhance the reactivity of nucleophilic reagents by complexing their cation components.[2]

1. G.M. Whitesides, E.E. Simanek, J.P. Mathias and C.T. Seto, *Acc. Chem. Res.*, **28**, 37 (1995).
2. G.W. Gokel, **Crown Ethers and Cryptands**, Royal Soc. Chem., Coral Gables, Florida, 1991.

While the binding between 18-crown-6 and potassium cation in this case is certainly non-covalent, it would be incorrect to call it weak.

Of course, Nature can perform similar feats. Recently it has been argued that the electron-rich faces of aromatic rings provide sites for cation binding in biological systems.[3]

Electrostatic interactions rely on the electric fields generated by a molecule's nuclei and electron cloud, and are easily explored using molecular models. Is it possible to investigate the structural preferences and energies of "weakly bonded molecules" by building and optimizing them, or to try to anticipate how a particular molecular might engage in electrostatic interactions by examining isosurfaces or electrostatic potential maps. Thus, molecular modeling is not only useful as a tool for rationalizing experimental observations, it can be used to design molecules with specific desired properties.

---

3.   D.A. Dougherty, *Science*, **271**, 163 (1996).

Structure of Water Dimer

Ab initio 6-31G* calculations are used to examine the dependence of hydrogen bond strength in water dimer as a function of the O–H••O angle, and to distinguish between possible dimer geometries.

Water and simple alcohols have boiling points much higher than their molecular weights would predict, an indication that these molecules associate in solution. Indeed, liquid water exhibits a highly structured network of hydrogen bonds, i.e.,

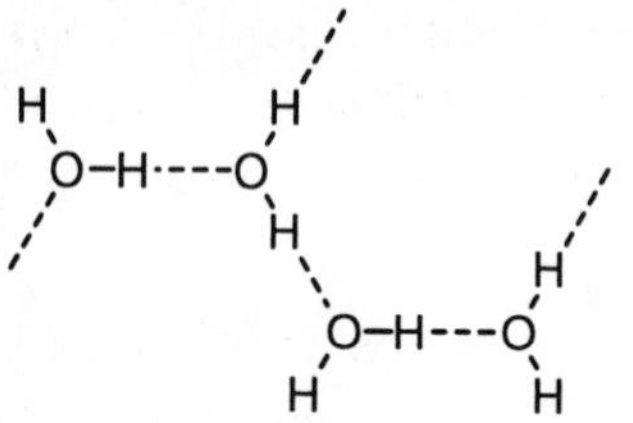

While individual hydrogen bonds are nowhere as strong as covalent or ionic linkages (see: **Weakly Bonded and Not-So-Weakly Bonded Molecules**), they are sufficiently strong to impart a degree of structure to the liquid. This has been observed experimentally using X-ray crystallography and neutron diffraction, which shows a relatively narrow distribution in the average separation between "nearest neighbor" oxygen atoms.

In this experiment, you will use *ab initio* 6-31G* calculations to determine the equilibrium geometry of the simplest, and best studied, hydrogen-bonded system, water dimer.[1] There are two plausible structures, linear and bifurcated (symmetry point groups are given in parenthesis).

linear (C$_s$)   bifurcated (C$_{2v}$)

These structures contain different numbers of hydrogen bonds, and require different OH••O bond angles. If a linear OH••O arrangement is strongly preferred then the linear structure may be more stable even though it contains but one hydrogen bond.

Procedure

Build linear water dimer (C$_s$ symmetry) in the conformation shown above, and optimize its geometry using *ab initio* 6-31G* calculations. Record its energy, OO distance, and OH••O angle. What is the strength of the hydrogen bond? (The 6-31G*

---

1. For a review, see: P.A. Kollman, in **Applications of Electronic Structure Theory**, H.F. Schaefer, III, Ed., vol. 4, Plenum Press, New York, 1977, p. 109. K. Liu, J.D. Cruzan and R.J. Saykally, *Science*, **271**, 929 (1996).

energy of water is -76.01075 hartrees.) How does the OO distance compare to the experimental distance of 2.97 Å? Does the dimer maintain a linear geometry?

Constrain the OH••O angle in your optimized dimer to a value that is 20° smaller than the equilibrium value, and reoptimize its 6-31G* geometry subject to this constraint. Record the energy of this dimer, and repeat the process three more times for dimers with OH••O angles 40° smaller, 60° smaller, and 80° smaller than the equilibrium value. Compute the energy of each constrained dimer relative to the optimal structure, and plot this vs. OH••O angle.

Qualitatively estimate the OH••O angles in the bifurcated structure, and then using the data obatined above, estimate its stability relative to the linear structure (assume that both hydrogen bonds in the bifurcated structure are equally strong). Now build the bifurcated dimer (C_{2v} symmetry) and optimize its 6-31G* geometry. Which is more stable, the linear or the bifurcated dimer? Decide if the bifurcated dimer is a true energy minimum by calculating its vibration frequencies (all of the frequencies of a true minimum-energy structure must be positive; see: **Finding Molecular Geometries**).

Optional

1. Intramolecular hydrogen bonds are believed to stabilize the keto-enol tautomers of 1,3-dicarbonyl compounds relative to their diketo forms. The strength of this bond is not likely to be optimal, however, because it is bent. Determine the strength of the hydrogen bond in 1,3-malonaldehyde by comparing the energies of "in" and "out" conformers.

    Optimize the AM1 geometry of each conformer, then use this geometry to calculate the 6-31G* energy. How strong is the intramolecular "hydrogen bond"? Is this value reasonable given the results obtained above for "bent" water dimers? What other factors might contribute to the apparent "bond strength"? Hint: Consider electrostatic interactions and the conformational preferences of vinyl alcohols (see: **Conformational Isomerism in Alkenes**).

2. The equilibrium structure of linear water dimer shows that there is also an optimal angle between the HOH plane of the "acceptor" molecule and the O••H bond axis. Build constrained linear dimers in which this angle is systematically varied, optimize their 6-31G* geometries (subject to the angle constraint), and determine the energetic penalty associated with distortions of this angle.

Molecular Recognition. Hydrogen-Bonded Base Pairs

Electrostatic potential maps obtained from AM1 semi-empirical calculations are used to anticipate the geometries of hydrogen-bonded nucleotide complexes. AM1 energies are then used to estimate the relative stabilities of various naturally-occurring complexes, as well as "unnatural" complexes of these nucleotides and some synthetic nucleotide mimics.

Each person's genetic code is contained in the nucleotide sequence of his or her DNA. The "message" in this code is "read" through the selective formation of hydrogen-bonded complexes, or Watson-Crick base pairs, of adenine (A) and thymine (T) (forms A-T base pair), and guanine (G) and cytosine (C) (forms G-C base pair). The importance of this hydrogen bond-based code lies in the fact that virtually all aspects of cell function are regulated by proper base pair formation, and many diseases can be traced to "reading" errors, i.e., the formation of incorrect base pairs. This kind of "molecular recognition", the formation of intermolecular complexes between specific molecules, and with specific geometries, is an important theme that runs throughout molecular biology, and control of molecular recognition is a new and rapidly developing branch of chemistry.

In this experiment, you will use semi-empirical AM1 calculations to examine several problems related to base pair stability. First, you will generate electrostatic potential maps of methylated derivatives of the four DNA nucleotides, and use these to identify potential hydrogen bonding sites, and potential base pair geometries. Then, you will compare the stabilities of various natural and unnatural base pair geometries involving these bases. You will also evaluate the hydrogen bonding capabilities of synthetic molecules that might mimic the base pairing properties of natural nucleotides, and might be used to direct drugs towards specific DNA and RNA targets.

Procedure

DNA nucleotides: Build **MeT**, **MeA**, **MeC**, and **MeG**, and optimize their geometries using semi-empirical AM1 calculations.

| 1-methylthymine | 9-methyladenine | 1-methylcytosine | 9-methylguanine |
| **(MeT)** | **(MeA)** | **(MeC)** | **(MeG)** |

Record their energies, and display their electrostatic potential maps. Identify electron-attracting and electron-repelling sites in the molecular plane that would be suitable for hydrogen bonding. Then use this information to draw all possible base pairs that involve two or three hydrogen bonds (consider all combinations; do not just limit yourself to Watson-Crick AT and GC pairs).

The AM1 geometries of the following base pairs have already been calculated, and their AM1 energies (ΔH_f(complex)), along with their experimental association constants (K_{assoc}),[1] are listed in the table below.

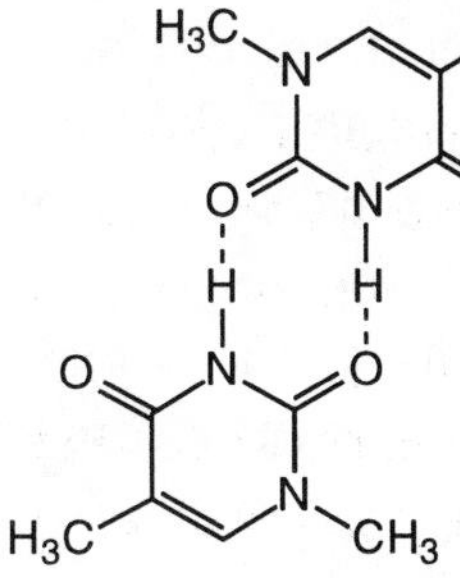

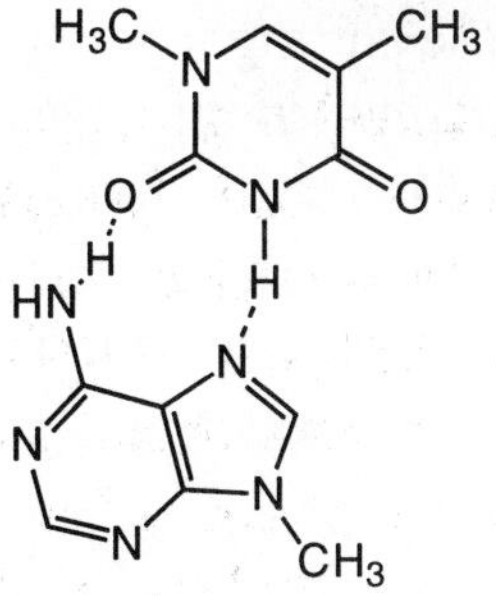

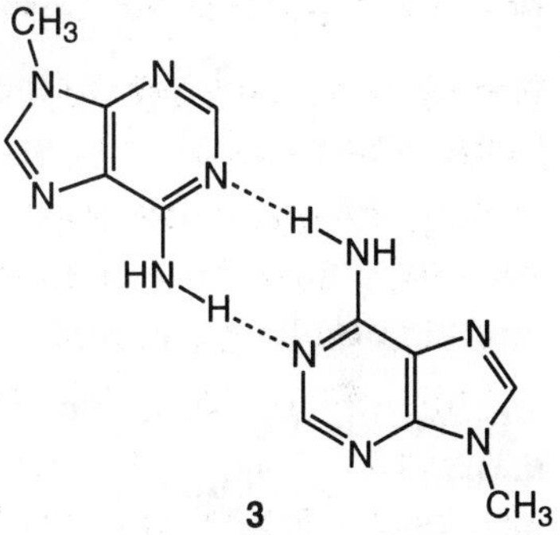

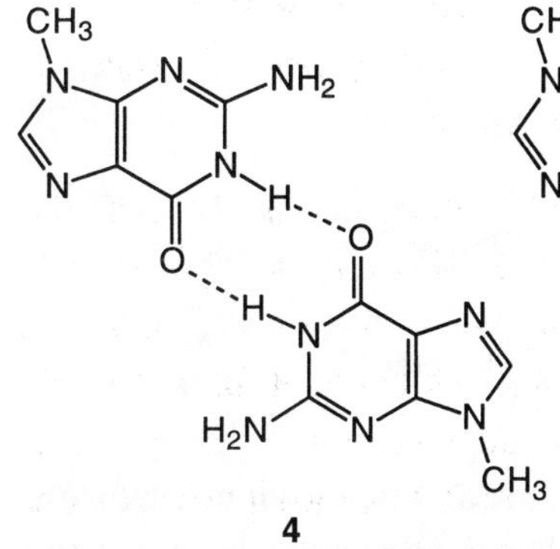

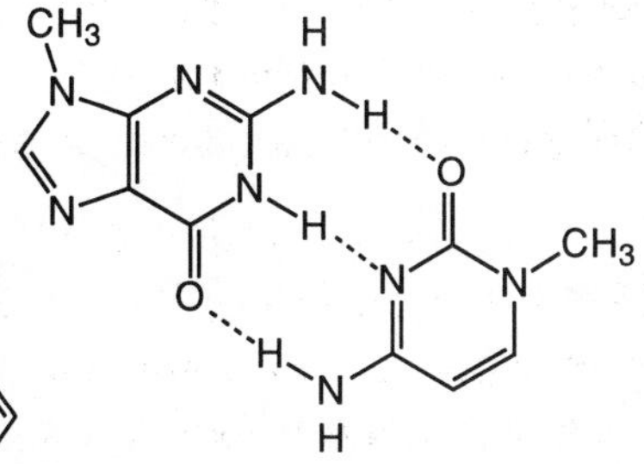

AM1 complex energies (kcal/mol) and experimental association constants		
	ΔH_f (complex)	K_{assoc}
1	-157.81	3.2
2	-28.81	1.3×10^2
3	105.46	3.1
4	1.88	5×10^3
5	-19.77	5×10^4
6	-18.55	1.7×10^4

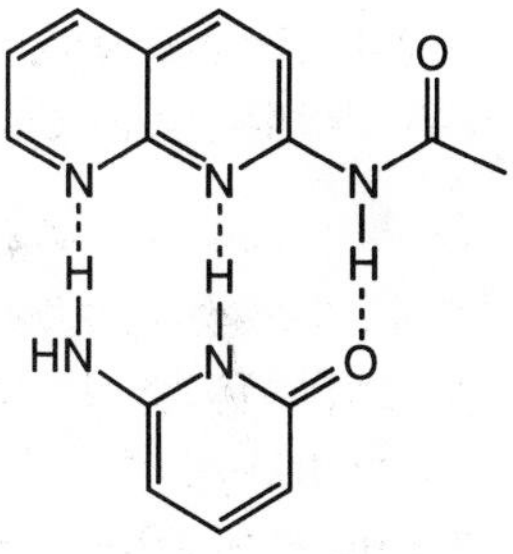

1. J. Pranata, S.G. Wierschke and W.L. Jorgensen, *J. Am. Chem. Soc.*, **113**, 2810 (1991); T.J. Murry and S.C. Zimmerman, *ibid.*, **114**, 4010 (1992).

Use your previously calculated monomer energies to calculate the combined energy of the isolated monomers ($\Sigma\Delta H_f$(monomer)) and, subtracting this from ΔH_f(complex), the energy of assocation (ΔH_{assoc}). Are the latter realistic, i.e., do they correspond to something like a hydrogen bond energy (or possibly a small multiple of this energy)? Are the association energies useful indicators of binding affinity, i.e., are $-\Delta H_{assoc}$ and log K_{assoc} correlated?

Other base pairing modes: Examine the base pairs in the figure above, and determine which, if any, correspond to Watson-Crick base pairs, i.e., the type of base pairing that naturally occurs in DNA. Do these base pairs have the largest K_{assoc}? Now examine the list of base pairs that you formulated using electrostatic potential maps. Which of these might form stronger complexes than the Watson-Crick base pairs? Test your ideas for three or four of these complexes (build the dimers, optimize their AM1 geometries, and calculate ΔH_{assoc}). Do your results confirm your expectations? What factors seem to favor formation of a strongly bound complex?

Nucleotide mimics: AP and **NA** are heterocyclic molecules that mimic the hydrogen bonding properties of DNA nucleotides by associating strongly in chloroform (see above).

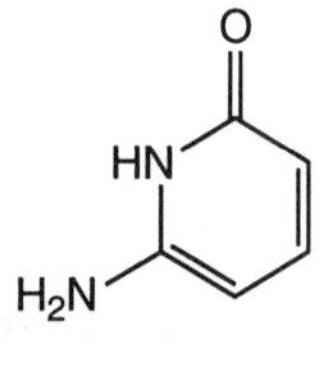

6-amino-2-pyridone
(AP)

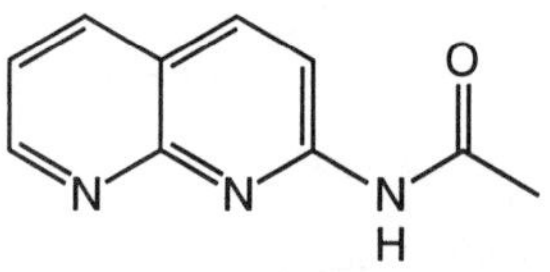

N-naphtharidinyl acetamide
(NA)

Build **AP** and **NA**, optimize their AM1 geometries, and calculate their electrostatic potential maps. Identify likely sites for hydrogen bond formation, and use this information to draw all possible base pairs between these mimics, and between the mimics and the DNA nucleotides listed above (only draw base pairs that involve two or three hydrogen bonds). Compute ΔH_{assoc} for **6**, an experimentally observed **AP-NA** base pair. Is ΔH_{assoc} a good indicator of binding affinity?

Optional

1. Try to account for the ΔH_{assoc} values calculated above using electrostatic potential maps for the individual nucleotides. Is binding energy correlated with the difference in electrostatic potentials between complementary hydrogen bonding sites? What types of atoms, O or N, appear to be the best hydrogen bond acceptors? What types of hydrogens, HO, HN_{ring}, H_2N, appear to be the best hydrogen bond donors?

2. DNA base pair recognition involves multiple hydrogen bonds within each base pair. A natural question to ask is whether ΔH_{assoc} depends on the placement of hydrogen bond acceptors (A) and donors (D) within a base pair (see first paper in ref. 1). For example, will an AA-DD or AD-AD complex be more stable, all other factors being equal? Design heterocyclic bases that will allow you to answer this question for base pairs containing two or three hydrogen bonds (all important base pair combinations are shown below).

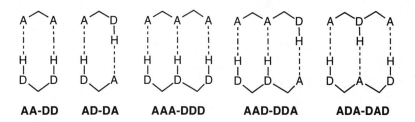

Calculate the AM1 energies of the individual bases and the base pairs, and use these energies to obtain the AM1 binding energies. What appears to be the best arrangement of A and D groups within each base? Can you account for these results using the electrostatic potential map of each base?

How Does Benzene Crystallize?

3-21G ab initio calculations together with electrostatic potential maps are used to anticipate the structure of benzene in the solid state.

What is the structure of benzene in the solid state? You might think this to be a "trick question". After all, benzene is a "flat" molecule seemingly ideal for efficient stacking.

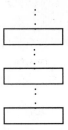

As there is no possibility for directed interactions, such as, hydrogen bonding, why shouldn't benzene crystallize in this way? As it happens, it doesn't.[1]

In this experiment, you will use 3-21G *ab initio* calculations to investigate two different structures for benzene dimer (as a model for crystalline benzene). You will first examine a parallel "stack" and then a perpendicular arrangement.

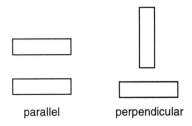

parallel   perpendicular

Calculated dimer energies as a function of separation should tell you which arrangement is favored, and electrostatic potential maps should help you to account for your results.

Procedure

Perform two series of single-point energy 3-21G *ab initio* calculations. For the first on parallel benzene dimer, constrain ring-ring separations to be 1.75, 2.00, ... 3.50 Å (8 calculations in total)[2], and for the second on perpendicular benzene

---

1. E.G. Cox, D.W.J. Cruickshank and J.A.S. Smith, *Proc. Roy. Soc.*, **A247**, 1 (1958); G.E. Bacon, N.A. Curry and S.A. Wilson, *ibid.*, **A279**, 98 (1964). See also: C.A. Hunter, *Angew. Chem. Int. Ed. Engl.*, **32**, 1584 (1993).
2. *SPARTAN* building instructions. Start with **Ligands: Benzene** from the *expert* tool kit and then add another benzene ligand to the "free valence" coming out of the ring. Set the ring-ring distance using **Geometry:Distance**.

dimer, constrain H-ring (measure separations from center of ring) separations to be 1.75, 2.00, ... 3.50 Å (8 calculations in total)[3]. Also calculate the 3-21G energy of benzene. (Make certain that the geometry of benzene is the same as that in benzene dimer.) Determine binding energies (E_{dimer} - $2E_{benzene}$) for the two sets of dimer calculations, and plot energy vs. separation. Which of the two dimer arrangements do you find to be more favored? See if you can find evidence for your conclusions in the known X-ray crystal structure of benzene.[1]

What is the origin of the preference which you observe? A hint comes from examining which regions on benzene are likely to be electron rich and which are likely to be electron poor. Calculate and display an electrostatic potential map for benzene. Describe what you observe insofar as the predicted charge distribution. How does it fit in with the results of your dimer calculations?

Optional

Perform a full geometry optimization using 3-21G *ab initio* calculations on whichever dimer (parallel or perpendicular) you find to be more stable. How does your structure relate to the known structure of crystalline benzene?

---

2.   *SPARTAN* building instructions. Start with **Ligands**: **Benzene** from the *expert* tool kit, add linear divalent hydrogen to the "free valence" coming out of the ring. Select **Ring**: **Benzene** and add to the free valence on hydrogen. Set the distance using **Geometry**: **Distance**.

Complexing Anions with Macrocycles

AM1 semi-empirical calculations together with electrostatic potential maps are used to investigate binding of halide anions to porphyrinogen macrocycles.

Macrocyclic ligands, among them crown ethers, cryptands and cyclodextrins, are known to form stable complexes with atomic and molecular cations, e.g.,

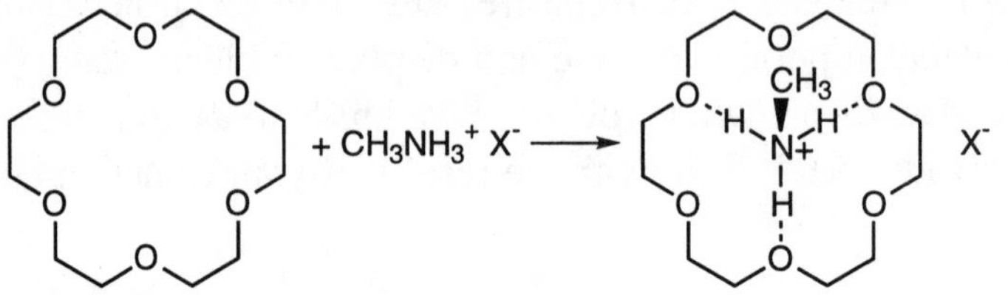

There are many uses for such systems. For example, the reactivity of anionic nucleophiles may be significantly enhanced by complexation of its associated counterion.

Far fewer macrocycles are available to complex anions. The reason for this is that it is much easier synthetically to incorporate electronegative atoms into a macrocycle, e.g., oxygens in the case of crown ethers, than it is to incorporate electropositive elements. One strategy for complexing anions is to build macrocycles containing "acidic hydrogens". For example, porphyrinogens, **1**, have been found to complex a variety of atomic and molecular anions.

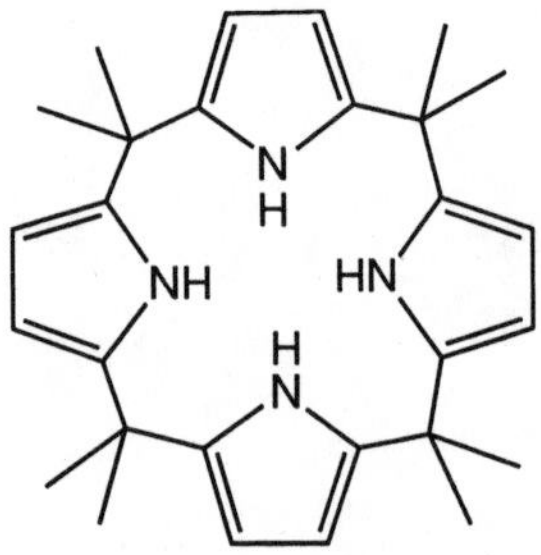

1

X-ray crystal structures have been obtained for several porphyrinogens, and these show the four NH bonds pointing in the same direction, thereby forming a "basket" into which the anion fits, e.g., **2X**, where X is an atomic or molecular anion.[1]

---

1.   P.A. Gale, J.L. Sessler, V. Kral and V. Lynch, *J. Am. Chem. Soc.*, **118**, 5140 (1996).

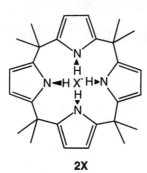

2X

In this experiment, you first will use AM1 semi-empirical calculations to examine the structures and predict binding energies for complexes **2X** (X=F⁻, Cl⁻, Br⁻). You will then calculate the structure (conformation) of the "free" macrocycle, **1**, to see if incorporation of the anion causes any significant changes.

Procedure

Build **2** (X⁻ = Cl⁻)[2] and optimize its AM1 geometry. Examine the resulting structure. Do the hydrogens form a "basket" in which to place the halide anion? Calculate atomic charges. How much charge has been removed from the anion as a result of complexation?

Use **2Cl** as a template for building **2F** and **2Br**, then optimize their AM1 geometries. Do you see a trend in structure, e.g., X⁻--H bond distances? Does this parallel any variation in halide charge (in the complexes) or ionic radii?

Build **1** (use **2Cl** as a template and remove the Cl) and optimize its AM1 geometry. Is the resulting structure similar to the macrocycle in the halide complexes or is it different? Try to rationalize your results. (Hint: What electrostatic interactions are likely to be important in **1**?)

Finally, calculate halide binding energies for the three complexes, i.e., E(**2X**) - E(**1**) - E(X⁻).[3] Do these parallel the structural changes and/or the halide charges or atomic radii noted previously?

Optional

Perform single-point energy calculations on **1** and **2X** (X=F, Cl, Br). using the AM1-SM2 model which accounts for solvent. Rationalize the changes in binding energies (relative to the gas-phase AM1 results) that you observe.[4]

---

2. *SPARTAN* building instructions. Select **Chelates**: **Porphyrin** from the *expert* tool kit. Add monovalent Cl to one of the (*axial*) free valences and then delete first the central atom, then the free valence on chlorine. Next, change the bond types for the eight linkages connecting the pyrrole rings from partial double (╌╌╌╌) to single (────). Switch to the *entry* tool kit, select sp³ carbon, and *double click* on the four carbons connecting the pyrrole rings; they will be changed to sp³ carbons. Add eight methyl groups. **Do not minimize.**
3. AM1 heats of formation (kcal/mol): F⁻, 3.4; Cl⁻, -37.7; Br⁻, -20.4.
4. AM1-SM2 heats of formation (kcal/mol): F⁻, -103.6; Cl⁻, -114.7 ; Br⁻, -92.4.

Proton Sponge and Other Super Bases

AM1 semi-empirical calculations are used to examine the geometry of protonated 1,8-bis(dimethylamino)naphthalene (proton sponge), and to predict structures and proton affinities of even stronger bases.

Among the known "super bases" are molecules like 1,8-bis(dimethylamino)-naphthalene, **1** ("proton sponge").[1]

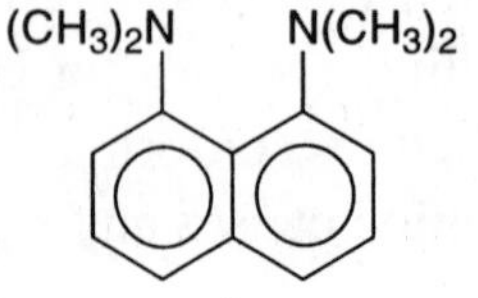

(CH₃)₂N   N(CH₃)₂

1

This owes its exceptional basicity to the fact that the two nitrogens are held in close proximity. This allows one nitrogen to stabilize the ammonium group by hydrogen bonding, **1H⁺hb**, or it might even allow formation of a symmetric chelate, **1H⁺c**.

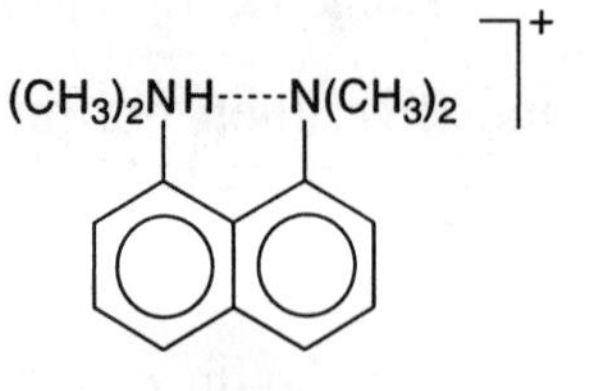

1 H⁺hb 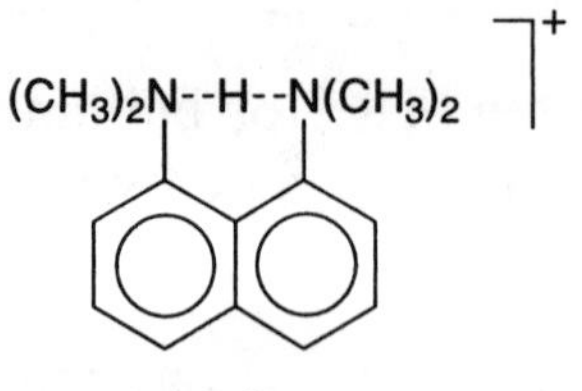   **1 H⁺c**

In either event, molecules such as proton sponge closely resemble macrocyclic systems, such as crown ethers and cryptands, both in geometry and in their ability to stabilize positively charged centers.

In this experiment, you will use semi-empirical AM1 calculations first to assign the equilibrium structure of the protonated form of **1**, either as "hydrogen bonded" or "chelated", and then go on to predict structures and basicities for molecules which offer the possibility of three and four nitrogens associated with a proton.

Procedure

Build **1H⁺hb** and **1H⁺c** and optimize their geometries using the AM1 method. (For **1H⁺c** you will need to maintain a plane of symmetry to keep it from falling back to **1H⁺hb** in case the molecule is not an energy minimum.) Which structure is preferred? If your calculations indicate this to be **1H⁺hb**, calculate the vibrational frequencies of **1H⁺c** to see whether it is an alternative energy minimum or a transition state for proton migration (see: **Finding Molecular Geometries** and **Finding Transition States**).

---

1.   H. Staab, *Angew. Chem. Int. Ed. Engl.*, **27**, 865 (1988), March, p.267.

Build **1**, its isomer, **2**, and conjugate acid, **2H⁺**.

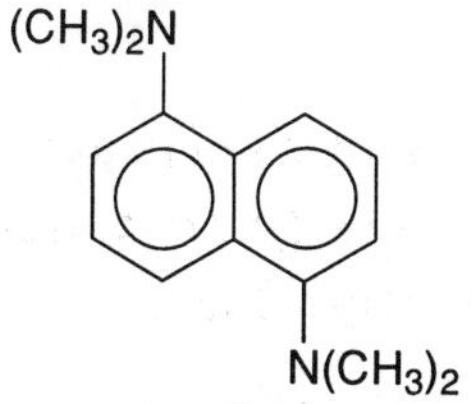

$$(CH_3)_2N$$

2

Optimize their AM1 geometries and use your results to determine the relative proton affinities of **1** and **2** (use the *isodesmic* reaction: **1H⁺ + 2 → 1 + 2H⁺**; see: **Isodesmic Reactions**). Are proximity effects important?

Build the conjugate acids of polyamines **3** and **4** and optimize their geometries.

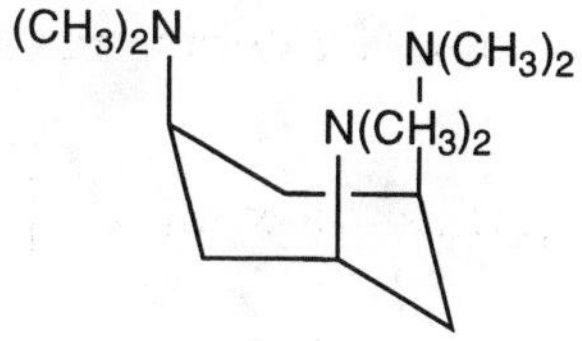

3 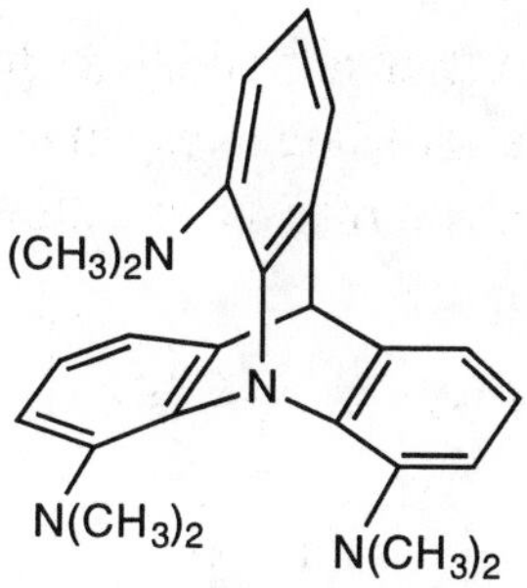   **4**

Do you find "delocalized" (chelated) protonated structures to be favored over "localized" structures in which the proton is associated primarily with only a single nitrogen? Which of the two localized structures for **4** is lower in energy? How do you explain this? (Hint: consider the relative proton affinities of N,N-dimethylaniline and quinuclidine).

Build **3** and **4** and optimize their AM1 geometries. Use your results to assess the proton affinities of **3** and **4** relative to that of proton sponge (use *isodesmic* reactions for this comparison). Does the possibility of additional interactions (or enhanced chelation) have a significant effect?

Optional

Design an amine that will act as an even stronger base than any considered in this experiment.

XIV

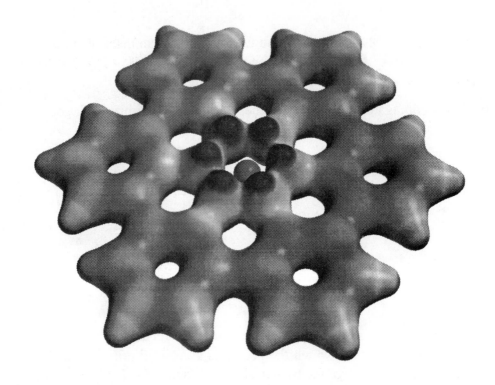

Isotope Effects

T he "electronic" Schrödinger equation as it is applied to chemical systems contains absolutely no mention of nuclear mass. This is due to the Born-Oppenheimer approximation, which in effect states that, from the "point of view" of the electrons, the nuclei are fixed (see: **Computational Tools**). One consequence of this approximation is that a molecule's potential energy surface does not depend on the molecule's isotopic constitution. How then is it possible to explain the different reactivity often observed for molecules containing different isotopes?

A simple explanation for many isotope effects may be found in the shape of a molecule's energy surface. The solid curves shown below depict a molecule's potential energy (PE) as a function of geometry.

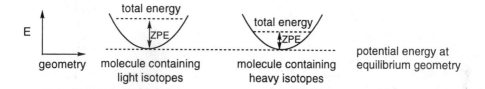

This energy is the one obtained from the electronic Schrödinger equation plus the component due to nuclear repulsion. It does not depend on nuclear mass, i.e., the same curve describes a molecule regardless of its isotopic composition. The diagram also shows each molecule's total energy, and this energy is seen to be: 1) greater than the potential energy, and 2) mass-dependent (the molecule that contains lighter isotopes has a higher total energy). Reactivity is determined not by potential energy, but by total energy; therefore, it is important to understand what factors make the two energies different. The most important factor turns out to be the zero-point energy (ZPE), the energy due to molecular vibrations.[1] This energy increases with increasing vibrational frequency, and vibrational frequencies are inversely related to nuclear mass. Or, to put it another way, lighter isotopes vibrate more rapidly and the molecules that contain them have higher energies.

Zero-point energies are actually quite large in chemical terms for typical organic molecules (on the order of tens of kcal/mol). However, zero-point energy differences between isomeric molecules, or between the reactants and products in a "balanced" chemical reaction, tend to be much smaller (often on the order of a few tenths of a kcal/mol). Therefore, changes in zero-point energy that might occur along a reaction coordinate are often ignored for the sake of convenience. Nevertheless, these changes are real, and can lead to measurable isotope effects on stability and reactivity, as the following examples illustrate.

---

1. The difference between the total and potential energies at 0K is exactly equal to the zero-point energy. Other factors (rotation, translation, and increased vibration) begin to contribute, however, as the temperature is increased. The latter, like the zero-point energy, are mass-dependent and easily calculated, but they will be ignored here for the sake of simplicity.

A particularly simple example of an isotope effect is the difference in the bond dissociation energy of H_2 and D_2. The same potential energy surface describes both molecules, but the zero-point energy of H_2 is much higher (the lighter atoms vibrate more rapidly) than the zero-point energy of D_2. Since the separated atoms do not vibrate, their zero-point energies are necessarily zero, and the bond dissociation energy of H_2 is less than that of D_2. The difference in bond dissociation energies is equal to the difference in zero-point energies, a value that has been measured to be 1.8 kcal/mol.

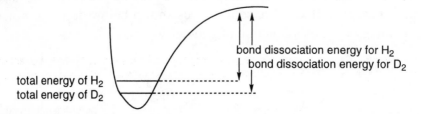

Similar arguments can be used to explain isotope effects on chemical equilibria and reaction rates. The equilibrium ratio of two isomeric molecules, **A** and **B**, is related to their energy difference, ΔE, by

$$\frac{[B]}{[A]} = e^{-\Delta E/k_B T}$$

where k_B is Boltzmann's constant and T is the absolute temperature, respectively. The energy difference between **B** and **A** depends on both the difference in their potential energies and the difference in their zero-point energies, and the latter is isotope-dependent.

$$\Delta E = E_B - E_A$$

$$= (PE_B + ZPE_B) - (PE_A + ZPE_A)$$

$$= (PE_B - PE_A) + (ZPE_B - ZPE_A)$$

Therefore, some equilibria may be sensitive to isotopic composition. This is referred to as an "equilibrium isotope effect".

A similar picture describes isotope effects on reaction rates. Reaction rates are determined, in part, by the energy barrier for the reaction, $\Delta H^\ddagger$, and the barrier may involve changes in both potential energy and zero-point energy.

$$\Delta H^\ddagger = E^\ddagger - E_{reactant} = (PE^\ddagger - PE_{reactant}) + (ZPE^\ddagger - ZPE_{reactant})$$

Since the latter are isotope dependent, the magnitude of the barrier may also be isotope-dependent, and a "kinetic isotope effect" may be observed.

Chemists distinguish between two types of kinetic isotope effects, primary and secondary. Primary kinetic isotope effects are defined as isotope effects on reactions that break (or make) a bond involving an atom for which a mass change has been made isotope. An example of this is hydrogen (or deuterium) atom abstraction from chloroform, i.e.,

$$Cl_3C\text{-}H + Cl\bullet \xrightarrow{k_H} Cl_3C\bullet + HCl$$

vs.

$$Cl_3C\text{-}D + Cl\bullet \xrightarrow{k_D} Cl_3C\bullet + DCl$$

the isotope effect for which follows from inspection of the potential energy diagram.

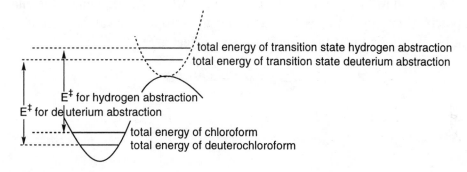

Secondary kinetic isotope effects are defined as isotope effects on reactions where the isotope acts as a spectator. This is illustrated by the following pair of solvolysis reactions.

Secondary kinetic isotope effects tend to be much smaller than primary isotope effects. This is because the change in the isotope's chemical environment is very small when it is a spectator atom. Consequently, the isotope-dependent change in zero-point energy is correspondingly small.

The computation of equilibrium and kinetic isotope effects is straightforward. Most programs that calculate vibrational frequencies automatically[2] provide the

---

2.   Two pieces of information are required in order to calculate vibrational frequencies: the molecule's force constant matrix (or Hessian), and the molecule's isotopic composition. The calculation of the force constant matrix is rather time-consuming for most molecules, however, the same matrix is used for all of the isotopic derivatives.

corresponding zero-point energy, and these energies can be used to calculate the isotope effect based on changes in zero-point energy directly. For example, the equilibrium isotope effect is given by the following relationships:

$$\frac{(K_{eq})_i}{(K_{eq})_j} = e^{-(\Delta E_i - \Delta E_j)/k_B T}$$

$$= e^{-(\Delta ZPE_i - \Delta ZPE_j)/k_B T}$$

where $(K_{eq})_i$ refers to the equilibrium constant for the ith isotopic derivative. The final equation follows from the fact that $\Delta E_i = \Delta PE + \Delta ZPE_i$, where the change in potential energy, ΔPE, is the same for all isotopic derivatives (see above). Similar reasoning gives the kinetic isotopic effect as:[3]

$$\frac{k_i}{k_j} = e^{-(\Delta ZPE^{\ddagger}_i - \Delta ZPE^{\ddagger}_j)/k_B T}$$

---

3. These equations for the equilibrium and kinetic isotope effects both assume that the entropy change for each process is the same for all isotopic derivatives.

Hydrogen vs. Deuterium Abstraction. Reaction of Chlorine Atom with d_1-Dichloromethane

Ab initio $3\text{-}21G^{(*)}$ calculations are used to obtain kinetic and equilibrium isotope effects for hydrogen (deuterium) abstraction from d_1-dichloromethane.

d_1-Dichloromethane can react with chlorine atoms in either of two ways: by hydrogen abstraction (producing HCl) or by deuterium abstraction (producing DCl), i.e.,

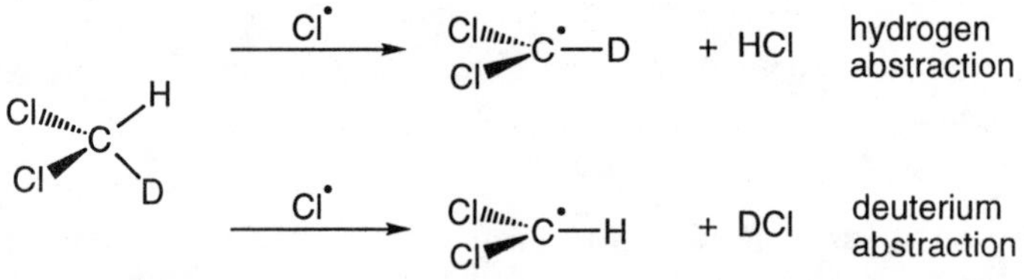

The two reactions follow the "same" reaction coordinate, and pass through transition states and products that are geometrically identical except for the disposition of the hydrogen and deuterium. The disposition of the isotopes is important, however, because it affects the zero-point energies of the transition state and products, and can produce both equilibrium and kinetic isotope effects (see: **Isotope Effects**).

In this two-part experiment, you will use *ab initio* $3\text{-}21G^{(*)}$ calculations to determine the kinetic and equilibrium isotope effects for hydrogen atom abstraction from d_1-dichloromethane. The kinetic isotope effect will be obtained by calculating the transition state for hydrogen abstraction from dichloromethane, **A**, and then calculating zero-point energies for the two monodeutero derivatives of this transition state, **B** and **C**.

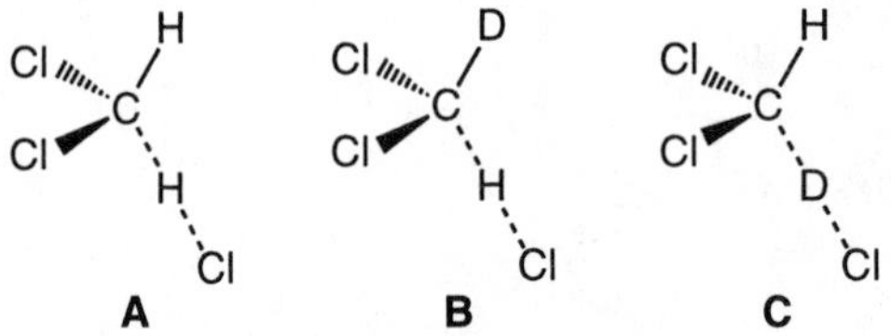

The equilibrium isotope effect on the product distribution will be determined by calculating zero-point energies for the products of abstraction.

Procedure

Kinetic isotope effect: Build a transition state for hydrogen abstraction from dichloromethane (**A**). Constrain the lengths of the reacting bonds to the values shown below, and optimize the geometry subject to these constraints using *ab initio* 3-21G$^{(*)}$ calculations.

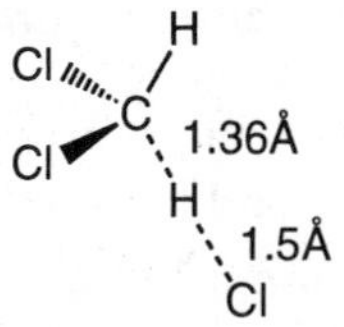

Then, use the resulting structure to search for the 3-21G$^{(*)}$ transition state (no constraints). Verify that the resulting structure is a true transition state by calculating its vibrational frequencies (a transition state will have exactly one imaginary frequency; see: **Finding Transition States**). Next, calculate the frequencies and zero-point energies for the monodeutero derivatives, **B** and **C**. Which transition state is preferred? Calculate the kinetic isotope effect (see: **Isotope Effects**; note that it will not be necessary to obtain the zero-point energy of the reactants). Can you find any experimental data that supports or refutes your results?

Equilibrium isotope effect: Build the two reaction products, dichloromethyl radical and HCl, and optimize each 3-21G$^{(*)}$ geometry. Calculate the vibrational frequencies and zero-point energy for each molecule and its monodeutero derivative, and use your results to calculate the equilibrium isotope effect at 25°C, i.e., calculate the equilibrium constant for the following equilibrium (see: **Isotope Effects**).

$$Cl_2\overset{\bullet}{C}H + DCl \rightleftharpoons Cl_2\overset{\bullet}{C}D + HCl$$

Can you find any experimental data that supports or refutes equilibrium control of this reaction?

Isotope Effects on the Solvolysis of Norbornyl Chloride

Semi-empirical AM1 calculations are used to obtain secondary kinetic isotope effects for solvolysis of norbonyl chloride, and to relate differences in isotope effects to the geometry of the carbocation intermediate.

Solvolysis of *endo*-2-norbornyl brosylate displays a variety of secondary kinetic isotope effects depending on the location of the isotopic label.[1]

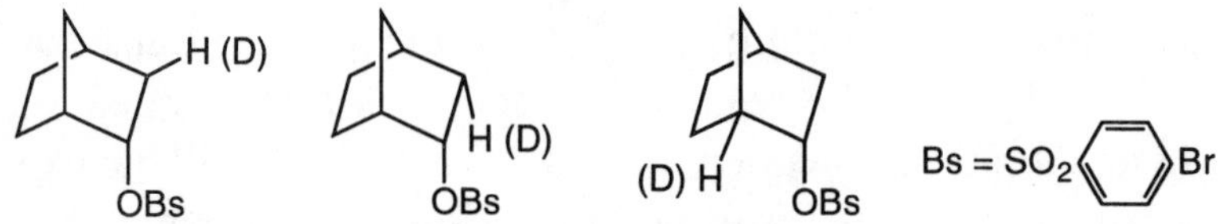

	1	**2**	**3**
$k_H/k_D =$	1.19	1.12	1.00

Assuming the transition state resembles a "classical" 2-norbornyl cation, the isotope effects are due to changes in zero-point energy from the reactant to the carbocation. In this experiment, you will use semi-empirical AM1 calculations to determine the kinetic isotope effects that accompany the following process (the chloride is used instead of the brosylate for the sake of computational simplicity).

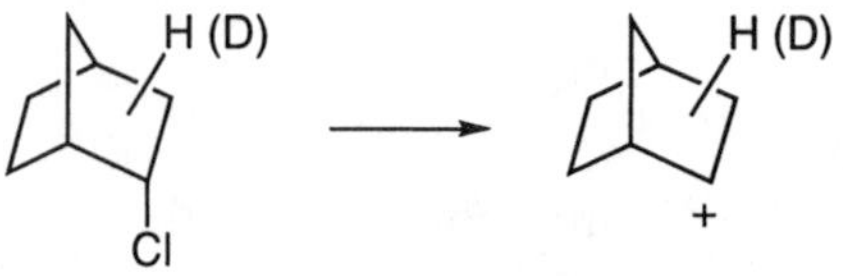

You will also examine the geometry of norbornyl cation, in particular, the angle which the CD bond makes with the carbocation center, to see if the different isotope effects are reflected in its geometry.

Procedure

Build *endo*-2-norbornyl chloride and 2-norbornyl cation, and optimize their geometries using AM1 semi-empirical calculations. Examine the optimized carbocation structure; does it appear to be a "classical" or "nonclassical" carbocation? (A nonclassical ion is not suitable for this exercise.) Next, calculate vibrational frequencies and zero-point energies for the chloride and carbocation, and then repeat this calculation for the three monodeuterated derivatives (**1-3**). Use the change in zero-point energy to calculate the secondary kinetic isotope effect of deuterium substition at each position (see: **Isotope Effects**).

---

1. For a review of experimental and theoretical literature on this and other related problems, see: D.E. Sunko and W.J. Hehre, *Prog. Phys. Org. Chem.*, **14**, 205 (1983).

What seems to make a larger contribution to the calculated isotope effects: changes in the chloride's zero-point energy or changes in the carbocation's zero-point energy? Are your results consistent with the experimental isotope effects, i.e., *exo* > *endo* >> bridgehead (no isotope effect)? The largest isotope effect should occur when deuterium is substituted for the hydrogen experiencing the largest decrease in vibrational frequency during ionization. Assuming that CH bond strengths (and vibrational frequencies) in the carbocation are inversely related to CH bond lengths, what are the relative strengths of the *exo*, *endo*, and bridgehead CH bonds? How do you account for these differences? Hint: see if you can relate relative CH bond strengths (and the magnitudes of the isotope effects) to the angle between the vacant orbital at the carbocation center and the CH bond (in the carbocation intermediate). This angle is close to 30° in the carbocations resulting from solvolysis of **1** and **2**, and close to 90° in the carbocation from solvolysis of **3**. (See: **Hyperconjugation and the Structures and Stabilities of Carbocations**.)

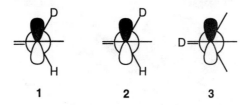

1	2	3

Optional

The relationship between hyperconjugation in carbocations and secondary kinetic isotope effects on solvolysis rates can be studied by examining the relationship between zero-point energy and conformation in a simple carbocation, such as "open" ethyl cation.

Build the eclipsed conformer of the "open" ion (C_s symmetry), and optimize its AM1 geometry subject to this symmetry constraint. Then, calculate its vibrational frequencies and zero-point energy. Next, increase the H*CCH* torsion angle in 15° increments (up to 60°), recalculate the zero-point energy at each geometry, and plot zero-point energy against torsion angle. Can your results be expressed by a simple mathematical formula? How do you account for your results?

XV

Finding Transition States

C hemists recognize a transition state as the structure that lies at the top of a potential energy surface connecting reactant and product.

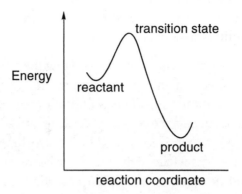

In more quantitative terms, a transition state is a point on the potential energy surface for which the gradient is zero (just as it is for an equilibrium structure; see: **Finding Equilibrium Geometries**), but for which the diagonal representation of the Hessian has one and only one negative element, corresponding to the "reaction coordinate" (see diagram above). All the other elements are positive. In other words, a transition state is a structure that is an energy minimum in all dimensions except one, for which it is an energy maximum.

The geometries of transition states on the pathway between reactants and products are not easily anticipated. This is not to say that they do not exhibit systematic properties as do "normal" (stable) molecules, but rather that there is not sufficient experience to identify what systematics do exist, and more importantly how to capitalize on structural similarities. It needs to be recognized that transition states cannot even be detected let alone characterized experimentally, at least not directly. While measured activation energies relate to the energies of transition states above reactants, and while activation entropies and activation volumes as well as kinetic isotope effects may be invoked to imply some aspects of transition state structure, no experiment can actually provide direct information about the detailed geometries and/or other physical properties of transition states. Quite simply, transition states do not exist in terms of a stable population of molecules on which experimental measurements may be made. Experimental activation parameters may act as a guide, although here too it needs to be pointed out that their interpretation is in terms of a theory ("transition state theory"), and even then, they tell little about what actually transpires in going from reactants to products.

The lack of experience about "what transition states look like" is only one of the reasons why their detailed geometries are more difficult to calculate than equilibrium geometries. Other important reasons include:

1)   The mathematical problem of finding a transition state is probably (but not necessarily) more difficult than that of finding a minimum. What is

certainly true, is that techniques for locating transition states are much less well developed than procedures for finding minima (or maxima). After all, minimization is an important task in many diverse fields of science and technology, whereas transition state location has few if any important applications outside of chemistry.

2)   Adding to the difficulty, is the likelihood that the potential energy surface in the vicinity of a transition state is more "flat" than the surface in the vicinity of a local minimum. (This is entirely reasonable; transition states represent a delicate balance of bond breaking and bond making, whereas overall bonding is maximized in equilibrium structures.) As a consequence, the potential energy surface in the vicinity of a transition state may be less well described in terms of a simple quadratic function (assumed in all common optimization procedures) than the surface in the vicinity of a local minimum.

3)   To the extent that transition states incorporate partially (or completely) broken bonds, it might be anticipated that low-level theoretical treatments will not provide entirely satisfactory descriptions.

In time, all of these problems will be overcome, and finding transition states will be as easy and routine as finding equilibrium geometries is today. Chemists can look forward to the day when reliable and quantitative tools become available for the elucidation of reaction mechanisms.

The same iterative procedure previously described for optimization of equilibrium geometry applies as well to transition states. However, the number of "steps" required for satisfactory completion is likely to be much larger. This is due to the factors discussed above. What is important to emphasize is that the task of transition state determination is even today completely automated and needs no more "human intervention" than that involved in locating equilibrium geometries.

Verifying Transition Structures

Having found a transition structure, two "tests" need to be performed in order to verify that it actually corresponds to a "proper" transition structure, and further that it actually corresponds to the transition structure for the process of interest, i.e., it smoothly connects energy minima corresponding to reactant and product:

1)   Verify that the Hessian yields one and only one imaginary frequency. This requires that a normal mode analysis be carried out on the proposed transition structure. The imaginary frequency will typically be in the range of 400-2000 cm^{-1}, quite similar to real vibrational frequencies. In the case of flexible rotors, e.g., methyl groups, or "floppy rings", the analysis may yield one or more additional imaginary frequencies with very small ($<100\ cm^{-1}$) values. These typically correspond to couplings of low energy

modes and can usually be ignored; make certain to verify what motions these small imaginary frequencies actually correspond to (see below) before doing so. Be wary of structures that yield only very small imaginary frequencies. This suggests a very low energy transition structure, which quite likely will not correspond to the particular reaction of interest.

2)   Verify that the normal coordinate corresponding to the imaginary frequency smoothly connects reactants and products. A simple qualitative way to do this is to "animate" the normal coordinate corresponding to the imaginary frequency, that is, to "walk along" this coordinate without any additional optimization. This does not require any additional calculations beyond the normal mode analysis already performed. "Incorrect" transition states located by calculation, that is, not linking the expected reactant to the expected product, may indicate new chemistry. Don't discard them so quickly!

Reactions Without Transition Structures

Not all chemical reactions have transition states (especially in the gas phase). In fact, reactions without energy barriers are quite common. Two radicals will typically combine without activation, for example, two methyl radicals to form ethane.

$$H_3C^\bullet + {}^\bullet CH_3 \longrightarrow H_3C-CH_3$$

Radicals will often add to paired-electron species with no (or very small) activation, for example, methyl radical and ethylene forming 1-propyl radical.

$$H_3C^\bullet + H_2C{=\!\!=}CH_2 \longrightarrow H_3C\text{-}CH_2\text{-}CH_2^\bullet$$

Exothermic ion-molecule reactions may require activation in solution, but not in the gas phase. Approach of a gas-phase ion and molecule is usually exothermic by itself, and any subsequent energy increases are likely to be small for an exothermic reaction. Therefore, the entire reaction coordinate might lie below the energy of the separated reactants e.g., nucleophilic attack by OH^- on CH_3Cl to give CH_3OH and Cl^-.

Failure to find a transition state, and location instead of what appears to be a stable intermediate or even the final product, does not necessarily mean failure of the computational model (nor does it rule this out). It may simply mean that there is no transition state! Unfortunately it is very difficult to tell which is the true situation.

The Pinacol Rearrangement

Semi-empirical AM1 calculations and ab initio 3-21G calculations are used to examine the structures and relative stabilities of carbocation intermediates involved in the pinacol rearrangement of 2,3-dimethylbutane-2,3-diol (pinacol), and the transition state connecting these intermediates.

One of the most common and synthetically useful rearrangements involving carbocations is the pinacol rearrangement, e.g., in 2,3-dimethylbutane-2,3-diol **1** (pinacol).[1]

$$\underset{\textbf{1}}{(CH_3)_2\overset{\overset{OH}{|}}{C}-\overset{\overset{OH}{|}}{C}(CH_3)_2} \xrightarrow{H^+} \underset{\textbf{2}}{CH_3\overset{\overset{O}{\|}}{C}-C(CH_3)_3} + H_3O^+$$

The mechanism involves protonation of **1** to give oxonium ion **3**, loss of water to give carbocation **4**, and 1,2-methyl migration to give α-hydroxycarbocation **5**. Deprotonation then yields ketone **2**.

$$\textbf{1} \xrightarrow{H^+} \underset{\textbf{3}}{(CH_3)_2\overset{\overset{OH}{|}}{C}-\overset{\overset{OH_2{}^+}{|}}{C}(CH_3)_2} \xrightarrow{-H_2O} \underset{\textbf{4}}{(CH_3)_2\overset{\overset{OH}{|}}{C}-\overset{+}{C}(CH_3)_2} \longrightarrow \underset{\overset{+}{\textbf{5}}}{CH_3\overset{\overset{OH}{|}}{C}-C(CH_3)_3} \xrightarrow{-H^+} \textbf{2}$$

In this experiment, you will use AM1 semi-empirical calculations to determine the structures of the two carbocation intermediates (**4** and **5**) involved in the rearrangement of pinacol, and then the transition state connecting the two. *Ab initio* 3-21G energy calculations, using AM1 geometries, will be employed to provide a better account of the relative energies of the two intermediates (the thermodynamics of the rearrangement) and the height of the reaction barrier (the kinetics of the rearrangement).

Procedure

Build **4** and **5** and optimize the geometry of each using AM1 semi-empirical calculations. Perform single-point energy calculations on the two cations using the 3-21G *ab initio* method. Is the rearrangement **4→5** predicted to be endothermic or exothermic? Is the ordering of carbocation stabilities in line with what you would expect based on the relative π donor abilities of CH₃ and OH groups? (See: **Substituent Effects on Reactive Intermediates.**)

---

1. Carey and Sundberg B, p. 499; Lowry and Richardson, p. 426; March, p. 1059.

Build a model for the transition state for **4** → **5**. The easiest way to do this is to start with one of the two carbocations, and then constrain the "migrating" methyl group, i.e.,

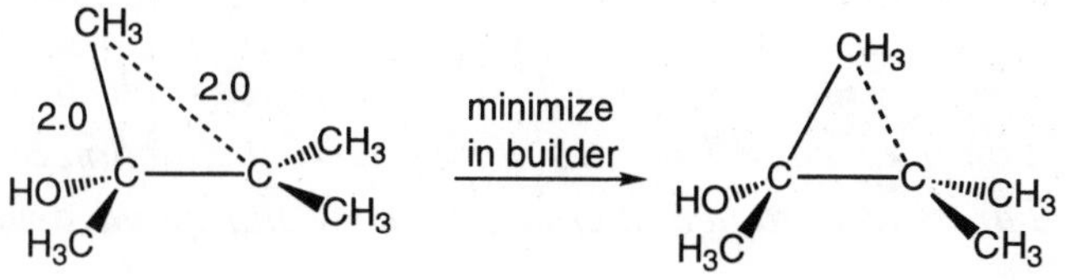

$$CH_3 \quad 2.0 \quad \xrightarrow[\text{in builder}]{\text{minimize}} \quad CH_3$$

Minimizing (using molecular mechanics) will produce a reasonable guess structure. Use this to search for the AM1 transition state. Verify that this is a true transition state by calculating its vibrational frequencies (a transition state should have one and only one imaginary frequency; see: **Finding Transition States**).

Is the structure you find for the transition state closer in geometry to the reactant or to the product? Given the thermochemistry of the rearrangement, is your result consistent with the Hammond postulate? (See: **Reactive Intermediates and the Hammond Postulate**.) Perform an energy calculation on the transition state using the 3-21G method and AM1 geometry. Is the calculated barrier consistent with the observation that the pinacol rearrangement is a facile process?

Compare the electron distribution in the transition state with that in **4** and **5** (compare atomic charges and/or electrostatic potential maps). Are these distributions consistent with the Lewis structures drawn above? Does the migrating CH_3 group gain or lose electron density in the transition state? Does electronic structure obey the Hammond postulate?

Optional

Does solvation affect either the thermodynamics and/or the kinetics of the pinacol rearrangement? Perform single energy calculations on **4**, **5** and the transition state separating the two using the AM1-SM2 model (which accounts for aqueous solvent) and AM1 geometries. Add the difference on AM1-SM2 and AM1 heats of formation to the (gas-phase) 3-21G energies.

Regioselectivity of Free Radical Allylic Bromination

Semi-empirical AM1 calculations are used to explore the regioselectivity of free radical allylic bromination of alkenes. The sites of hydrogen abstraction and bromination are rationalized in terms of radical thermochemistry and spin distributions.

Terminal alkenes containing allylic hydrogens undergo reaction with N-bromosuccinimide (NBS) to give a mixture of allylic bromides.[1]

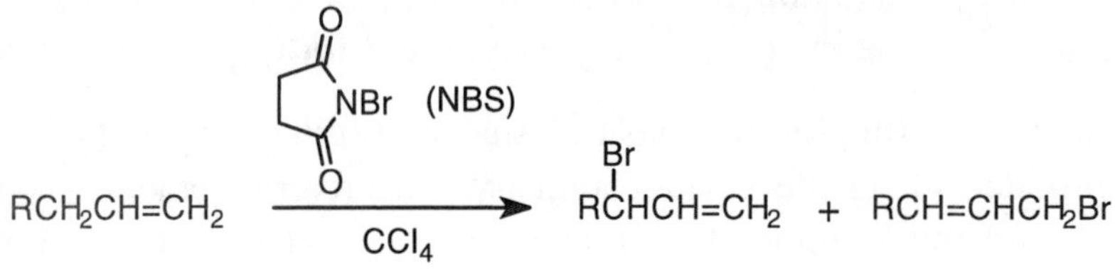

$$RCH_2CH=CH_2 \xrightarrow[CCl_4]{\text{(NBS)}} \underset{\overset{|}{Br}}{R\overset{}{C}HCH=CH_2} + RCH=CHCH_2Br$$

The reaction mechanism involves a radical chain process in which Br_2 is the actual bromination agent. The mechanism involves three steps: 1) allylic hydrogen abstraction by Br• (addition of Br• is reversible, and is not observed because $[Br_2]$ is low[2]), 2) bromine abstraction by an allylic radical, and 3) regeneration of Br_2 from NBS and HBr.

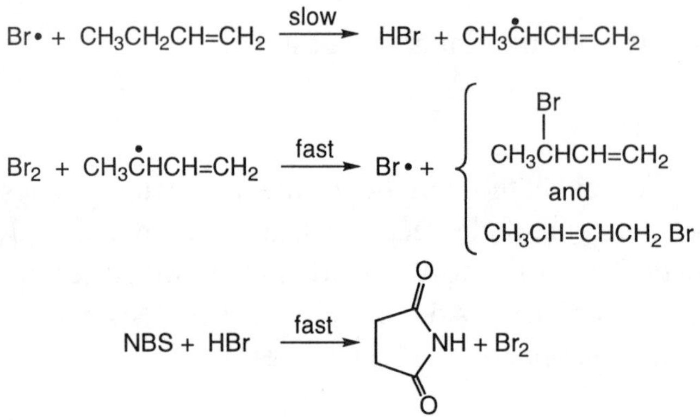

$$Br\bullet + CH_3CH_2CH=CH_2 \xrightarrow{\text{slow}} HBr + CH_3\overset{\bullet}{C}HCH=CH_2$$

$$Br_2 + CH_3\overset{\bullet}{C}HCH=CH_2 \xrightarrow{\text{fast}} Br\bullet + \begin{cases} \underset{\overset{|}{Br}}{CH_3\overset{}{C}HCH=CH_2} \\ \text{and} \\ CH_3CH=CHCH_2Br \end{cases}$$

$$NBS + HBr \xrightarrow{\text{fast}} \text{(succinimide)} NH + Br_2$$

In this experiment, you will examine the regioselectivity of hydrogen abstraction and allylic bromination. There are five possible hydrogen abstraction sites in 1-butene (note that radicals **1** and **2** might be stereoisomers, depending on radical geometry).

---

1. Carey and Sundberg A, p. 688; March, p. 615, 624.
2. C.C. Wamser and L.T. Scott, *J. Chem. Ed.*, **62**, 650 (1985).

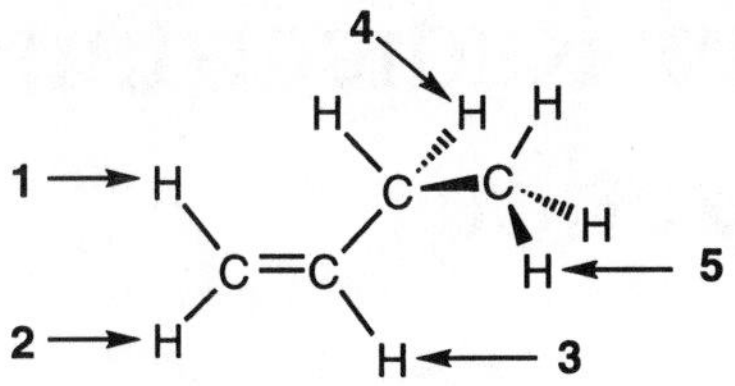

The underlying assumption is that the most stable radical will be formed in greatest abundance. Assuming that this is allylic radical, **4**, then the distribution of spin density in **4** should indicate which site is better able to abstract bromine from Br_2.

Procedure

Build the five radicals, **1-5**, obtained by removing hydrogen atoms from *gauche* 1-butene (see above), and optimize their geometries using semi-empirical AM1 calculations. Do you obtain five distinct radicals (are **1** and **2** distinguishable)? What other conformations of each radical might be possible? Record the energy of each radical, and determine which radical will form most rapidly. Is this consistent with the bromination mechanism shown on the facing page?

Next, calculate and display a spin density map of allylic radical **4**. Where are the regions of highest spin density? Assuming that the carbon with the greatest spin density preferentially attacks Br_2, which allylic bromide will be the preferred product?

Finally, build two of the allylic bromides derived from 1-butene, i.e., 3-bromo-1-butene and *trans*-1-bromo-2-butene, and optimize their AM1 geometries. Which product is more stable? Would the same product mixture be predicted if this reaction were thermodynamically controlled instead of kinetically controlled?

Selectivity in Radical-Initiated Alkene Copolymerization

Semi-empirical AM1 calculations and ab initio 3-21G calculations are used to examine the electronic character of alkenes and radical intermediates involved in the copolymerization of acrylonitrile and styrene.

Radicals are typically classified as nucleophilic (electron-rich) or electrophilic (electron-poor) according to their reactivity toward different types of alkenes. Thus, a nucleophilic radical adds more rapidly to electron-poor alkenes, while an electrophilic radical adds more rapidly to electron-rich alkenes.

One way to explain the electronic selectivity of radical-alkene reactions is to use Frontier Molecular Orbital (FMO) theory. FMO theory states that the barrier to radical addition will be reduced by a stabilizing three-electron interaction between the radical's singly-occupied molecular orbital (SOMO) and the alkene π orbital, or by a stabilizing one-electron interaction between the SOMO and the alkene π^* orbital. The lowest barrier results when there is a good match between the energies of the interacting orbitals. Therefore, a nucleophilic radical (high-energy SOMO) is expected to react preferentially with electron-poor alkenes (low-energy π^*orbital), while an electrophilic radical (low-energy SOMO) should favor an electron-rich alkene (high-energy π orbital).

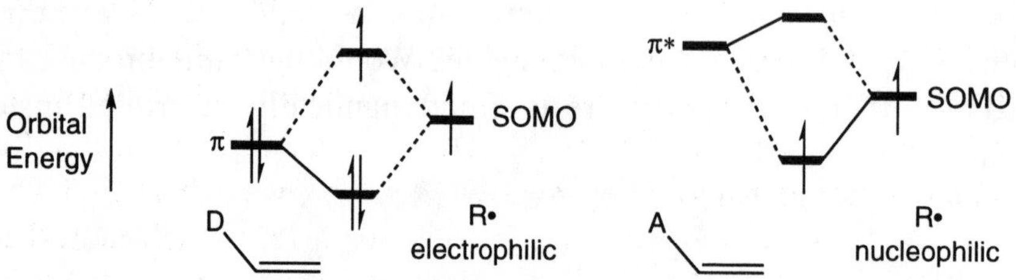

These types of radical-alkene interactions explain the formation of alternating copolymers of styrene and acrylonitrile. The reaction mixture contains benzylic radicals ($\bullet$C-Ph), cyano-substituted radicals ($\bullet$C-CN), styrene, and acrylonitrile, yet the resulting copolymer has an alternating structure. This suggests that benzylic radicals add selectively to acrylonitrile, while cyano-substituted radicals add selectively to styrene, i.e.,

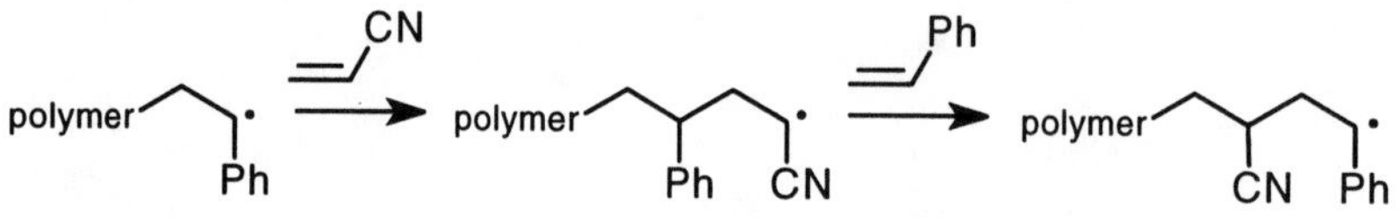

In this experiment, you will develop an FMO analysis of this copolymerization by using semi-empirical AM1 and *ab initio* 3-21G calculations to evaluate SOMO energies of model radicals, and π and π^* energies of acrylonitrile and styrene. You will also decide whether the FMO analysis or an alternative analysis based on electrostatic interactions, provides a clearer explanation of the observed selectivity.

Procedure

Build acrylonitrile and styrene, optimize their geometries using semi-empirical AM1 calculations, and use these geometries to calculate *ab initio* 3-21G wavefunctions. Record the 3-21G energies of the π and π^* orbitals for each alkene. (Note: these orbitals may or may not correspond to the alkene's HOMO and LUMO; therefore, you should display and identify the desired orbitals first.) Compare the energies of the corresponding orbitals in each alkene. Which alkene is the better donor? Which alkene is the better acceptor?

Next, build PhĊHCH$_3$ and NCĊHCH$_3$ radicals (the methyl group simulates the growing polymer chain), optimize their AM1 geometries, and calculate their 3-21G wavefunctions. Record the 3-21G energy of the highest occupied orbital of α spin in each radical (this is the SOMO energy). Which radical is more nucleophilic? Consider the structure of the alternating styrene-acrylonitrile copolymer. Is this structure consistent with the FMO analysis developed above? That is, does the more nucleophilic radical react with the better acceptor alkene? Does the more electrophilic radical react with the better donor alkene?

An alternative way to explain the observed selectivity in copolymerization is to look at electrostatic interactions between radicals and alkenes. The fastest reaction should occur between the molecule (radical or alkene) with the most negative potential and a partner with the most positive potential. Calculate and display electrostatic potential maps of the two alkenes and the two radicals. Which alkene has the more negative potential over its reactive face? Which radical has the more negative potential? What radical-alkene combinations should be favored, and are these consistent with experiment?

Cycloaddition of Singlet Difluorocarbene and Ethylene

Semi-empirical AM1 calculations are used to determine the electronic structure of singlet difluorocarbene, and to generate a transition state for its cycloaddition with ethylene.

Singlet carbenes add to alkenes to yield cyclopropanes. The reaction is stereospecific in that *cis* and *trans* alkenes give *cis* and *trans* cyclopropanes, respectively.[1]

Although experimental evidence is consistent with a one-step concerted process, a simple theoretical analysis suggests otherwise.[2] The electron configuration of singlet dihalocarbenes involves a filled hybrid orbital in the carbene plane, and an empty p orbital perpendicular to this plane, i.e.,

When this electronic configuration is combined with a Woodward-Hoffmann analysis of the "least-motion" transition state, **1**, a thermally forbidden $_\pi 2_s + _\sigma 2_s$ process seems to be required. This in turn suggests that the reaction is either stepwise or proceeds via a less symmetric transition state, **2** or **3** (these structures are consistent with a thermally allowed $_\pi 2_s + _\omega 0_s$ reaction). Both the latter involve interaction of an empty carbene orbital and a filled alkene orbital, and these transition states should be characterized by electron transfer from the alkene to the carbene.

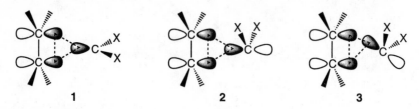

| 1 | 2 | 3 |

1.   Carey and Sundberg A, p. 92.
2.   Lowry and Richardson, p. 556.

In this experiment, you will use semi-empirical AM1 calculations to examine the electronic structure of singlet difluorocarbene, CF_2. You will then obtain the AM1 transition state for the reaction and examine its geometry and electronic structure.

Procedure

Build CF_2 and optimize the geometry of the singlet species using semi-empirical AM1 calculations. Examine the highest-occupied and lowest-unoccupied molecular orbitals (HOMO and LUMO). Are these consistent with the qualitative description given above? Calculate and display the carbene's electrostatic potential map. What regions are electrophilic and nucleophilic, respectively?

Next, build difluorocyclopropane; constrain the CC bond distances to the values shown below, and optimize the AM1 geometry subject to these constraints.

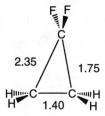

Use the resulting model (remove constraints) to search for the AM1 transition state, and verify that the resulting structure is a transition state by calculating its vibrational frequencies (a transition state must have one and only one imaginary frequency; see: **Finding Transition States**).

What are the CC distances of the forming bonds? Are these more consistent with a concerted or stepwise transition state? If the former, which transition state seems to provide the most accurate description, **1**, **2**, or **3**? The transition state geometry can also be interpreted in terms of the Hammond postulate ("reactant-like" and "product-like" transition states are expected for highly exothermic and endothermic reactions, respectively; see: **Reactive Intermediates and the Hammond Postulate**). Build ethylene and 1,1-difluorocyclopropane, optimize their AM1 geometries, and calculate ΔH_{rxn} and $\Delta H^{\ddagger}$ for the cycloaddition. Are the transition state geometry and energy consistent with the Hammond postulate?

Animate the motion along the normal coordinate. Does this connect the transition state with the separated reactants and cyclopropane product? Rationalize what you see, in particular with regard to the orientation of reactants.

Finally, calculate and compare electrostatic potential maps of the transition state, reactants, and product. Describe the electronic changes that occur along the reaction coordinate. How do you account for these changes?

Selectivity in Carbene Insertion Reactions

Semi-empirical AM1 calculations and ab initio 3-21G$^{(*)}$ calculations are used to obtain transition states for carbene insertion into primary, secondary, and tertiary CH bonds in simple alkanes.

Carbenes readily insert into CH bonds.

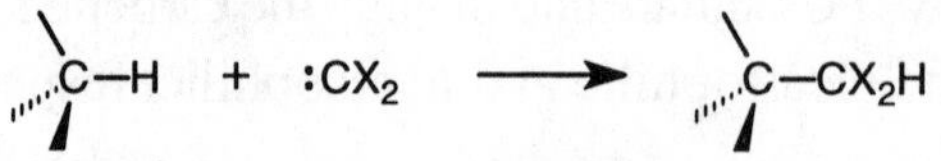

For example, singlet methylene, CH_2, inserts rapidly and indiscriminantly into primary, secondary, and tertiary CH bonds.[1] Stabilized carbenes, such as $C(CO_2R)_2$, are less reactive and more selective than CH_2 in their insertions into CH bonds: tertiary > secondary > primary, e.g.,

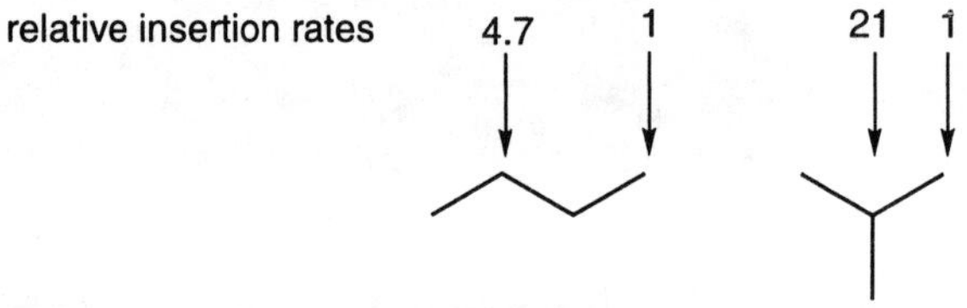

relative insertion rates   4.7   1   21   1

Dihalocarbenes are also less reactive toward CH bonds than methylene.

In this experiment, you will use semi-empirical AM1 and *ab initio* 3-21G$^{(*)}$ calculations to obtain transition states for insertion of singlet dichlorocarbene, CCl_2, into CH bonds in three simple hydrocarbons: methane, propane (1° and 2° CH), and isobutane (1° and 3° CH). You will then use the models to assess the selectivity of this carbene, the electronic structure of the transition states, and the roles that steric and electronic factors may play in controlling selectivity.

Procedure

Build a transition state for CCl_2 insertion into a CH bond of methane by constraining the CC and CH distances to the values shown below.

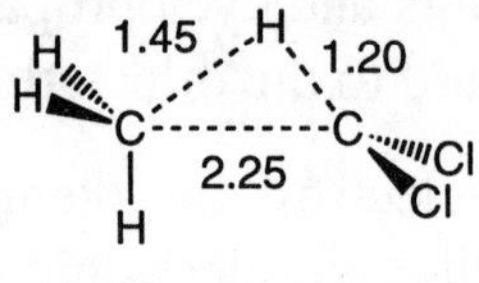

---

1. Lowry and Richardson, p. 554.

Optimize the geometry using semi-empirical AM1 calculations (use constraints), and then use the resulting structure to search for the AM1 transition state (remove constraints). Verify that you have obtained a true transition state by calculating its vibrational frequencies (a transition state should have one imaginary frequency; see: **Finding Transition States**).

Build the reactants, CH_4 and singlet CCl_2, optimize their AM1 geometries, and record their energies. Perform single-point 3-21G$^{(*)}$ *ab initio* calculations on the reactants and on the transition state (use AM1 geometries). What is $\Delta H^{\ddagger}$ for this reaction? Calculate and compare 3-21G$^{(*)}$ electrostatic potential maps of the reactants and transition state on the same color scale. Is there significant charge development in the transition state? Based on these maps, and the lengths of the forming/breaking bonds, what resonance structure(s) would you use to characterize the transition state?

Build, in turn, transition states for CCl_2 insertion into a 1° CH bond of propane, a 2° CH bond of propane, a 1° CH bond of isobutane, and the 3° CH bond of isobutane.[2] Verify that these structures are true transition states by calculating their vibrational frequencies. Also obtain AM1 structures for propane and isobutane, and perform single-point 3-21G$^{(*)}$ calculation on the reactants (you already have done so for CCl_2) and on the four transition states (use the AM1 geometries). Finally, calculate the 3-21G$^{(*)}$ barrier for each type of insertion. What would the product ratio be for each reaction if it were run at room temperature? (Assume that $\Delta(\Delta G^{\ddagger}) = \Delta(\Delta H^{\ddagger})$ and remember to take into account the different number of hydrogens of each type within each molecule.)

Try to account for the relative magnitudes of the different insertion barriers. Can the barriers be correlated with the electronic model developed above, i.e., do changes in the alkyl substitution pattern facilitate or inhibit charge development? Can the barriers be correlated with CH bond strengths? Might steric factors play a role? Cite specific data that support your analysis, such as experimental CH bond strengths, or the appearance of space-filling models of the various transition states.

---

2. The easiest way to build these transition states is to use the transition state for CCl_2 insertion into methane (the "methane transition state") as a template. That is, add alkyl groups to the appropriate positions in the AM1 methane transition state, and use these structures to search for the desired transition states directly (do not perform a constrained optimization).

The McLafferty Rearrangement

Ab initio 3-21G calculations are used to obtain the transition state for the first step in the McLafferty rearrangement, and to describe the radical cation species involved.

Radical cations that are generated in mass spectrometers rapidly undergo a variety of rearrangements and fragmentation reactions. This is partly due to excess energy delivered to the radical cation during ionization, and partly due to the intrinsic reactivity of these unstable, and poorly understood, intermediates.

One fragmentation that is routinely observed for radical cations derived from aldehydes (and ketones) bearing γ hydrogens is the so-called McLafferty rearrangement.[1] The process involves transfer of a γ hydrogen and cleavage of the $\alpha\beta$ CC bond.

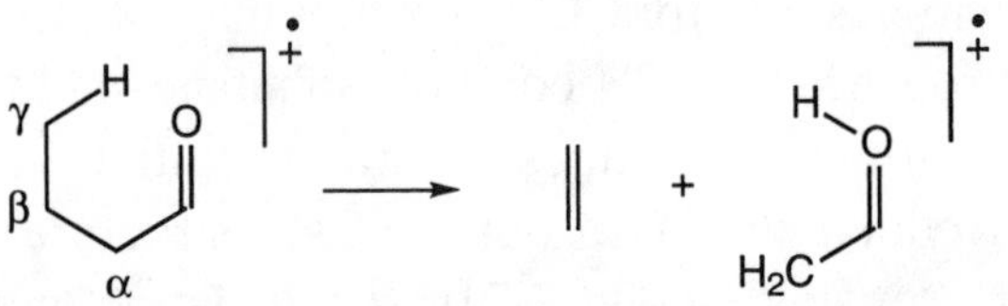

The McLafferty rearrangement is useful for identifying the structures of unknown compounds. For example, all aldehydes with the formula $C_5H_{10}O$ display a peak for the molecular radical cation, $C_5H_{10}O^+$, at m/e 86 in their mass spectra. 2-Methylbutanal and 3-methylbutanal, however, give different fragment ions. The 2-methylbutanal radical cation loses ethylene, and can be identified from a large peak at m/e 58 (M-28), while 3-methylbutanal loses propene and gives a peak at m/e 44 (M-42).

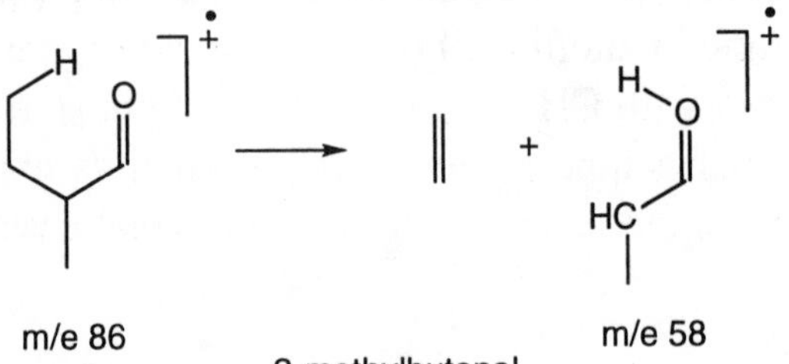

m/e 86   m/e 58

2-methylbutanal

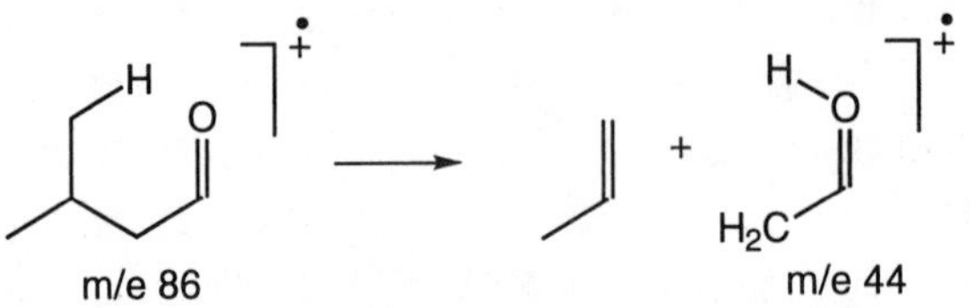

m/e 86   m/e 44

3-methylbutanal

1. F.W. McLafferty and F. Turecek, **Interpretation of Mass Spectra**, 4th ed., University Science Books, Mill Valley, California, 1993.

Although it was originally believed that the McLafferty rearrangement was a concerted process, it is now believed to be a stepwise process in which hydrogen transfer occurs to give an intermediate, and CC bond cleavage occurs in a second step. In this experiment, you will use *ab initio* 3-21G calculations to locate the hydrogen-transfer transition state, **TS**, for the rearrangement of butanal radical cation, **B**, and you will examine the electronic structure of **B** and the intermediate molecular ion, **I**, in order to better understand the nature of the rearrangement.

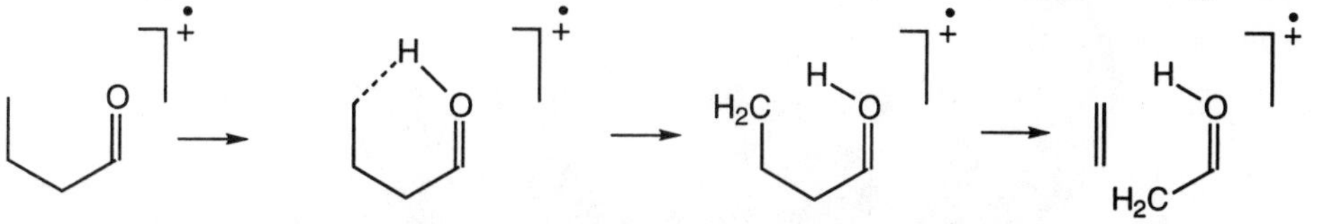

Procedure

Build butanal, **B**, and the structure resulting from γ hydrogen transfer, **I**, in the *cisoid* conformations shown above, and optimize the geometries of their radical cations using *ab initio* UHF/3-21G calculations. Record the energy of each ion. Is hydrogen transfer exo or endothermic? Is **I** an isomer of the butanal radical cation or does it fragment into ethylene and a radical cation? Cite specific data that support your conclusion.

Calculate and display spin density maps for each species. Is the unpaired electron localized on one atom in each molecule or is it delocalized? Which atom(s) carry the unpaired electron? What does the position(s) of the unpaired electron in **B** suggest about the nature of the ionization process, i.e., which electron, bonding or nonbonding, is ionized? Is this consistent with the bond distances in **B**?

Build **TS** with the distances shown below, and use molecular mechanics to optimize its geometry subject to these distance constraints.

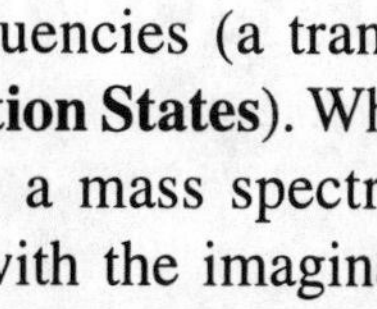

Use the resulting structure to search for the UHF/3-21G radical cation transition state (no constraints), and verify that the new structure is a transition state by calculating its vibrational frequencies (a transition state has one imaginary frequency; see: **Finding Transition States**). What is the barrier for the reaction? Would this reaction be fast in a mass spectrometer? Examine the reaction coordinate, i.e., the vibration with the imaginary frequency. How would you describe this motion, as a hydrogen transfer, CC bond cleavage, or simultaneous transfer and bond cleavage? Calculate and display the transition state's spin density map. Is the unpaired electron localized or delocalized, and on which atom(s)? Does this result support or contradict the notion that a transition state is a resonance hybrid of the resonance structures describing the reactant and product.

230

XVI

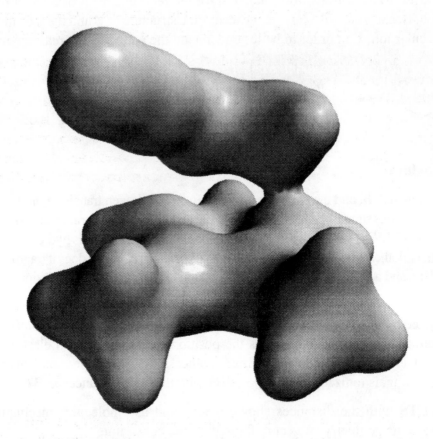

Thermodynamic
vs.
Kinetic Control

C hemical reactions can display complex, and even paradoxical, behavior. A certain experiment might yield one mixture of products under one set of conditions and a completely different mixture of products under a different set of conditions. In addition, many factors (temperature, reagent concentrations, solvent and reaction time) can affect the outcome of an experiment. Thus, understanding the relationship between these experimental factors and the outcome of chemical processes is a major goal of chemistry.

The way in which experimental conditions might favor one chemical reaction over another can be explained by invoking the notions of "kinetic control" and "thermodynamic control". A kinetically-controlled system is one where the outcome is determined by the relative rates of competing reactions. A thermodynamically-controlled system is one where the outcome is determined by the relative stability of competing reaction products. The difference between these reaction types can be more easily understood using a reaction coordinate diagram.

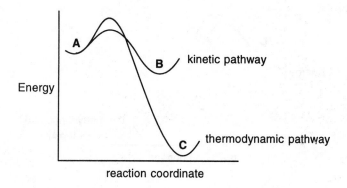

This shows the change in energies for two different competing reactions: **A** → **B** and **A** → **C**. Both reactions are exothermic, i.e., the energies of **B** and **C** are both lower than the energy of **A**. However, the energy of **C** is much lower than the energy of **B**. There is also an energy barrier for both reactions, and the size of the barrier(s) will determine how much heat must be added before a "reasonable" reaction rate can be achieved. The barrier for formation of **C** is higher than the barrier for formation of **B**, and **B** will form more rapidly than **C** during the first stages of reaction regardless of how much heat is applied.

It is easy to see how the reaction conditions, in this case, temperature and reaction time, can affect the outcome of the experiment. At relatively low temperatures, more **A** molecules can cross the barrier leading to **B** than can cross the barrier leading to **C**. Since it takes less time to equilibrate **A** and **B** than it does **A** and **C**, a short reaction time will yield [**B**]>>[**C**] at the end of the experiment. In other words, the combination of low temperature and short reaction time favors the faster process, **A** → **B**, over the slower process, **A** → **C**, and the system is kinetically controlled. **B** is often referred to as the "kinetic product" for just this reason.

If the experiment is carried out at a higher temperature, the formation of **B** will still be favored. However, if the reaction time is extended beyond the time required for equilibration of **A** and **B**, then **C** will gradually accumulate. This occurs because there is enough energy for some **A** molecules to cross the barrier leading to **C**, and because there is enough energy for some **B** molecules to return to **A** (remember that all chemical equilibria are dynamic). Given a long enough reaction time, nearly all of **A** and **B** will be converted to **C**. The combination of higher temperatures and longer reaction times favors the more stable product, **C**, which is referred to as the "thermodynamic product", over the less stable product, **B**. (Of course, intermediate reaction times and reaction temperatures might provide only partial equilibration of **A**, **B**, and **C**.)

A commonly encountered situation is one where the kinetic and thermodynamic products are the same. This is described by the following reaction coordinate diagram.

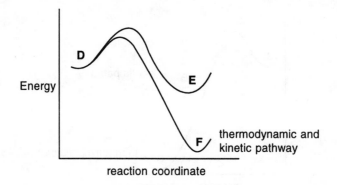

Although **F** is both the kinetic and thermodynamic product, reaction conditions will still influence the exact product distribution. High temperatures and long reaction times will cause complete equilibration of **E** and **F**, and their ratio will be determined by their relative energies. Low temperatures and short reaction times will give a different product ratio, one that is determined by the relative barrier heights. Even here, different conditions will lead to different product distributions.

Thermodynamic vs. Kinetic Control of Intramolecular Radical Addition to Multiple Bonds

Semi-empirical AM1 calculations and ab initio 3-21G calculations are used to explain why cyclization of hex-5-enyl radical leads preferentially to cyclopentylmethyl radical instead of cyclohexyl radical.

Cyclization of unsaturated radicals is of considerable use synthetically, providing that the regioselectivity of the reaction can be anticipated. For example, cyclization of hex-5-enyl radical, **1**, can yield cyclopentylmethyl radical, **2**, or cyclohexyl radical, **3**, i.e.,

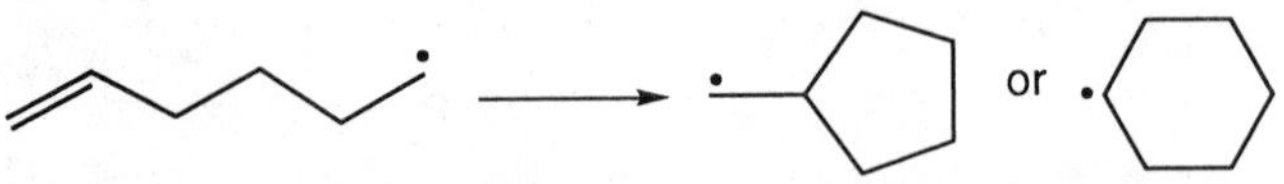

While the latter might be expected (**3** should be less strained than **2**, and 2° radicals are generally more stable than 1° radicals), the opposite is normally observed, e.g.,

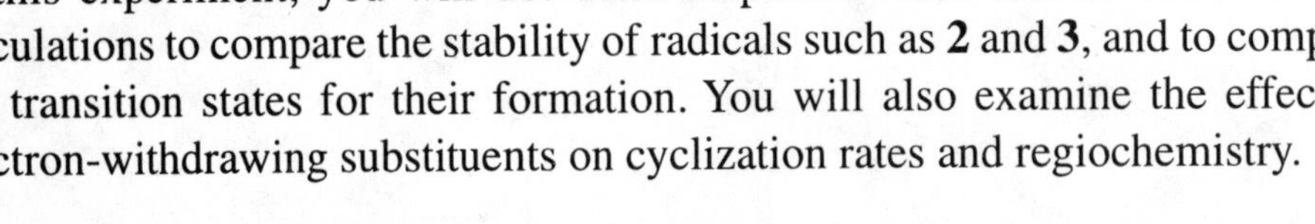

In this experiment, you will use semi-empirical AM1 and *ab initio* 3-21G calculations to compare the stability of radicals such as **2** and **3**, and to compare the transition states for their formation. You will also examine the effect of electron-withdrawing substituents on cyclization rates and regiochemistry.

Procedure

Build radicals, **1-3**, and optimize their geometries using semi-empirical AM1 calculations. Then, use these geometries to calculate their *ab initio* 3-21G energies. Is cyclization an exothermic process for each pathway? What would be the major product if cyclization were thermodynamically controlled? Build cyclization transition states, **1→2** and **1→3**, by reoptimizing the AM1 geometries of **2** and **3** subject to the constraint that the "forming" CC bond length is 2.1 Å. Use the resulting geometries to search for the corresponding AM1 transition states (no

235

constraints). Verify that these are true transition states by calculating their vibrational frequencies (see: **Finding Transition States**), and calculate their 3-21G energies. What is the barrier for each cyclization? Should these reactions be fast at room temperature? What would be the major product if cyclization were kinetically controlled?

An "electronic" interpretation of the cyclization transition state suggests that a primary alkyl radical is "nucleophilic", and should add preferentially to the most electron-poor carbon of the alkene (see: **Selectivity in Radical-Initiated Alkene Copolymerization**). The best electron-acceptor site on an alkene can be determined by examining the shape of its π* orbital. Build propene (a model for the alkene "tail" of radical **1**), and optimize its AM1 geometry. Calculate the 3-21G wavefunction, and use this to obtain the LUMO map. Is the LUMO polarized towards one alkene carbon or the other? What, if anything, does this model suggest about the kinetically preferred mode of cyclization for **1**?

A stronger case for this electronic model can be made when the alkene is substituted with a strong electron-withdrawing group, such as CN. Build 1- and 2-cyanopropene, and optimize their AM1 geometries. Calculate their 3-21G wavefunctions (use the AM1 geometries), and examine their LUMO maps. Which alkene carbon is the better electron-acceptor site in each alkene? Now consider cyclization of radicals **4** and **7**. Which cyclization products are predicted by your models?

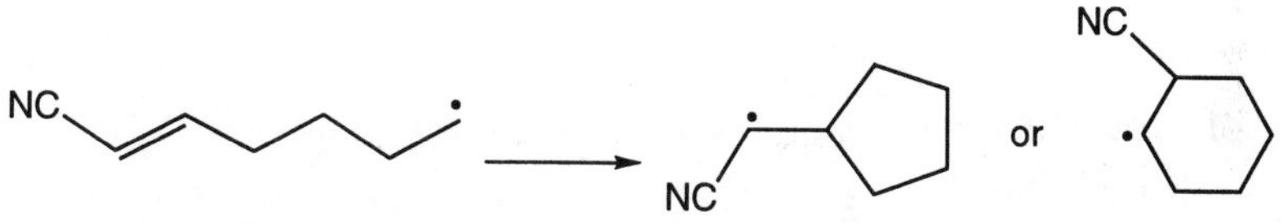

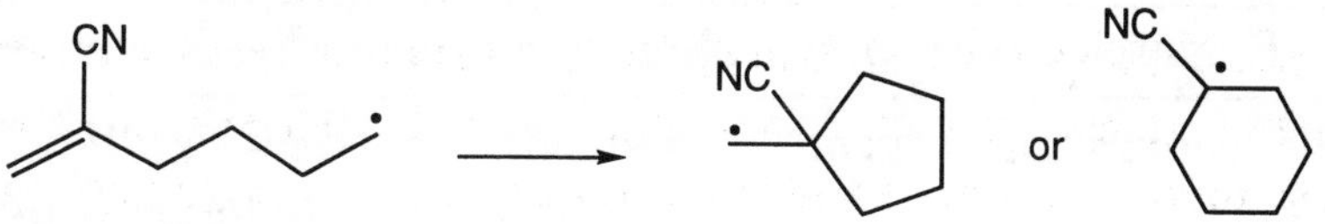

Verify your predictions by finding the AM1 transition states for **4→5** (**6**) and **7→8** (**9**) (use the procedure described above), and calculate the 3-21G energies for these transition states. What is the barrier for each cyclization? What effect does CN have on cyclization rate and regioselectivity? Are these effects consistent with the electronic model examined above?

Regioselectivity in Hydroboration of Alkenes[1]

AM1 semi-empirical calculations are used to investigate regioselectivity in hydroboration of 4-methyl-2-pentene, and to interpret this selectivity in terms of steric and/or electronic effects.

Hydroboration of 4-methyl-2-pentene, **1**, may lead to two different alcohols, **2**, and **3**, depending on whether boron attaches itself to the more or less substituted olefin carbon.

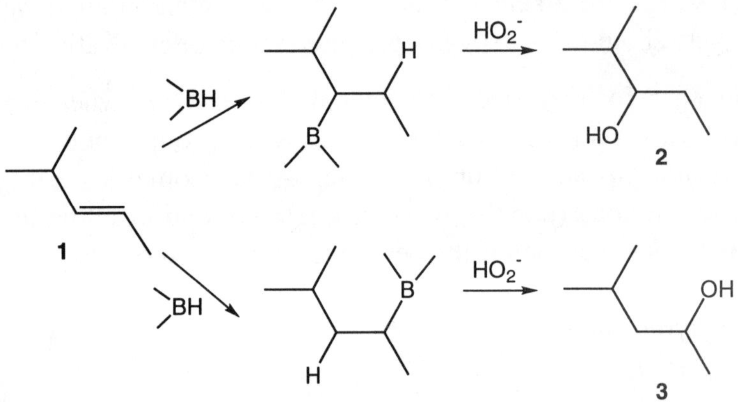

The ratio of products depends on the hydroborating agent ⟩BH. As seen below, reagents with two and three hydrogens bonded to boron result in only modest regioselectivity, while those with only a single hydrogen bonded to boron result in good regioselectivity.

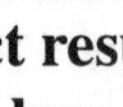

Regioselectivity in hydroboration of 4-methyl-2-pentene	
hydroborating reagent 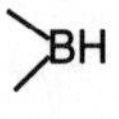 ⟩BH	**percent of product resulting from boron addition to less substituted alkene carbon**
diborane[2]	57
thexylborane	66
disiamylborane	97
9-BBN	99.8

---

1. Carey and Sundberg B, p. 200; Lowry and Richardson, p. 584; March, 702; for a review, see: H.C. Brown, **Organic Synthesis Via Boranes**, Wiley, New York, 1975.
2. The active reagent is diborane resulting from the equilibrium: diborane ⇌ 2 BH₃ .

diborane   thexylborane   disiamylborane   9-BBN

$(CH_3)_2CHC\!-\!BH_2$ with CH_3 top and CH_3 bottom (thexylborane)

$((CH_3)_2CHCH\!-\!)_2BH$ (disiamylborane)

In this experiment, semi-empirical AM1 molecular orbital calculations will be used to investigate the regiochemistry of hydroboration of 4-methyl-2-pentene by 9-BBN (9-borabicyclo[3.3.1]nonane). It is assumed that the reaction is under kinetic control, meaning that the distribution of products should be anticipated by the relative heights of the energy barriers leading to them. The product ratio of >99:1 corresponds to a difference in barrier heights on the order of 3 kcal/mol at room temperature. (See: **Thermodynamic vs. Kinetic Control.**)

Procedure

Build a transition structure corresponding to the addition of 9-BBN to ethylene. Start with a structure with the following bond lengths,

and optimize (constrained geometry optimization) using AM1 semi-empirical calculations. Next, use this geometry to search for the AM1 transition state (no constraints).[3] Finally, use the 9-BBN/ethylene transition state as a "template" for the two regioisomeric transition states for addition to 4-methyl-2-pentene,[4] and search for the corresponding AM1 transition states.

Do the calculations show a clear preference for one transition state over the other? If so, is the result consistent with the experimental preference? What ratio of major:minor products is predicted by your data (assume 298K)? What do you think is the reason behind the preference? Is it purely steric (examine molecular structure models of the two transition states) or is there an electronic component as well (examine atomic charges and/or electrostatic potential maps for 4-methyl-2-pentene)?

---

3. This is a common procedure for establishing transition states for complex reactions. Even so, the final (transition state) optimization is likely to require several hundred optimization cycles.

4. Make 2 copies of the 9-BBN/ethylene transition state. For each, add methyl and isopropyl groups appropriately to "ethylene". "Freeze" the molecule except for the groups you have just added, and minimize using molecular mechanics. This will relieve obvious steric interactions but will not alter the geometry of the underlying transition state.

Activated Dienophiles in Diels-Alder Cycloadditions

Semi-empirical AM1 calculations and ab initio 3-21G calculations are used to relate reactivity and regioselectivity in Diels-Alder cycloadditions, to the properties of dienophiles derived from Frontier Molecular Orbital theory and electrostatic potential maps.

The most common (and synthetically most useful) Diels-Alder reactions involve electron-rich dienes and electron-deficient dienophiles.

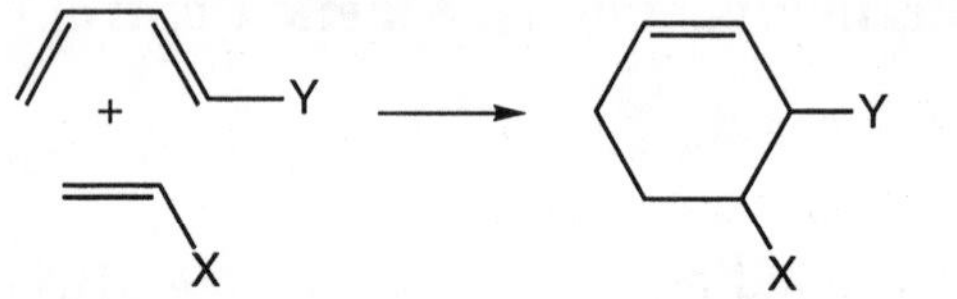

Y = R, OR

X = CN, CHO, CO_2H

The rate of these reactions generally increases with the π donor ability of the diene substituent, Y, and with the π acceptor ability of the dienophile substituent, X. This behavior can be rationalized using Frontier Molecular Orbital (FMO) theory by examining the interaction of the diene's highest-occupied molecular orbital (HOMO) and dienophile's lowest-unoccupied molecular orbital (LUMO).

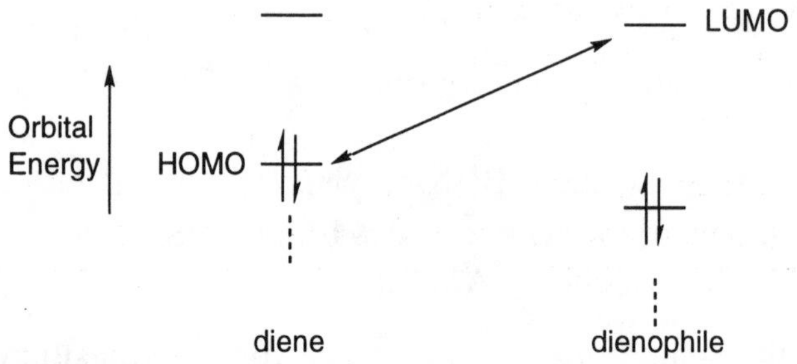

The HOMO-LUMO interaction lowers the reaction barrier by stabilizing the electrons in the diene HOMO, and FMO theory suggests that this effect will be largest when the $HOMO_{diene}$-$LUMO_{dienophile}$ energy difference is smallest. In other words, diene reactivity should be positively correlated with HOMO energy, and dienophile reactivity should be negatively correlated with LUMO energy.

Electrostatic interactions between the diene and dienophile may also play a significant role in determining the reaction rate.[1] According to this view, rapid reactions occur when electrostatic interactions between the diene and dienophile are most attractive (least repulsive). Thus, diene reactivity should increase as the electrostatic potential over the π face of the diene becomes more negative, and dienophile reactivity should increase as the electrostatic potential over the π face of the dienophile is made more positive.

---

1. S.D. Kahn, C.F. Pau, L.E. Overman and W.J. Hehre, *J. Am. Chem. Soc.*, **108**, 7381 (1986).

The FMO and electrostatic models can also be used to predict the regiochemistry of a cycloaddition reaction. According to the FMO model, the best HOMO-LUMO interaction for a given diene-dienophile pair results when there is good overlap between the interacting orbitals. Thus, the best orientation for two unsymmetric reactants is usually the one that aligns the HOMO and LUMO so that the atomic orbitals that contribute most to each molecular orbital are able to overlap.[2]

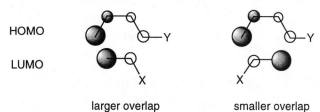

HOMO

LUMO

larger overlap   smaller overlap

The electrostatic potential model suggests that the best orientation is the one that brings together the region of most negative potential on the diene with the region of most positive potential on the dienophile.

In this experiment, you will explore to what extent the behavior of a series of dienophiles (cyanoalkenes) can be correlated with either their frontier molecular orbitals and/or their electrostatic potentials as provided by *ab initio* 3-21G calculations.

Procedure

Rates of Diels-Alder cycloadditions: Build the dienophiles listed in the table below, optimize their geometries using semi-empirical AM1 calculations, then use the resulting structures to calculate their *ab initio* 3-21G energies.

Experimental relative rates for Diels-Alder cycloadditions of cyclopentadiene and various dienophiles	
dienophile	**relative rate (k_{rel})[3]**
acrylonitrile	1
trans-1,2-dicyanoethylene	78
cis-1,2-dicyanoethylene	88
1,1-dicyanoethylene	4.4×10^4
tricyanoethylene	4.6×10^5
tetracyanoethylene	4.1×10^7

Record the 3-21G LUMO energy of each dienophile, and plot these energies against the log of the experimental k_{rel} values. Is there a relationship between LUMO energy and reaction rate? If there is, is this relationship consistent with the FMO model?

2. I. Fleming, **Frontier Orbitals and Organic Chemical Reactions**, Wiley, New York, 1976.
3. J. Sauer, H. Wiest and A. Mielert, *Chem. Ber.*, **97**, 3183 (1964); see also: Isaacs, p.713.

The FMO model actually depends on two factors: the HOMO-LUMO energy difference and the degree of HOMO-LUMO overlap. You can assess the "availability" of the LUMO for HOMO-LUMO overlap by comparing LUMO maps of the various dienophiles on the same color scale. Is there a relationship between the maximum value of the LUMO on this map and dienophile reactivity?

Now consider the electrostatic argument. Calculate and display electrostatic potential maps of the dienophiles, and record the most positive potential found over the π face of each. Plot this quantity against $\log(k_{rel})$. Are the two correlated? If so, is the direction of the correlation consistent with the electrostatic model?

Regioselectivity in Diels-Alder cycloadditions: You will use the reaction of 1-methylcyclopentadiene with various unsymmetrical dienophiles (acrylonitrile, 1,1-dicyanoethylene, and tricyanoethylene) to examine FMO and electrostatic potential-based analyses of regioselectivity.

Build 1-methylcyclopentadiene, optimize its AM1 geometry, and calculate its 3-21G wavefunction. Calculate and display a HOMO map. Which end of the diene is best suited for overlap with an unsymmetrical dienophile? Next, examine LUMO maps of the three unsymmetrical dienophiles. Which alkene carbon is best suited for overlap? What product regioisomers are predicted by the FMO model? Do you arrive at the same predictions using resonance arguments? (Your answer should include drawings of the specific resonance structures that your predictions are based on.)

Next examine electrostatic potential maps of the diene and the three dienophiles. Which end of the diene has the most negative potential? Which end of each dienophile has the most positive potential? What product regioisomers are predicted by the electrostatic model? Are the FMO and electrostatic predictions the same?

Optional

1. Lewis acids such as BF_3 generally increase both the rate and regioselectivity of Diels-Alder reactions. These effects might be due to enhanced and more selective frontier orbital interactions, or they might be due to changes in the nature of key electrostatic interactions. Examine the effect of a simple "Lewis" acid, H^+, on the reaction of acrylonitrile and 1-methylcyclopentadiene. Lewis acids preferentially bind to the strongest Lewis base present, the nitrogen lone pair of acrylonitrile.

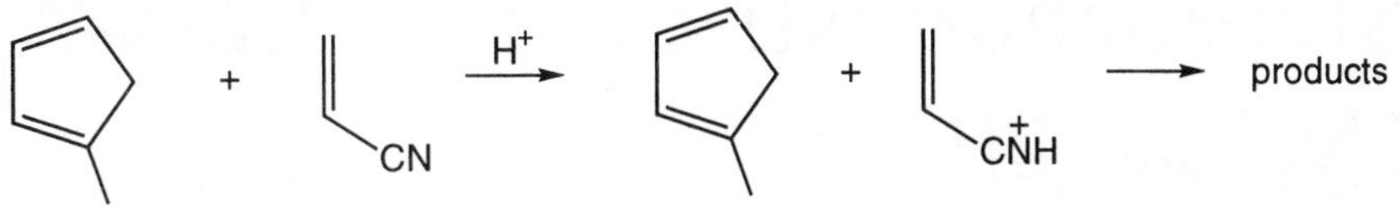

Therefore, build *N*-protonated acrylonitrile cation, optimize its AM1 geometry, and use this geometry to calculate its 3-21G energy as well as the various quantities needed to evaluate the reactivity and regioselectivity of this dienophile (see above). Which factors appear to contribute to this dienophile's enhanced reactivity and selectivity?

2. The most direct way to evaluate substituent effects on Diels-Alder reactivity is to calculate energy barriers for reactions involving differently substituted dienophiles. Build transition states for the reaction of cyclopentadiene and acrylonitrile, 1,1-dicyanoethylene, and tetracyanoethylene (TCNE) using the "template" method. That is, add cyano groups to a simpler transition state "template": 1) build norbornene (the cycloaddition product obtained from cyclopentadiene and ethylene), 2) constrain the two "forming" CC bonds to the distances shown below and optimize the AM1 geometry of this constrained structure,

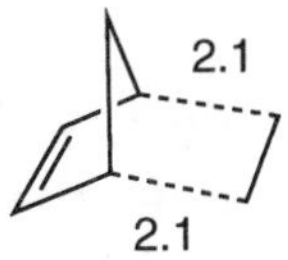

"Norbornene" transition state template

3) use this structure to search for the AM1 transition state (remove constraints), and finally, 4) add cyano substituents as needed to generate the desired transition state models.

Use AM1 calculations to search for each of the desired transition states, and verify that these are transition states by calculating their vibrational frequencies (a transition state has one imaginary frequency; see: **Finding Transition States**). Next, calculate the 3-21G energy of each transition state, and use these energies along with those of the reactants[4] to obtain the 3-21G energy barriers for each cycloaddition. Are the energy barriers correlated with $\log(k_{rel})$?

---

4. Total energies (in hartrees): cyclopentadiene, -191.7122; acrylonitrile, -168.8167; 1,1-dicyanoethylene, -260.0231; tetracyanoethylene, -442.4226.

Stereochemistry in Diels-Alder Cycloadditions

Semi-empirical AM1 calculations and ab initio 3-21G calculations are used to determine transition states for Diels-Alder cycloaddition of 5-substituted cyclopentadienes and acrylonitrile. The transition states are used to establish the facial and endo/exo stereoselectivity of these reactions, and the steric and electronic factors responsible for this selectivity.

Diels-Alder cycloadditions can lead to an array of stereoisomeric products, as shown below (*syn* and *anti* refer to the orientation of X and the dienophile).

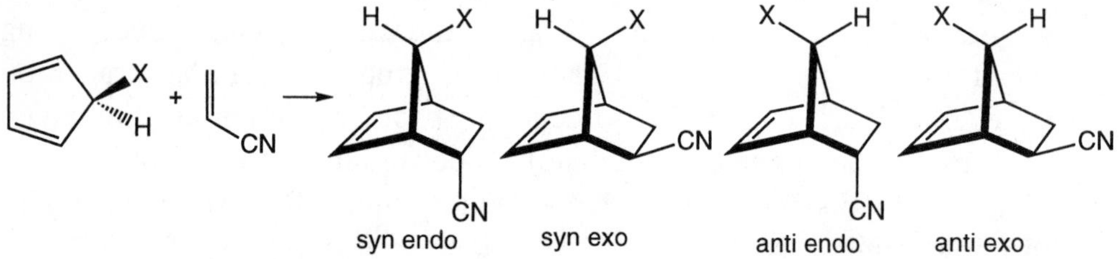

syn endo   syn exo   anti endo   anti exo

Experimental work has shown that *syn/anti* selectivity depends on the substituent X. In particular, if X=R, *anti* products form preferentially, while *syn* products are favored for X=OR.[1] Diels-Alder reactions also display a general preference for formation of *endo* adducts (see: ***Endo/Exo* Selectivity in Diels-Alder Cycloadditions**). Since these reactions occur in a single step, it is interesting to inquire whether these preferences are due to "product-control", i.e., the relative stability of different products is reflected in the stability of their transition states, or if unique steric and/or electronic factors exist in the transition state.

In this experiment, you will use semi-empirical AM1 calculations to obtain the four stereoisomeric transition states that describe the reaction of acrylonitrile with 5-methylcyclopentadiene (X=CH$_3$), and with 5-fluorocyclopentadiene (X=F). You will determine whether the transition state energies duplicate the experimental results cited above, and you will examine the transition states for steric and electronic factors that might account for the observed preferences.

Procedure

Since the eight required transition states have similar geometries, it is advantageous to first build a transition state for the cycloaddition of cyclopentadiene and ethylene (do this by building norbornene, then constrain the forming CC bond distances to 2.1 Å, and optimize the constrained geometry using semi-empirical AM1 calculations).

---

1.   For a review of the experimental and theoretical literature, see: S.D. Kahn and W.J. Hehre, *J. Am. Chem. Soc.*, **109**, 663 (1987); T. Chao, Ph.D. dissertation, University of California, Irvine, 1992.

"Norbornene" transition state template

Then, use this transition state as a template for each of the eight desired structures, i.e., make copies of this structure (remove constraints) and then add fluorines, methyl groups, and cyano groups where needed. Search for the AM1 transition states, and verify that these are true transition states by calculating their vibrational frequencies and displaying the "reaction coordinate"; a transition state should have one imaginary frequency, and the normal mode associated with this frequency should resemble the reaction coordinate for the reaction (see: **Finding Transition States**).

What is the lowest energy transition state for each reaction? Are these results consistent with the experimental observations? (Note: AM1 energies often do not provide a good account of energy differences in these systems. Calculate the *ab initio* 3-21G energy of each AM1 transition state. What qualitative and quantitative differences exist between the two sets of results?)

Try to determine the factors that affect transition state stability. Compare space-filling models of the different transition states. Which transition states appear to offer the least (most) steric hindrance? Can you find supporting evidence for this in terms of specific geometrical parameters (interatomic distances, angles)? Compare the polarity of isomeric transition states by calculating and comparing their dipole moments and atomic charges. Is there a correlation between transition state energy and polarity? Calculate and compare electrostatic potential maps of isomeric transition states on the same color scale. Is the variation in potential consistent with the variation in dipole moment?

Optional

Changes in transition state energy may be due to "product control", i.e., the factors that affect product stability may be reflected in the transition state because the latter already has some product-like characteristics. Build the eight reaction products i.e., norbornene substituted *syn* and *anti* by CH_3 and F, and *endo* and *exo* by CN (eight structures in total).

Optimize their AM1 geometries. Is transition state energy correlated with product energy? If it is, which energy, transition state or product, is more sensitive to stereochemistry and why?

244

Endo/Exo Selectivity in Diels-Alder Cycloadditions

Semi-empirical AM1 calculations and ab initio 3-21G calculations are used to determine the kinetic and thermodynamic products resulting from Diels-Alder cycloaddition of cyclopentadiene and maleic anhydride. The transition states are examined for evidence of Frontier Molecular Orbital and electrostatic interactions.

The Diels-Alder reaction of cyclopentadiene and maleic anhydride gives a mixture of *endo* and *exo* products, with the *endo* stereoisomer being the major product.

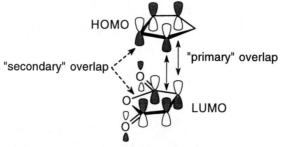

endo (99%)   *exo* (1%)

It has been established experimentally that the *endo* product is the kinetic product, and not the thermodynamic product, and that *endo* products are generally preferred for conjugated dienophiles (this is referred to as the Alder *endo* rule).[1]

The *endo* rule can be accounted for in several ways, and different factors may be at work in different Diels-Alder reactions. Frontier Molecular Orbital (FMO) theory proposes that the favored transition state will be the one that has the best HOMO$_{\text{diene}}$-LUMO$_{\text{dienophile}}$ overlap. The "primary" overlap in the transition state, i.e., the overlap leading to bond formation, is assumed to be the same for the *endo* and *exo* transition states. However, there is the possibility of bonding "secondary" overlap between the carbonyl groups on the dienophile and the forming π orbital in the *endo* transition state (one of the two "secondary" overlaps is identified below).

HOMO

"secondary" overlap   "primary" overlap

LUMO

Electrostatic interactions may also contribute to an *endo* preference. It might be expected that the forming π bond will be the most electron-rich site in the diene, and the carbonyl carbons to be the most electron-poor sites in the dienophile.

---

1. Carey and Sundberg B., p. 284; Lowry and Richardson, p.925; March, p. 851.

The *endo* transition state appears to place these sites in close proximity (see facing page), and might be stabilized by the resulting electrostatic interactions.

In this experiment, you will examine *endo* and *exo* transition states for the cycloaddition of cyclopentadiene and maleic anhydride. Kinetic selectivity will be established using AM1 and 3-21G energies, and the operation of favorable frontier orbital and/or electrostatic interactions will be determined using 3-21G. You will also verify that the *exo* product is thermodynamically preferred by comparing 3-21G energies of the isomeric products.

Procedure

Build *endo* and *exo* transition states by building the corresponding *endo* and *exo* products, and constraining the distances of the "forming" CC bonds to 2.1Å. Optimize each constrained structure using semi-empirical AM1 calculations, and use the resulting structures (remove constraints) to search for the AM1 transition states. Verify that each structure is a transition state by calculating its vibrational frequencies (see: **Finding Transition States**), then calculate its *ab initio* 3-21G energy. Compare the AM1 energies of the two transition states. Which pathway is preferred? Now compare the 3-21G energies. Which pathway is preferred? Do the two calculations lead to the same conclusions?

Calculate and display the several highest-energy occupied orbitals of the *endo* transition state. Which of these describe "forming" bonds, i.e., the two new σ bonds or the new π bond? Do you see evidence of secondary orbital overlap?

Calculate and display an electrostatic potential map for each transition state. What are the most electron-rich and electron-poor sites in each transition state? Does this agree with the qualitative analysis given above? Which transition state places these sites in closest proximity? Is this transition state less polar? (Note: you can assess relative polarity by displaying the two maps on the same color scale, or by comparing the dipole moments for each transition state.) Might electrostatic interactions contribute to the observed *endo* preference?

It is also interesting to compare the electron distribution and atomic charges in the *endo* transition state with those in the isolated reactants (this provides information about how charge shifts during the reaction). Build maleic anhydride and cyclopentadiene, optimize their AM1 geometries, and calculate their 3-21G wavefunctions and electrostatic potential maps. Compare the transition state and reactant maps on the same color scale. Also calculate and compare atomic charges. Do you see evidence of charge transfer? If so, in what direction does this charge transfer occur?

Finally, determine the preferred product by building each isomer, optimizing its AM1 geometry, and calculating its 3-21G energy. Which isomer is the AM1 thermodynamic product? The 3-21G thermodynamic product? Are these results consistent with each other and with experiment?

246

The Origin of the Barrier in Diels-Alder Cycloadditions

Semi-empirical AM1 calculations and ab initio 3-21G calculations are used to dissect a Diels-Alder energy barrier. In particular, the energy required to distort the reactant geometries into that of the transition state is compared to the reaction's energy barrier.

Why do chemical reactions have barriers? One reason is that bond breaking, whether it involves a covalent bond or an ion-solvent bond, requires energy. It is less obvious, however, why concerted reactions, i.e., reactions where bond making accompanies bond breaking, have barriers. One might suppose that the energy released by bond making would compensate for the energy required by bond breaking, but this is rarely the case.

A useful model for describing this problem has been developed by Shaik and Pross.[1] They begin by noting that the wavefunction of a reacting system (Ψ_{true}) can be thought of as a superposition of the valence bond wavefunctions that describe the isolated reactants (Ψ_R) and products (Ψ_P), i.e., $\Psi_{true} = c_R\Psi_R + c_P\Psi_P$.[2] Thus, the behavior of a system at each point along the reaction coordinate can be attributed to three factors: the behavior of Ψ_R, the behavior of Ψ_P, and the stabilization that results from their interaction.

The behavior of each wavefunction for a typical reaction is shown below.

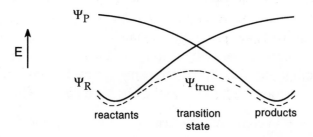

1. This model is referred to colloquially as the "curve crossing model". S.S. Shaik, H.B. Schlegel and S. Wolfe, **Theoretical Aspects of Physical Organic Chemistry**, Wiley, New York, 1992; A. Pross, **Theoretical and Physical Principles of Organic Reactivity**, Wiley, New York, 1995; A. Ioffe and S. Shaik, *J. Chem. Soc., Perkin Trans. 2*, 2101 (1992).
2. Readers who are not familiar with the distinctive properties of valence bond (VB) wavefunctions might find it helpful to view them as the quantum mechanical equivalents of Lewis structures. Thus, just as a Lewis structure describes a particular bonding pattern, and resonance between Lewis structures stabilizes a molecule, so it is with VB wavefunctions. Each wavefunction corresponds to a particular bonding pattern, the superposition of several wavefunctions is stabilizing, and this "resonance" stabilization is greatest when two VB wavefunctions have identical energies.

The Ψ_R and Ψ_P curves rise and fall at different rates, and their "crossing point", which roughly corresponds to the transition state, has an energy significantly higher than that of either the reactants or products. On the other hand, the energy of the transition state is much less than the energy of the crossing point because maximum resonance stabilization occurs when Ψ_R and Ψ_P have identical energies. Therefore, reaction barriers find their origin in a competition of two opposing factors: the energy required to reach the Ψ_R-Ψ_P crossing point, and the energy released by resonance stabilization at the crossing point.

Reaction barriers have their origin, then, in the various destabilizing factors that lead to the high-energy crossing points. One way to approach this problem is to examine what happens to Ψ_R as the reacting molecules move from "reactants" to "transition state". Two factors turn out to be important: 1) the molecules change their geometry and their bonds become weaker, and 2) repulsion builds up between reactant bonds, i.e., between the bond pairs defined by Ψ_R, wherever these bonds begin to approach one another. Referring to these two energy changes and the resonance stabilization energy between Ψ_R and Ψ_P (at the crossing point) as $E_{distort}$, $E_{repulsion}$, and RE, respectively, the energy barrier for a reaction, $\Delta E^{\ddagger}$, can be written as.[3]

$$\Delta E^{\ddagger} \approx E_{distort} + E_{repulsion} - RE$$

Another way to describe the destabilizing factors that contribute to reaction barriers is to examine the energy gap between Ψ_R and Ψ_P at the reactant geometry, i.e., for well-separated reactants. This initial energy gap turns out to be very large for reactions like Diels-Alder. To convert Ψ_R into Ψ_P, three π bonds in the diene and dienophile must be broken (3 x 65 kcal/mol), and only one product π bond is made in return (the σ bonds do not lower the energy of Ψ_P because the reactants are not yet within bonding distance).

Moving the reactants together and changing their geometries ultimately erases the Ψ_R-Ψ_P energy gap, but this occurs at a geometry where both wavefunctions have higher energies than the reactants or products.

Examining the initial energy gap also allows the prediction of reactivity of different systems. For example, all other things being equal, it is to be expected that reactions with larger initial energy gaps (R→P2) will have higher energy crossing points and larger barriers than reactions with smaller gaps (R→P1).

---

3. This formula ignores the small resonance stabilization experienced at the reactant geometry.

Thus, an intramolecular Diels-Alder reaction, in which the diene and dienophile are properly pre-positioned to begin σ bond formation, might have a smaller initial energy gap and will proceed more rapidly.

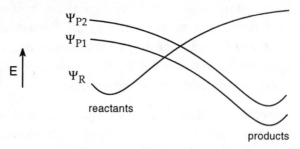

In this experiment, you will use semi-empirical AM1 and *ab initio* 3-21G calculations to examine the relationship between the Diels-Alder energy barrier, $\Delta E^{\ddagger}$, the geometrical distortion energy, $E_{distort}$, and the size of the Ψ_R-Ψ_P energy gap, for two reactions: the addition of ethylene and butadiene, and the addition of ethylene and *o*-xylylene.

$E_{distort}$ will be estimated by finding the transition state for each cycloaddition reaction, calculating the energy of each reactant at its transition state geometry, and then comparing this to the energy of the isolated, undistorted reactant. This method necessarily underestimates $E_{distort}$ because the wavefunction of the distorted molecule does not correspond purely to Ψ_R.

As shown previously, the initial energy gap in the Diels-Alder reaction is roughly equal to the sum of the singlet-triplet excitation energies for the diene and dienophile, i.e., ΔE_{ST}(diene) + ΔE_{ST}(dienophile). Since you are only interested in the change in the energy gap, and since both reactions utilize the same dienophile (ethylene), you will calculate ΔE_{ST}(diene) only. The triplet state of *o*-xylylene is expected to have some aromatic character (see below), therefore, this diene should have a smaller excitation energy and should be more reactive than butadiene.

singlet diene ΔE_{ST} excited triplet diene

ΔE_{ST}

Procedure

Barriers: Build ethylene, butadiene and *o*-xylylene, and optimize their geometries using semi-empirical AM1 calculations. Then, use *ab initio* 3-21G calculations to obtain the energies of these structures, and record these energies.

Build the ethylene + butadiene transition state using a "constrained cyclohexene" template: 1) build cyclohexene, 2) constrain the "forming" CC bonds to 2.1Å and constrain the dihedral angle between these bonds to 0°, and 3) optimize the AM1 geometry of this constrained structure. Use this structure to search for the AM1 transition state. Use an analogous procedure to build a template for the ethylene + *o*-xylylene transition state, and search for the corresponding AM1 transition state. Verify that each structure is a transition state by calculating its vibrational frequencies (see: **Finding Transition States**), and then calculate their respective 3-21G energies. Combine your results to obtain barriers for the two Diels-Alder reactions. Which cycloaddition has a lower barrier?

Distortion energies: Make two copies of each transition state, and delete atoms from each one so that you are left with just the atoms of either the diene or dienophile. Calculate 3-21G energies for the "distorted" diene and dienophile, and compare these to the energies of the isolated, optimized reactants to get $E_{distort}$ for each molecule (remember that these values underestimate $E_{distort}$). What is the total distortion energy for the two reactants in each reaction? Which is larger, the total distortion energy or the reaction energy barrier? Use the difference between the two energies to obtain an estimate of RE, the resonance energy in the transition state. Is this estimate a lower or upper bound?

Initial energy gap: Make a copy of each optimized diene and calculate the AM1 energy of its triplet state. Use this energy and the AM1 energy of the optimized diene to estimate ΔE_{ST}(diene). Which energy is larger? What does this imply about the relative size of the initial Ψ_R-Ψ_P energy gaps? Do the excitation energies correctly identify the reaction with the smaller barrier?

It was suggested above that the triplet state of the diene corresponds to a resonance structure in which there is enhanced π bonding between the central two carbons, and diminished bonding in the original π bonds. Calculate the AM1 bond orders of the two optimized dienes, and of their triplet excited states. Do the bond orders support the simple resonance picture given above? Do the bond orders support the notion that the *o*-xylylene triplet has aromatic character (you will need to examine the CC bond orders around the ring)?

1,3-Dipolar Cycloadditions of Azomethine Ylides

Semi-empirical AM1 calculations and ab initio 3-21G^(*) calculations are used in conjunction with Frontier Molecular Orbital theory to examine the regioselectivity of a 1,3-dipolar cycloaddition between an azomethine ylide and an electron-poor alkene.

1,3-Dipoles, such as azomethine ylides, undergo cycloadditions with alkenes to give five-membered ring products. This is demonstrated by the following synthesis of the biologically active amino acid, α-allokainic acid.[1] Azomethine ylide, **1**, generated *in situ* with base, combines with the unsaturated ketone to give a single regioisomer as a ~10:1 ratio of *endo* and *exo* molecules (*endo* shown). Subsequent removal of sulfur and epimerization then gives the desired amino acid.

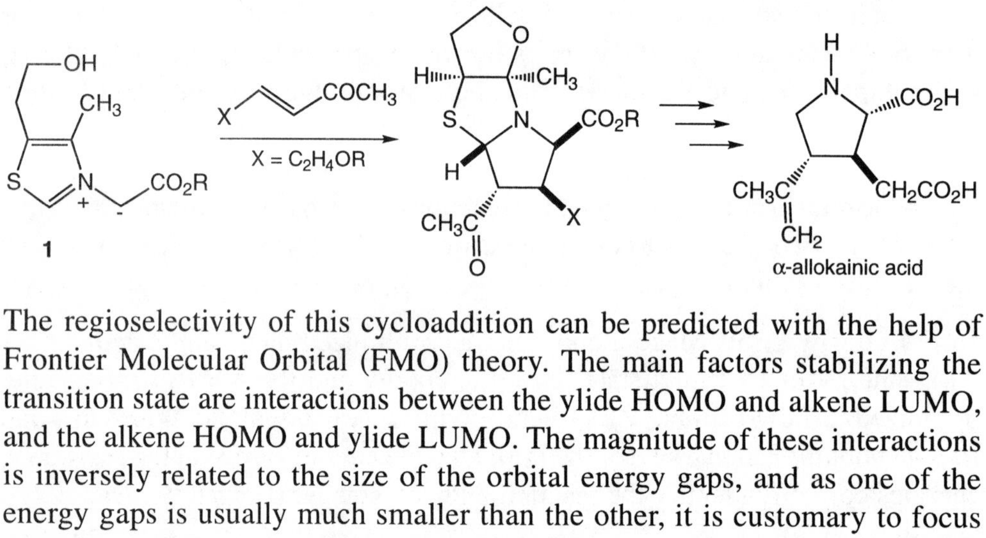

The regioselectivity of this cycloaddition can be predicted with the help of Frontier Molecular Orbital (FMO) theory. The main factors stabilizing the transition state are interactions between the ylide HOMO and alkene LUMO, and the alkene HOMO and ylide LUMO. The magnitude of these interactions is inversely related to the size of the orbital energy gaps, and as one of the energy gaps is usually much smaller than the other, it is customary to focus attention on the interaction with the smaller energy gap. The preferred regiochemistry is then identified by aligning the ylide and alkene so that the interacting orbitals produce the best overlap.

In this experiment, you will use semi-empirical AM1 calculations to identify the most stable conformer of a simple model azomethine ylide, **2**, and *ab initio* 3-21G^(*) calculations to generate the frontier orbitals of **2** and a model alkene, **3**. You will also use AM1 and 3-21G^(*) calculations to compare the energies of the regioisomeric cycloaddition transition states, **4** and **5**.

---

1.   G.A. Kraus and J.O. Nagy, *Tetrahedron Lett.*, **22**, 2727 (1981); *ibid*, **24**, 3427 (1983).

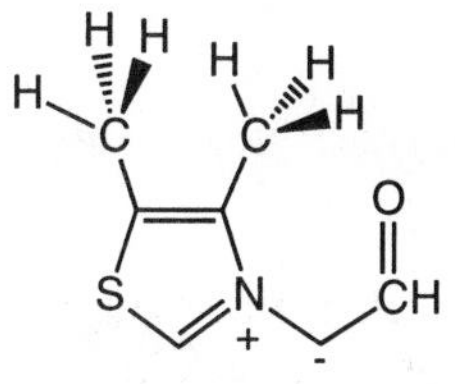

2 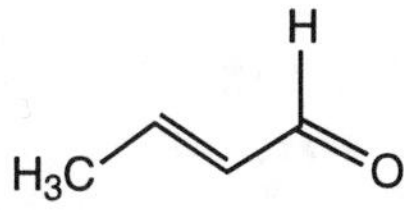 **3** 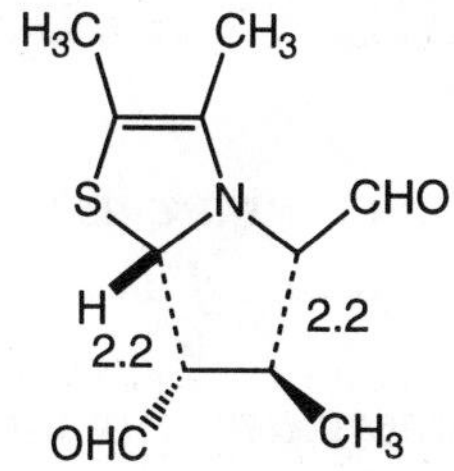 **4** 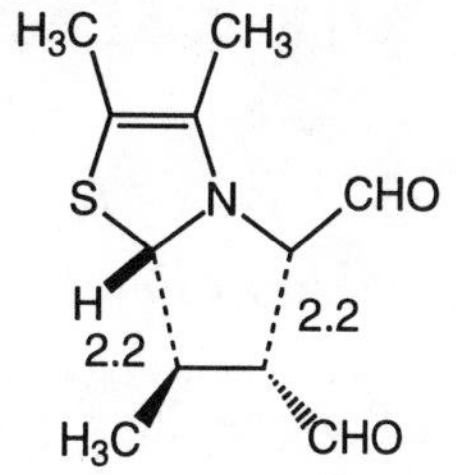 **5**

Procedure

Conformational analysis of 2: Build azomethine ylide, **2**, and arrange the methyl hydrogens in the conformation shown above. Generate the four different "planar" conformers resulting from internal rotation of the CHCHO substituent, i.e., keep the substituent coplanar with the ring. Optimize each conformer using semi-empirical AM1 calculations, record the energy of each, and use the most stable conformer for the rest of the experiment (this will be referred to as "**2**").

FMO analysis: Calculate the *ab initio* 3-21G$^{(*)}$ wavefunction of **2** (use the AM1 geometry), and record the energies of its frontier orbitals. Next, build unsaturated ketone **3** in the *trans* conformation, and optimize its AM1 geometry. Calculate the 3-21G$^{(*)}$ wavefunction of **3** (AM1 geometry), and record the energy of its frontier orbitals. Which is the dominant frontier orbital interaction, HOMO$_{ylide}$-LUMO$_{alkene}$ or HOMO$_{alkene}$-LUMO$_{ylide}$? Calculate and display the key orbitals and decide which regiochemistry gives the best orbital overlap.[2] Which reactant appears to be more "electronically" unsymmetric?

Transition states: Build **4** and **5** with the bond distances shown above (note: the ring and alkene substituents should have the same orientations used in **2** and **3**). Optimize the AM1 geometry of **4** and **5** subject to these distance constraints, then use the resulting structures to search for the AM1 transition states (no constraints). Verify that the structures are true transition states by calculating their vibrational frequencies (a transition state has one imaginary frequency; see: **Finding Transition States**). Then, calculate their 3-21G$^{(*)}$ energies (AM1 geometries). Which transition state is preferred? Which transition state is more (less) symmetric? Calculate and compare electrostatic potential maps of the two transition states. Is there a correlation between bond distances, the degree of charge transfer, and frontier orbital interactions? Why?

---

2. Alternatively, calculate and display orbital maps of the key orbitals. The best regiochemistry can be predicted by obtaining the values of the orbitals directly over the reacting carbon atoms.

Dimerization of Ketene

Ab initio 3-21G calculations are used to investigate the dimerization of ketene, and to rationalize formation of an unsymmetric dimer.

Ethylene does not dimerize readily upon heating. This fact is readily explained by Frontier Molecular Orbital (FMO) theory which notes that the highest-occupied molecular orbital (HOMO) and lowest-unoccupied molecular orbital (LUMO) of ethylene have opposite symmetries, and do not produce net overlap during a face-to-face encounter. In other words, this sterically reasonable pathway is "forbidden" by orbital symmetry.

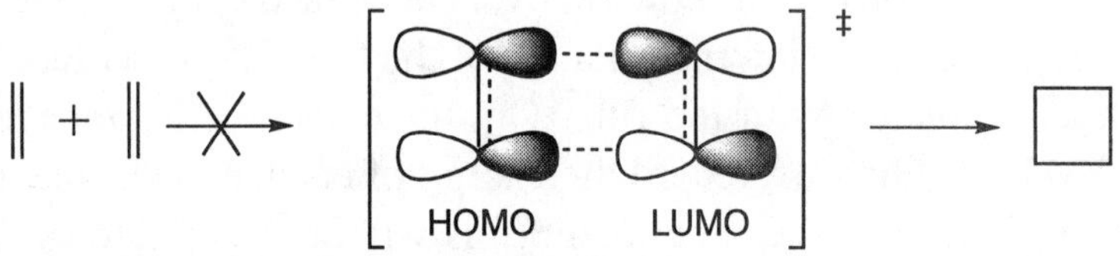

FMO analysis suggests that cycloadditions involving pairs of alkenes will be rare because the frontier orbitals of most alkenes are similar to those of ethylene.[1] An apparent exception to this rule is ketene, $CH_2=C=O$, which undergoes cycloaddition both with alkenes, to form cyclobutanones, e.g., with ethylene,

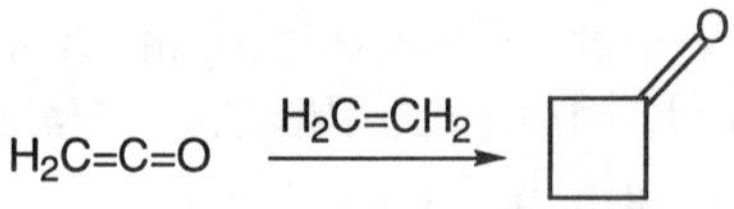

and with itself to form an unsymmetrical dimer, **1**.

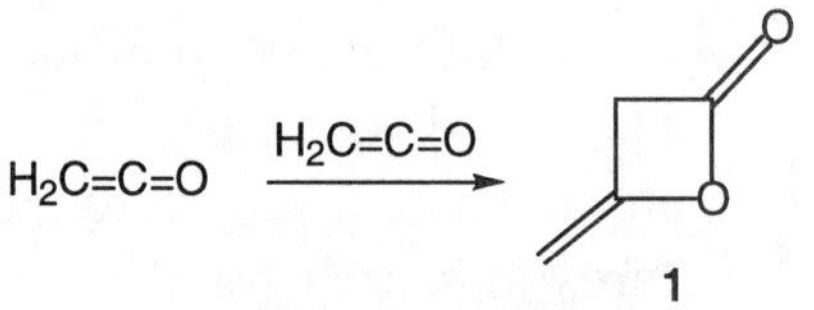

In this experiment, you will use *ab initio* 3-21G calculations to determine the frontier orbitals of ketene in order to rationalize why it behaves differently from ethylene. Since ketene dimerization might be expected to produce 1,3-cyclobutadione, **2**,[2]

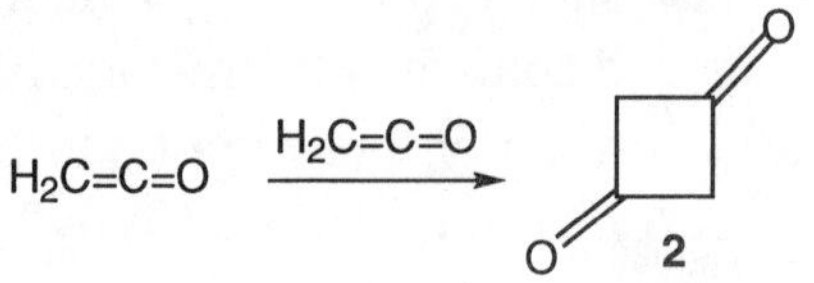

---

1. There is an alternative mechanism to account for the observed dimerization of ethylene and other alkenes at high temperatures. Rather than proceed in a concerted manner, the reaction may actually go in two steps, one CC bond forming well in advance of the other, i.e.,

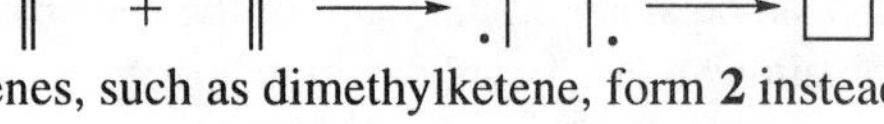

2. Certain substituted ketenes, such as dimethylketene, form **2** instead of **1**.

you will also determine whether the observed formation of **1** is controlled by kinetic or thermodynamic factors.[3]

Procedure

Build ethylene and ketene, and optimize their geometries using *ab initio* 3-21G calculations. Calculate and display the HOMO and LUMO of each molecule. What differences are there in the two molecules? Is there an orbital-symmetry-allowed path leading from ketene to either **1** or **2**? Unsymmetrical cycloadditions[4] often involve unsymmetric or asynchronous transition states, i.e., one CC bond (usually the one that gives the best HOMO-LUMO overlap) forms more rapidly than the other. Applying this logic to the formation of **1** and **2**, and assuming asynchronous transition states for each pathway, which bonds would form more rapidly, and which would form more slowly?

Thermodynamic control: Build **1** and **2**, and optimize their 3-21G geometries. Record their energies and calculate their relative energies. Which molecule is more stable? Is ketene dimerization thermodynamically controlled? Explain.

Kinetic control: Build the following "guesses" for the asynchronous transition states leading to **1** and **2**, respectively.

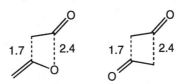

Optimize the AM1 geometry of each (constrain distances, but not the symmetry) and calculate their vibrational frequencies. Use the resulting structures and frequencies to search for the 3-21G transition states (no constraints). Verify that these are true transition states by calculating their vibrational frequencies (a transition state should have one and only one imaginary frequency; see: **Finding Transition States**). Record their energies and calculate their relative energies. Is ketene dimerization kinetically controlled? Explain. Are the transition state structures consistent with the FMO analysis? Calculate electrostatic potential maps of ketene and the two transition states, and compare them on the same color scale. Does electron transfer occur during each cycloaddition? Use these results along with the calculated bond distances to draw the principal Lewis structure(s) for each transition state. Are these structures consistent with the FMO analysis?

3.   E.T. Seidl and H.F. Schaefer, III, *J. Am. Chem. Soc.*, **112**, 1493 (1990); *ibid.*, **113**, 5195 (1991).
4.   "Symmetry" here can refer either to molecular structure or electronic structure, such as electron-rich and electron-poor.

Ylides

Ab initio 3-21G$^{(*)}$ calculations are used to compare the proton affinities of ylides containing nitrogen, oxygen, phosphorus, and sulfur, and to better understand the electronic structure of these synthetic reagents.

Phosphorus and sulfur ylides find many uses in synthetic organic chemistry.[1] Both types of ylides can be generated by deprotonation of an appropriate salt precursor, and react with ketones and aldehydes, but that is where their similarity ends. Phosphorus ylides, **1**, convert ketones and aldehydes into alkenes, while sulfur ylides, **2** and **3**, yield epoxides.

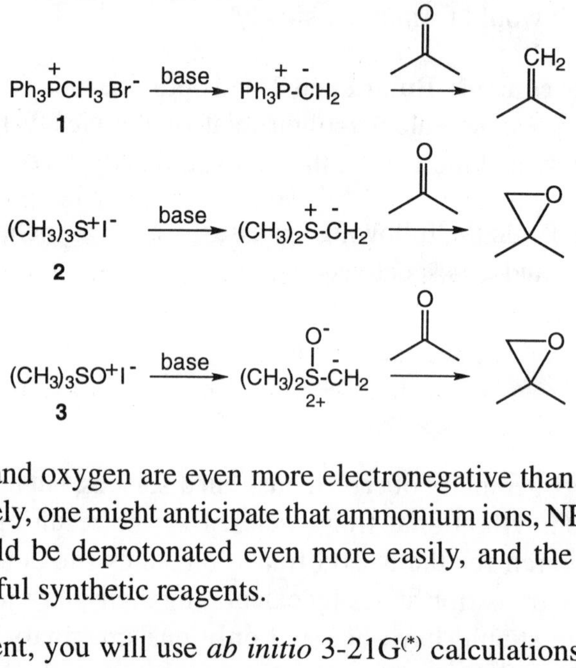

Since nitrogen and oxygen are even more electronegative than phosphorus and sulfur, respectively, one might anticipate that ammonium ions, NR_4^+, and oxonium ions, OR_3^+, could be deprotonated even more easily, and the resulting ylides might make useful synthetic reagents.

In this experiment, you will use *ab initio* 3-21G$^{(*)}$ calculations to compare the acidities of various "onium" ions. Specifically, you will use the following equation to calculate acidities of different ions relative to tetramethylammonium ion, $(CH_3)_4N^+$, where the "onium" ions are: $(CH_3)_4P^+$, $(CH_3)_3O^+$, $(CH_3)_3S^+$, $(CH_3)_3SO^+$

$$(CH_3)_4N^+ + \text{ylide} \rightleftharpoons (CH_3)_3\overset{+}{N}\text{-}\overset{-}{C}H_2 + \text{"onium" ion}$$

Procedure

Build $(CH_3)_4N^+$, $(CH_3)_4P^+$, $(CH_3)_3O^+$, $(CH_3)_3S^+$, and $(CH_3)_3SO^+$, and optimize their geometries using 3-21G$^{(*)}$ calculations. Record their energies and the XC bond distances (X = N, P, O, S). Next, build the ylides derived from each of the onium

---

1.   Carey and Sundberg A, p. 95.

ions listed above, and optimize their 3-21G$^{(*)}$ geometries. Record their energies and XC bond distances. For each ylide/onium-ion combination, calculate the reaction energy of the proton transfer reaction shown above. Are there are large differences in the ability of N, O, P, and S at stabilizing adjacent anions? Order the onium ions according to their relative acidities. Which can be generated most easily?

What effect does deprotonation have on XC bond distance? Is there a correlation between the anion-stabilizing ability of X and the change in the X–CH$_2$ bond distance? Bond shortening is usually associated with multiple bonding. Based on your energy and geometry data, which ylides appear to be stabilized by the following type of resonance? Multiple bonding also requires XC-p overlap. Examine the HCH and XCH bond angles around the CH$_2$ groups in each of the ylides; are these consistent with your conclusions? Explain.

$$R_n\overset{+}{X}-\overset{-}{C}H_2 \longleftrightarrow R_nX{=}CH_2$$

Although chemists have traditionally attributed multiple bonding in ylides to the overlap of d and p orbitals on X and C, respectively, considerable evidence suggests that this does not occur.[2] Another way multiple bonding might occur is by π overlap of a p orbital on C$_{methylene}$ and a π-type XC$_{methyl}$ antibonding orbital (see: **π Interactions Involving σ Electrons**). However, for this type of interaction to be important, the antibonding orbital must contain a large contribution from X. Is this more likely to be the case with X = N or O, or X = P or S? Why? Are the changes in X-CH$_3$ bond distance caused by deprotonation consistent with this explanation?

---

2. D.C. Gilheany, *Chem. Rev.*, **94**, 1339 (1994).

Appendix A

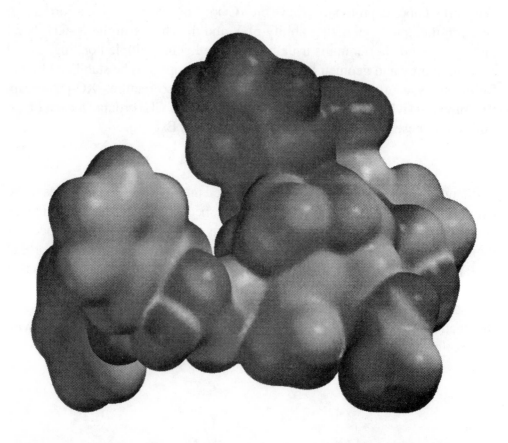

Tutorials

T he three tutorials which follow describe the use of the SPARTAN molecular modeling program.[1] This actually comprises a whole suite of computer programs which perform different functions: a program for building and modifying models, several programs for calculating molecular structure and energy, a program for deriving molecular properties from previously calculated structures, and a program to display properties using three-dimensional graphs. All of the programs communicate with one another automatically as needed. Therefore, you do not need to concern yourself with their operation. From your point of view, it will always seem as though you are using a single program.

Although SPARTAN's large and varied assortment of modeling tools may at first seem a bit intimidating, there is a natural relationship between them. Consider a "typical" modeling session.

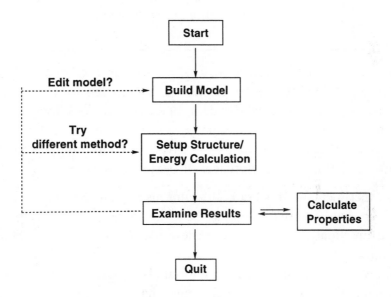

The starting point for most molecular modeling is the user's need to learn something about a molecule. Thus, the user typically: 1) builds a trial model of the molecule, 2) optimizes the molecule's structure and calculates its energy, 3) examines the results to see if they are reasonable, and then 4) repeats one of the earlier steps with a new model if needed, or proceeds to calculate interesting properties. Each tool, therefore, plays a logical role in the modeling process, and the user quickly becomes familiar with each tool after just a small amount of practice.

---

1.   Available from: Wavefunction, Inc. 18401 Von Karman Avenue, Suite 370, Irvine, CA 92612. The tutorials may also be used with MacSPARTAN, MacSPARTAN *Plus* and PC SPARTAN, the operation of which are similar to that of SPARTAN. Consult the user's guides for these programs for further information.

The tutorials included here attempt to illustrate this sequence of steps required for a complete computational investigation. They follow the order of topics found in a typical organic chemistry textbook, and should be performed in sequence.

tutorial	related organic chemistry topics
1. Electron Distributions and Electrostatic Potential Maps	Chemical Bonds (covalent, polar covalent, multiple)
2. Energy	Acid-Base Behavior, Solvation Energy, Resonance Stabilization
3. Conformation	Conformation (acyclic, cyclic molecules, effect of hydrogen bonding)

The first tutorial emphasizes electron distributions, partial charges and resonance structures. It describes how to use calculated electron densities and electrostatic potential maps to characterize the bonding and charge distributions in molecules. The second tutorial continues these themes, and shows how to calculate reaction energies for gas phase and aqueous processes. The third tutorial shows how to predict the conformational preferences of flexible molecules.

Like the experiments in this book, the tutorials need to be "read" in front of a computer so that each operation can be tried immediately. It is important to follow the directions exactly as they are described (note, however, that modeling errors, unlike errors in the laboratory, are usually "safe" and easy to correct), and to allow sufficient time for each tutorial (1-2 hours is typical, but this will depend on the user and the type of computer that is being used).

Tutorial 1: Electron Distributions and Electrostatic Potential Maps

Chemists use structural models to learn about overall molecular size and shape, and to identify arrangements that are likely to be too crowded to be stable. They use resonance models both to pick out systems that should be especially stable, and to identify especially reactive molecules in terms of a buildup of charge. Such simple models are neither quantitative nor are they general. Conventional structural models require detailed knowledge about bonding ("what atoms are connected to what atoms") before they can be used, and resonance models are limited to very simple systems, and even here are qualitative at best.

Models based on quantum mechanics do not suffer from such deficiencies. Calculations can directly provide information about electron distributions, which in turn relates to overall molecular dimensions. Electrostatic potential maps are able to show which regions of a molecule are especially electron rich, hence subject to electrophilic attack, and those which are especially electron poor, hence subject to nucleophilic attack. The models are completely general, applicable to "unusual molecules" and even transition states, as well as "normal molecules". They require no preconceptions about bonding and are inherently quantitative. Small differences in chemical environments are reflected in different electron distributions and different electrostatic potentials.

This tutorial comprises a number of simple exercises relating to the calculation of electron densities and electrostatic potential maps, and to their use in interpreting molecular structure, thermochemical stability and chemical reactivity.

Oct-8-en-1-ol

Oct-8-en-1-ol is an unsaturated 8-carbon alcohol.

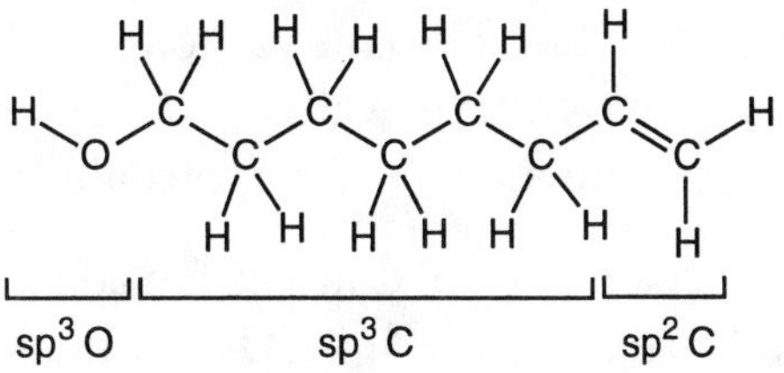

oct-8-en-1-ol

The techniques used to model this molecule can be applied to nearly any neutral, acyclic molecule. The following instructions show how to construct a model using one-atom fragments, how to manipulate the model on the computer screen, and how to calculate and display its charge distribution using an electrostatic potential map.

260

Creating a new model: To create a new model, select **New** from the **File** menu. This causes a model kit to appear. This contains several different building blocks: a library of one-atom fragments, a library of functional groups (**Groups**), and a library of rings (**Rings**). Each atom fragment specifies a fixed and unique combination of element, bonding pattern, and geometry, and is represented by a button which conveys this information (the five N buttons are shown below).

Button	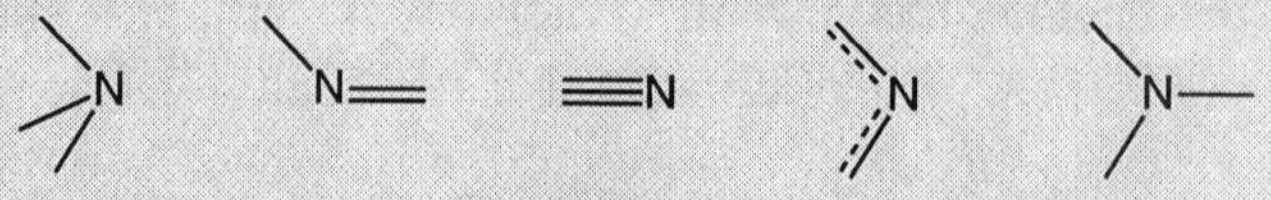 $\geq$N	N=	≡N	$\geq$N	N—
Hybridization	sp^3	sp^2	sp	sp^2 (aromatic)	sp^2
Geometry (bond angle)	pyramidal (109.5°)	bent (120°)	terminal	bent (120°)	trigonal (120°)
Possible Neighbors	3	2	1	2	3
Possible Bonds (unfilled valences)	3 single	1 single 1 double	1 triple	2 aromatic	3 single

Before you can build a model, you must decide which atom fragments are needed for its construction. This process is straightforward for most molecules. For example, to build octenol you need two different types of C atoms (6 sp^3 and 2 sp^2) and one kind of O atom (sp^3). H atoms are not needed since these are added automatically (see below).

Building the model (Stage 1): To begin, *click* (left mouse button) on the sp^3 C button, and then *click* anywhere on the screen. A "wire" model of a tetrahedral carbon atom appears. (If you make a mistake, select either **Undo Add** or **Clear** from the **Edit** menu.) The gray line segments intersect at the carbon nucleus. The yellow line segments are "unfilled valences", and identify places where new atom fragments can be attached to the carbon atom. The unfilled valences are automatically changed into hydrogen atoms when you save your model, so it is unnecessary to add hydrogen atoms to the molecule.

Oct-8-en-1-ol contains a chain of 6 sp^3 C atoms. To add one C atom, *click* on the end of the unfilled valence that points "up". The unfilled valence is replaced by a second C atom. Continue adding C atoms to the unfilled valences at the "top" of the model until you have a chain of 6 C atoms.

Next, *click* on the sp^3 O button, then *click* on the "top" unfilled valence of the model. The O is colored red. At this point the model appears to be very crowded with several overlapping unfilled valences. Before you try to add the two sp^2 C atoms, read the following instructions for moving the model into other, more easily studied, positions.

Moving models: The center and right mouse buttons are used to rotate and translate molecules. To rotate your molecule, move the mouse with the center button depressed; to translate it, move the mouse with the right button depressed. The **Shift key** in conjunction with the center mouse button allows rotation in the plane of the screen, and in conjuction with the right mouse button allows zooming (in and out of screen).

mouse button	mouse movement	model movement
center		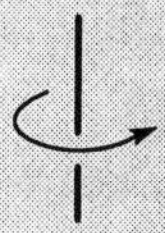
center		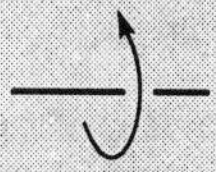
center (press **Shift key** too)		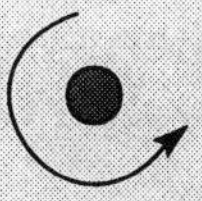
right		
right		
right (press **Shift key** too)		ZOOM Zoom

Building the model (Stage 2): To finish building the model, *click* on the sp^2 C button. Then, *click* on an unfilled valence of the terminal C atom (the C that does not carry O). Even though the sp^2 C atom can make both single and double bonds, a single bond between the sp^2 and sp^3 C atoms will be formed because this is the only type of unfilled valence that both fragments have in common.

If necessary, rotate the model again so that you can see both of the unfilled valences on the sp^2 C. One unfilled valence will be used to make a double bond, while the other will be used to make a single bond. *Click* on the double bond valence. Make sure that the two sp^2 C atoms are connected by a double bond (rotate the model if necessary). If it is not, select **Undo Add** from the **Edit** menu and try adding the second sp^2 C atom again.

Saving your model and quitting the builder: You have finished building a trial structure of your model, but you must save it and exit build mode before you can refine your model or calculate its properties. Both of these operations can be accomplished at the same time. Select **Quit** from the **File** menu (*click* on **Yes**, type "octenol" for your model name, and *click* on **Save**).[1]

At this point the model kit will disappear and your model will appear in the main window. Notice that the name of your model, "octenol" appears in the title bar above the window. You can still rotate and translate your model using the same mouse operations described above. You can also change the appearance of your model using the various menu items listed in the **Model** menu. (Go ahead and try the different representations in the **Model** menu, then select the **Tube** model before proceeding.)

Calculating the AM1 wavefunction: The charge distribution for octenol is calculated in two steps. First, an "approximate" wavefunction of the molecule is calculated (a "wavefunction" is just a mathematical formula that describes the movements and energies of subatomic particles), and then the wavefunction is converted into a three-dimensional graph or "map" of the charge distribution. This tutorial relies on the AM1 method to calculate wavefunctions. While this method is very fast, its reliability is relatively modest. More reliable methods for calculating wavefunctions exist, but they are much slower than AM1. Regardless of the method that is used, the wavefunction must be calculated before the charge distribution can be mapped.

Select **Semi-Empirical** from the **Setup** menu. Type "oct-8-en-1-ol" in the **Title** text box. Next, select **Task: Single Point Energy**, **Model: AM1**, **Solvent: None**, **Charge: 0**, and **Multiplicity: 1**. Do not press any of the pushbuttons: **Constraints**, **Freeze**, **Converge**, etc. (if you accidentally press a pushbutton, press it again so that it returns to the "up" position).

Single Point Energy means that the wavefunction (and the molecular energy) of the molecule will be calculated using the trial model you have built. No refinement of the geometry will occur. The remaining settings should be self-explanatory, except for **Multiplicity**. This refers to the "spin state" of the molecule. A "closed shell" molecule, i.e., one in which all molecular orbitals are doubly occupied

---

1. Note: Spartan windows never close on their own. You must close each dialog window yourself before you can use the rest of the program. In most cases, a window can be closed by *clicking* on a button in the window: **Save**, **OK**, **Done**, etc., or by pressing the **Return** key.

always has **Multiplicity: 1**. A molecule or atom with one unpaired electron, e.g., neutral CH_3 or H, always has **Multiplicity: 2**. Carbenes, e.g., CH_2, may be closed-shell molecules but also may have two unpaired electrons (**Multiplicity: 3**). Higher spin states are rarely encountered in organic chemistry.

Next, *click* on **Save**, and then select **Submit** from the **Setup** menu. Two dialog boxes will appear in succession. The first one announces that your calculation ("job") has been submitted (*click* on **OK**), and, after a while, the second announces it has been completed (*click* on **OK** again). Your model, of course, looks exactly the same, but the AM1 wavefunction for your molecule has been stored inside the computer, and is ready to be converted into a map of the charge distribution.

Calculating and displaying the electrostatic potential map: Select **Surfaces** from the **Setup** menu. Then, select **Surface: density**, **Property: elpot**. Finally, *click* on **Add** to place your request in the scroll box and *click* on **Save** to store your request and close the window.

Each surface request has two parts: **Surface** identifies which variable will be used to create a three-dimensional "isovalue" surface, e.g., **Surface: density** indicates that the graph will correspond to a surface of constant electron density. This is roughly equivalent in size and shape to a conventional space-filling (CPK) model. **Property** identifies which variable will be mapped onto the isovalue surface, e.g., **Property: elpot** indicates the electrostatic potential, i.e., the potential energy a positive unit point charge would feel. Since the electrons and nuclei in the molecule attract and repel a positive point charge, respectively, the electrostatic potential equals zero at any point where these forces are in balance. If the local region is unusually electron rich, then the attractive force of these electrons will dominate and the electrostatic potential will be negative. The same reasoning indicates that electron poor regions will have positive (repulsive) electrostatic potential values.

You need to submit the job (**Submit** from the **Setup** menu) again. When the surface calculation has finished, select **Surfaces** from the **Display** window. The surface is listed in a scroll box and should say "completed" (if it still says "pending" then your job is still running). *Click* on the surface in the scroll box (it becomes highlighted in reverse video), and then *click* on the **Display Surface** and **Map Property** pushbuttons. Note that the drawing **Style** and the property **Range** each become active as you click on the appropriate pushbutton (the latter identifies the most negative and most positive electrostatic potential values that were found).

Click on **OK** to close the window, and then look at the map. The map colors identify the value of the electrostatic potential according to the following color code: red (most negative) < orange < yellow < green < blue (most positive). In this case, the property range is very unsymmetric (red corresponds to a very negative electrostatic potential), and green regions are actually slightly electron rich.

Rotate the molecule and its surface, identify the most electron-rich (red) and electron-poor (blue) regions, and try to determine which two atoms these regions correspond to. If you are having trouble identifying the atoms, select **Display: Surfaces**, select the surface in the scroll box, and select a different **Style** (even if you have identified the atoms, experiment with different **Styles**). Since the molecule is neutral overall, it is reasonable to conclude that these electron-rich and electron-poor atoms carry partial charges. What property of these atoms is responsible for their partial charges? Would you describe the bond between these atoms as covalent? ionic? polar covalent? How would you describe the other bonds in the molecule?

The polar OH group dominates the chemistry of this molecule, and most other alcohols. The nonpolar $-CH_2-$ groups do not produce much "interesting chemistry". Close inspection of the $-CH=CH_2$ group shows that it too has an unsymmetric charge distribution. The group is relatively electron rich on its two faces, but relatively electron poor in the atomic plane containing the atoms. Can you produce a bonding model for this group that will account for its unsymmetric charge distribution?

The colors on the graph can be modified to give a more symmetric color scale. Once again, select **Display: Surfaces**, and select the surface in the scroll box. Now, *click* inside the **Range From:** and **Range To:** text boxes and replace the displayed values with "-15" and "15", respectively. Then *click* on **OK**. The map is now redrawn to fit the new range; any region where electrostatic potential is less than -15 is red, and any region where electrostatic potential is greater than +15 is blue. Note that this primarily affects the OH and $-CH=CH_2$ groups; the $-CH_2-$ groups are almost completely nonpolar (green).

When you are finished experimenting with the different options in the **Display: Surfaces** window, select **Close** from the **File** menu.

Glycine (Zwitterion)

Glycine is an α-amino acid and one of the building blocks that proteins are made from. Glycine prefers the "neutral" form in the gas phase and the doubly-charged "zwitterion" form in water, the solvent found in biological systems.

The zwitterion illustrates a general rule in organic chemistry, namely, that most organic ions can be derived from a neutral molecule by either adding or removing a proton (H^+). Comparing the neutral and zwitterion forms shows that the positive N has an extra proton and the negative O is missing a proton.

Building the zwitterion: Select **File: New**, then *click* on the sp^3 C button in the model kit and, finally *click* anywhere on screen. This C will become the methylene group. The next step is to add the charged NH_3 group. Note, however, that none of the N atom fragments in the entry model kit have four bonds. One way to add this fragment to a model is to use the expert model kit. Switch to the expert model kit by *clicking* on **Expert**. The atom fragments in the expert model kit differ from the fragments in the entry model kit in that: they 1) must be "assembled" (you choose the element and the geometry separately), 2) there are more element/geometry combinations to choose from, and 3) bond types are handled more loosely, e.g., it is possible for an atom to have "too many" bonds. In this case you will need a N with four valences arranged in a tetrahedral geometry. *Click* on **N** in the Periodic Table, then *click* on the "tetrahedral X fragment", and finally *click* on one of the unfilled valences of the C on screen. The element and geometry selections are automatically combined when you click on the model.

The "deprotonated" CO_2^- group can be built from the normal atom fragments in the entry model kit. However, we will illustrate instead a shortcut based on SPARTAN*'s* library of functional groups. *Click* on **Groups**, then select **Ester** from the menu underneath the group fragment button (it doesn't matter whether you are using the entry or expert model kit at this point). A drawing of the -CO_2- fragment should appear on the pushbutton. Note that this fragment has two unfilled valences, one on C and one on O, and that one valence is marked by a small circle "o". The "o" indicates which valence will be used to connect the group to the model. *Click* on the pushbutton several times and watch the "o" alternate between the C and O. Stop when the "o" is on C. Then *click* on the appropriate unfilled valence in the model.

At this point, one O still has an unfilled valence, and this valence will be converted into a hydrogen atom when you save the model. To "deprotonate" this O the unfilled valence must be deleted. *Click* on **Delete Atom**; instructions for using the tool appear at the top of the Builder window. *Click* on O's unfilled valence, then type "." to abort. Your model should now look like the glycine zwitterion. Rotate and examine your model carefully. If it has been built correctly, save the model by selecting **File: Quit** and naming the model: "glycine_single_point".

Calculating the AM1 wavefunction and the electrostatic potential map (trial model): Calculate the AM1 wavefunction of the trial model (same setup as "oct-8-en-1-ol" above, except for the title: "glycine (trial model)"; note **Charge: 0** is still used since this parameter refers to the overall charge). Next, calculate an electrostatic potential map (same setup as "oct-8-en-1-ol" above). Display the map and note the zwitterion's polar character. Which region of the molecule is electron poor? Which is electron rich? Did you expect both O atoms to be charged? Are there any differences in their charges?

Turn off the electrostatic potential map by selecting **Display: Surfaces**, and then the surface in the scroll box. Then *click* on the **Display Surface** pushbutton and on **OK**. The two O atoms occupy different chemical environments, and this can be seen by comparing the CO bond distances. Select **Distance** from the **Geometry** menu, follow the instructions in the title bar to measure the two CO bond distances (record them), and then *click* on **Done**. The longer distance should correspond to the single bond (you can see which CO bond was built as a single bond and which was built as a double bond by selecting **Wire** from the **Model** menu).

Calculating the AM1 wavefunction/geometry and the electrostatic potential map (optimized model): While for some molecules, the trial model's geometry is sufficiently reliable that it can be used without further refinement, as a general rule, the model's geometry should be recalculated at the same time that its wavefunction is calculated (refinement of the geometry is necessary because trial geometries are based on rather drastic assumptions about what "normal molecules" look like, and very few molecules are truly normal). The geometry refinement process is called "geometry optimization", and involves moving the atom positions, calculating the wavefunction and molecular energy after each move, and repeating this "move-calculate" cycle until the best possible (lowest energy) geometry is obtained.

To see that geometry optimization can alter molecular structure and properties, you will first need a trial model that can be optimized (you will want to save your original trial model for comparison with the optimized model). Select **File: Save As** and save the model under the name "glycine_optimized" (note that the model name in the title bar of the main window is updated). All of the setup information is saved with the new model, so only a few changes need to be made. Select **Setup: Semi-Empirical** and change **Task:** to **Geometry Optimization**, and change the title to "glycine (AM1 optimized model)" (the remaining parameters should be left unchanged); save the setup (this causes all of the previously saved results to be deleted) and submit the calculation. Both the AM1 wavefunction and geometry will be calculated (this takes a little longer), after which a new electrostatic potential map will be calculated (note that it is possible to combine wavefunction/geometry calculations with surface calculations). The structure on the screen will then be updated automatically.

When the calculation has finished, measure the CO bond distances (use **Model: Wire** to identify the "single" and "double" CO bonds; note that the drawing of these bonds is not affected by the geometry optimization process). Optimization makes the two distances very similar. Also measure the O••H(N) distance (choose the closest pair of O and H atoms). Any distance less than 2.0 Å is considered very short, and suggests a special attraction between these atoms. In this case, the attraction is electrostatic; the atoms have opposite partial charges and this produces a weak attraction known as a "hydrogen bond". Another way to detect a close approach between nonbonded atoms is to display a model as a space-filling model (select **Model: Space Filling**). Each atom is represented by a sphere, and the spheres of nonbonded atoms do not overlap much unless they are forced together. The unusually large overlap of the O and H spheres is typical of a "hydrogen bond". Select **Model: Tube** before proceeding.

A molecule's charge distribution and geometry are always closely related. Since optimization makes the two CO bond distances nearly identical, we might also expect the charge distribution to become more symmetric. Verify this by displaying the electrostatic potential map. On the basis of this geometry and charge information, what resonance structures should you use to describe the glycine zwitterion?

Finally, to reinforce the effect geometry optimization had, open the trial model of glycine (select **File: Open** and "glycine_single_point"; note that the model name in the title bar of the SPARTAN window has been updated) and place it alongside the optimized model. Try to put the two molecules into the same orientation (to work with a given model you must first *click* on the model to make it "active"; the name of the "active" model is always displayed in the title bar). When both models have similar orientations, simultaneously press the **Ctrl** key and rotate the models (note that they both move as long as the **Ctrl** key is pressed). How did optimization affect the geometry? When you are finished, close both models (**File: Close**).

Tutorial 2: Energy

Energy is one of the most important quantities to come from a quantum chemical calculation. It relates overall stability to the positions of the nuclei, and as such, is the criterion for determining molecular equilibrium structure (by locating local energy minima) and transition state structure (by finding lowest-energy pathways connecting equilibrium structures). The difference in energy between products and reactants for a chemical reaction determines whether the reaction is thermodynamically favored (exothermic) or disfavored (endothermic). The difference in energy between reactant and transition state directly relates to the rate of reaction.

Different types of calculations express "energy" in different ways. So called *ab initio* calculations typically give the energy of a molecule relative to fully separated nuclei and electrons. Such energies are very large in relation to typical chemical energy differences, e.g., bond energies. By convention, semi-empirical calculations give "energy" as a heat of formation, that is, the heat of a specific reaction that forms a molecule from a defined set of "standard" reactants. Finally, molecular mechanics calculations give "strain energies"; this measures structural distortion (strain) from an idealized system.

This tutorial is made up of several exercises describing the calculation of heats of formation (ΔH_f) using the AM1 semi-empirical method. They are intended to illustrate some of the many ways to use energies to anticipate molecular structure and chemical behavior.

Base Strength: Ammonia versus Pyridine

Which is the stronger base, ammonia or pyridine? This question, like many questions regarding chemical reactivity, can be better understood if it is expressed in the form of a chemical equilibrium. In this case, the equilibrium involves transfer of a proton between ammonia and pyridine, i.e.,

This equilibrium describes a "competition" by ammonia and pyridine for the proton. If ammonia is the stronger base it will "win" the proton, and the right-side of the equilibrium will be favored ($\Delta G_{rxn} < 0$). If pyridine is the stronger base, the left-side will be favored ($\Delta G_{rxn} > 0$). Obviously, the stronger base can be determined simply by calculating ΔG_{rxn}. Unfortunately, ΔG_{rxn} is very difficult to calculate. In many cases, however, you can use ΔH_{rxn}, the reaction enthalpy or "heat of reaction", in place of ΔG_{rxn} (ΔG and ΔH are related by $\Delta G = \Delta H - T\Delta S$ where ΔS is the entropy change; if ΔS is relatively small, $\Delta G \approx \Delta H$). That is the strategy that you will use here.

ΔH_{rxn} can be calculated as long as the heat of formation, ΔH_f, is known for each molecule appearing in an equilibrium. The relationship between ΔH_{rxn} and ΔH_f is given by $\Delta H_{rxn} = \Delta H_f(\text{products}) - \Delta H_f(\text{reactants})$, i.e., for the proton transfer equilibrium above, $\Delta H_{rxn} = \Delta H_f(\text{pyridine}) + \Delta H_f(\text{ammonium cation}) - \Delta H_f(\text{pyridinium cation}) - \Delta H_f(\text{ammonia})$. The required heats of formation are easily obtained because they are generated by the same AM1 calculations used to calculate wavefunctions and optimize geometries (note: geometry optimization of each molecule appearing in an equilibrium is often essential for a reliable prediction of ΔH_{rxn}).

Build and optimize (AM1) models of ammonia and ammonium cation: To build ammonia, select **File: New**, *click* on the sp^3 N fragment in the model kit and then *click* anywhere on screen. Use **File: Quit** to exit build mode and save your model ("ammonia"). Next, optimize geometry using the AM1 method (**Setup: Semi-Empirical, Task: Geometry Optimization, Model: AM1, Solvent: None, Charge: 0, Multiplicity: 1**, then **Setup: Submit**). When completed, *click* on **Properties** in the **Display** menu, then *click* on **Energy** and record ΔH_f (*click* on **OK** to close the window). Do not close the model.

Now repeat the entire process for the ammonium cation. Since the entry tool kit lacks the necessary tetrahedral N fragment, *click* on expert (in the entry model kit), select N from the Periodic Table and "tetrahedral X" from the list of geometries, and *click* anywhere on screen. Save your structure ("ammonium").[1] Optimize geometry using the AM1 method (same setup as above, except **Charge: 1**). When completed, record ΔH_f (**Display: Properties: Energy**). Do not close the model.

Build and optimize (AM1) models of pyridine and pyridinium cation: The next two calculations proceed in exactly the same way as above: build and save the models as "pyridine" and "pyridinium", respectively (special building instructions follow), setup and submit an AM1 geometry optimization (**Charge: 0** for pyridine, **Charge: 1** for pyridinium cation), and if all goes well, record the ΔH_f values. Do not close either of the models. Finally, calculate ΔH_{rxn} and decide which base is stronger, pyridine or ammonia.[2]

---

1. *Spartan* does not recognize "ammonium" as a cation until you set **Charge: 1** in **Setup: Semi-Empirical**.
2. Your values should agree with the following values to within ± 0.1 kcal/mol: $\Delta H_f(NH_3) = -7.3$, $\Delta H_f(NH_4 \text{ cation}) = 150.6$, $\Delta H_f(\text{pyridine}) = 32.0$, $\Delta H_f(\text{pyridinium cation}) = 184.2$, $\Delta H_{rxn} = +5.7$.

Special building instructions: Models of pyridine and pyridinium cation can be built very quickly using two new tools: the **Rings** tool and "atom replacement". The **Rings** tool works the same way with both the entry and expert model kits, but "atom replacement" gives different results for the two model kits. To build pyridine, *click* on **Rings** (a **Phenyl** fragment or benzene ring appears) and *click* anywhere on screen. The benzene ring is identical to pyridine except one aromatic C must be replaced by an aromatic N. *Click* on the aromatic N icon (entry model kit), then *double click* on a ring C. The C and its unfilled valence will be replaced by N, i.e.,

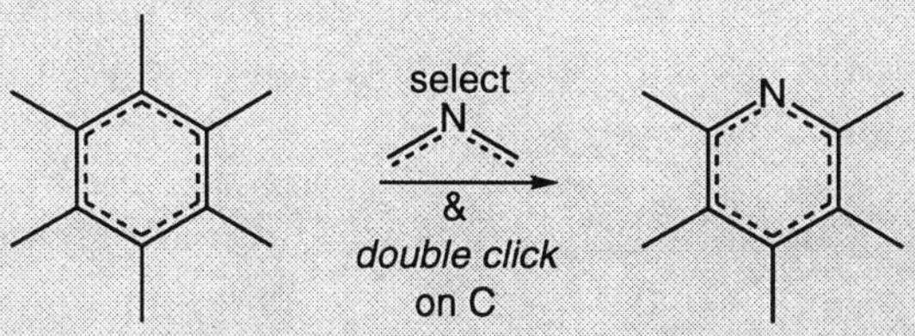

The two essential features of entry-level atom replacement are: 1) the bond properties of the new atom must match the bond properties of the atom it replaces (for those valences that are being used as bonds), and 2) the number of unfilled valences is adjusted to fit the new atom.

Expert-level atom replacement is used to build pyridinium cation. Use **Rings**: **Phenyl** to place a benzene ring on screen. This time you want to replace C with N, but without changing the number of unfilled valences. *Click* on **Expert**. Then select N from the Periodic Table and *double click* on any ring C. This gives pyridinium cation, i.e.,

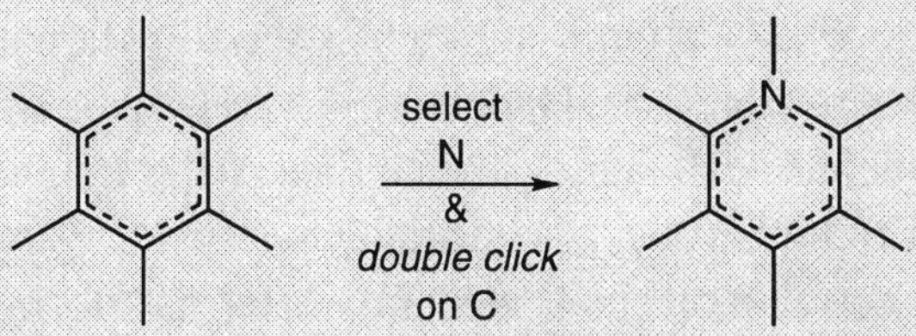

The two essential features of expert-level atom replacement are: 1) no match is required between the bond properties or geometry of the new atom and the atom it replaces, and 2) the new atom adopts the same number of unfilled valences, bond properties, and geometry as the old atom.

What's going on? If you calculate everything correctly, you should obtain ΔH_{rxn} = 5.7 kcal/mol. Pyridine is calculated to be the stronger base. This conclusion is qualitatively correct for the gas phase (the experimental ΔH_{rxn}(gas) is 15 kcal/mol), and this brings up an important point: each of the molecular models consists only of the atoms that are part of the molecule itself, i.e., each model simulates an isolated molecule. Therefore, the energy calculations apply to gas-phase behavior, and extrapolating them to other environments is risky. For example, in aqueous media, pyridine is known to be a weaker base than ammonia. The interaction of the solvent, water, with the various bases and conjugate acids makes protonation of ammonia more favorable.

Ammonia(aq) vs. pyridine(aq). Calculating AM1-SM2 energies: A variety of molecular properties (wavefunction, charge distribution, geometry, ΔH_f) for molecules dissolved in water (at infinite dilution) can be calculated using an extension of the AM1 method known as "AM1-SM2". The influence of the aqueous environment is reproduced using various empirical approximations, and no water molecules are actually added to the model. The following instructions show how to calculate ΔH_{rxn}(aqueous) and determine which base is stronger, ammonia or pyridine, in an aqueous environment.

For each model in turn (all four models should be open; remember that you make a model "active" by *clicking* on it): 1) setup an AM1-SM2 energy calculation, 2) submit the calculation, and 3) if all goes well, use **Display**: **Properties**: **Energy** to get ΔH_f(aqueous) for the model. To setup an AM1-SM2 energy calculation, select **Setup**: **Semi-Empirical**, set **Task**: **Single Point Energy**, i.e., use the current AM1-optimized geometry, set **Solvent**: **Water (C-T)**, *click* on the **Restart Using**: **Wavefunction** pushbutton (this is optional but it speeds up the calculation), and finally *click* on **Save** two times.

Once you finish getting all of the necessary ΔH_f(aqueous) values, combine them to get ΔH_{rxn}(aqueous). Your calculations should now show ammonia to be the stronger base, i.e., ΔH_{rxn}(aqueous) < 0.[3] The change in base strength is due to the fact that different molecules interact with water differently. The strength of this interaction is measured by a quantity known as the solvation enthalpy, or heat of solvation, ΔH_{solv}, where $\Delta H_{solv} = \Delta H_f$(aqueous) $- \Delta H_f$(gas). Use your AM1 (gas) and AM1-SM2 (aqueous) ΔH_f values to calculate ΔH_{solv} for each molecule. What do you notice? First, although ΔH_{solv} is invariably negative, it is much more negative for ions than for neutral molecules. Second, ΔH_{solv} is much more negative for smaller ions and for ions which can make more hydrogen bonds to the solvent. Ammonium is smaller than pyridinium and it is capable of making four hydrogen bonds (pyridinium can only make one), so its solvation enthalpy (-79.0 kcal/mol) is much more negative than pyridinium's (-56.8 kcal/mol). Ammonia is a stronger base than pyridine in water because its conjugate acid, ammonium cation, interacts more strongly with water molecules.

When you are finished, close all of the models.

---

3. Your values should agree with the following values to within ±0.1 kcal/mol: $\Delta H_f(NH_3) = -11.6$, $\Delta H_f(NH_4 \text{ cation}) = 71.6$, $\Delta H_f(\text{pyridine}) = 27.7$, $\Delta H_f(\text{pyridinium cation}) = 127.4$, $\Delta H_{rxn} = -16.6$.

Selectivity in an Acid-Base Reaction: 4-Aminopyridine

Another type of reactivity question that often arises is to find the most reactive site in a polyfunctional molecule. 4-Aminopyridine (**4AP**) contains two amine-type groups. Which is more basic: the amino N (**4AP→A** favored) or the ring N (**4AP→B** favored)?

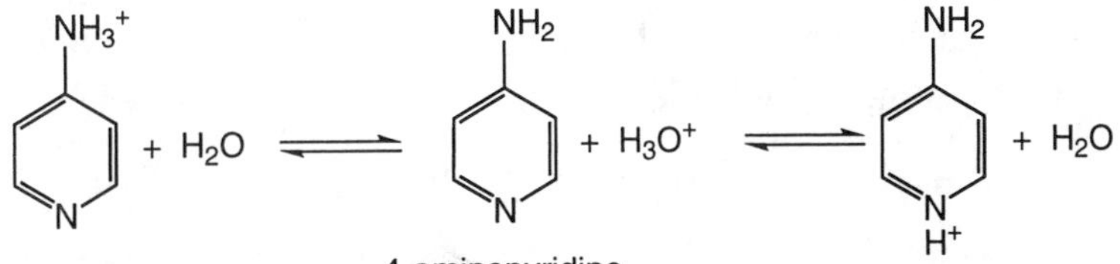

A   4AP   B

One way to answer this question would be to calculate ΔH_{rxn} for each equilibrium. This would require calculating ΔH_f for the five molecules appearing in the two equilibria. However, since the products of the competing reactions, **A** and **B**, are isomers, all we need to know are their relative energies. To put it another way, the difference between ΔH_{rxn} for the two equilibria is the same as ΔH_{rxn} for the single equilibrium shown below.

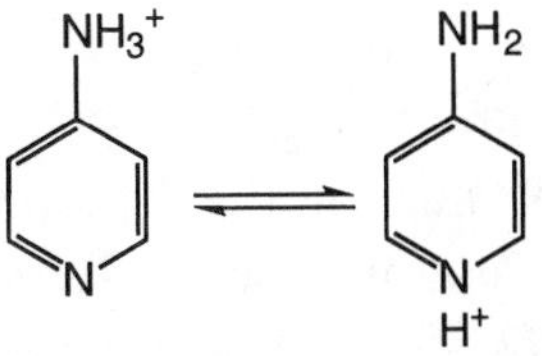

A   B

If $\Delta H_{rxn} < 0$ for the latter, then **B** is more stable than **A** and the ring N is more basic. Otherwise, the amino N is more basic. Before you build models of **A** and **B** and calculate ΔH_{rxn} (gas and aqueous), you may want to see how the charge distribution in the base, **4AP**, can provide qualitative clues regarding reactivity.

OPTIONAL. Qualitative reactivity prediction: Build and save a model of **4AP** using the **Rings** tool and atom replacement. Optimize its geometry using the AM1 method (**Task: Geometry Optimization, Method: AM1, Solvent: None, Charge: 0, Multiplicity: 1**). Then calculate and display its electrostatic potential map (**Surface: Density, Property: elpot, Resolution: med**). According to the electrostatic potential map, which N is more electron rich? The more electron rich N is likely to be more basic.

Quantitative reactivity prediction: Build and save models of **A** and **B** using the **Rings** tool and atom replacement. Optimize the geometry of each model using the AM1 method (**Charge: 1**). If the optimization proceeds satisfactorily, record ΔH_f(gas) for each molecule (**Display: Properties: Energy**). What is $\Delta H_{rxn}(\mathbf{A}\rightarrow\mathbf{B})$?

Which molecule is more stable? Which N of **4AP** is predicted to be more basic?[4]

OPTIONAL. What is going on? What makes the ring N more basic than the amino N? The answer lies in the electronic structure of **A** and **B**. The positive charge in each ion is delocalized to a different degree, and this affects the stability of each (generally, the more delocalized an ion, the more stable it is likely to be). You can decide which Lewis structures contribute most to **A** and **B** by examining carefully the geometry of each molecule and its charge distribution. Since both molecules contain a pyridine ring, two Kekule structures can be drawn for each, i.e.,

A   **B**

The distinguishing factor, then, is whether π-electron sharing occurs between the amino N and the ring. Effective π overlap requires a more planar sp^2-type N, and this should be reflected in a larger H-N-H bond angle and a shorter C_{ring}-N_{amino} bond distance. Use **Geometry: Distance** and **Geometry: Angle** to determine the CN distance and HNH angles in **A**, **B**, and **4AP**. What do these values suggest about the degree of π overlap in the two ions? π–electron sharing should also result in electron-transfer, either: ring$\rightarrow$N or N$\rightarrow$ring. Calculate and display electrostatic potential maps of **A** and **B**. For the ion which exhibits more π overlap, in which direction does electron-transfer occur? Reset the **Range: From** and **Range: To** values for the two electrostatic potential maps so that they are the same. In which ion is the positive charge more delocalized? Use these results (geometry, direction of electron transfer, and degree of delocalization) to determine the most important resonance contributors to **A** and **B**.

OPTIONAL. Solvent effect: You have previously noted the fact that AM1 ΔH_f calculations apply only to gas-phase molecules, and that solvent can exert a strong effect on ΔH_{rxn}. To assign the more basic N in **4AP** in water, recalculate $\Delta H_f(\mathbf{A})$ and $\Delta H_f(\mathbf{B})$ using the AM1-SM2 method, i.e., set **Task: Single Point Energy** (uses the current AM1-optimized geometry), set **Solvent: Water (C-T)**, *click* on the **Restart Using: Wavefunction** pushbutton (this is optional but it speeds up the calculation), then *click* on **Save** two times.

Which ion is more stable? What is ΔH_{rxn}(aqueous)?[5] Has your conclusion regarding the more basic N changed? Although you should reach the same conclusion, you should note that placing the ions in water reduces their energy

---

4. Your values should agree with the following values to within ±0.1 kcal/mol: $\Delta H_f(\mathbf{A}) = 193.6$, $\Delta H_f(\mathbf{B}) = 169.1$, $\Delta H_{rxn} = -24.5$.
5. Your values should agree with the following values to within ±0.1 kcal/mol: $\Delta H_f(\mathbf{A}) = 122.7$, $\Delta H_f(\mathbf{B}) = 111.5$, $\Delta H_{rxn} = -11.2$.

difference considerably. As before, examination of the solvation enthalpy, ΔH_{solv}, is informative. The more delocalized ion (which is more stable in the gas phase) is less stabilized by solvation: $\Delta H_{solv}(\mathbf{B}) = -57.6$ kcal/mol. The more localized ion is more stabilized by solvation: $\Delta H_{solv}(\mathbf{A}) = -70.9$ kcal/mol. You can explain this behavior in terms of ion "size", if you are willing to think of "size" as being defined by the region of the molecule that actually carries a partial charge. A delocalized ion is necessarily "larger" because the charge is spread over more atoms, and this reduces its solvation energy.

Close all of the models.

Resonance Stabilization: Pyrrole, Dihydropyrrole, Tetrahydropyrrole

Resonance and electron delocalization are very important factors to consider when evaluating the properties of ions, but they can also play an important role in neutral molecules as well. This is illustrated by a comparison of pyrrole and its 2,3-dihydro and 2,3,4,5-tetrahydro derivatives (each "hydro" refers to a hydrogen atom that has been added to the parent molecule). For each of these three molecules, build and save a trial model (special building instructions follow), calculate the AM1-optimized geometry and wavefunction, and calculate and display the electrostatic potential map.

Assuming that all of the calculations went satisfactorily, display the electrostatic potential maps of dihydro- and tetrahydropyrrole. The red region around each N shows the location of the lone pair of electrons on N (note the roughly tetrahedral arrangement of the lone pair, the NH bond, and the two NC bonds). Note that the double bond in dihydropyrrole is also electron rich. Now display the electrostatic potential map of pyrrole. Is this N more or less electron rich than the N atoms in dihydro- and tetrahydropyrrole? (Set **Range: From** and **Range: To** in each map to the values given for dihydropyrrole.)

One way you might explain the unusual characteristics of pyrrole is to invoke electron delocalization, i.e., to describe pyrrole as a hybrid of several significant resonance contributors. Draw the resonance contributors needed to describe pyrrole and its two derivatives (before you begin, record the ring CC and CN bond distances and the orientation of the NH bond in each molecule; your resonance contributors must be consistent with these geometrical parameters as well as with the charge distribution). Are you satisfied with your picture?

Frankly, the rules normally used to construct and evaluate Lewis structures suggest that each molecule should be described primarily by one neutral Lewis structure. Charged structures should play a very minor role because separation of charge is required, and also because N is more electronegative than C, so structures with N^+ and C^- should be minor contributors at best. Resonance theory also fails to provide any clue as to why pyrrole should differ significantly from dihydropyrrole,

Special building instructions: Each of these cyclic molecules can be built using the **Rings** tool and atom replacement to introduce N into the ring. The most convenient way to construct these models is to use tetrahydropyrrole as the starting point. The sequence of building operations is shown below.

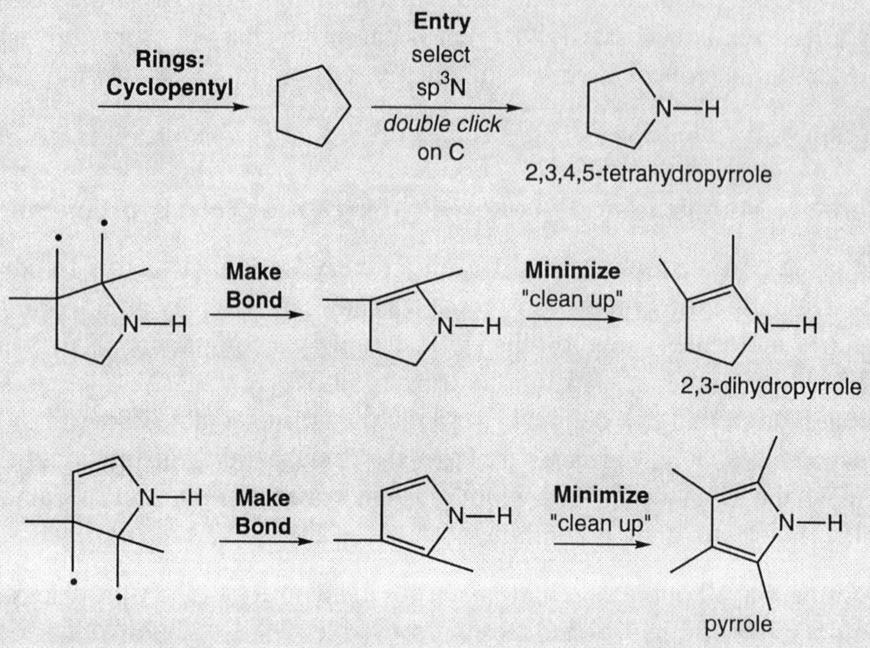

Tetrahydropyrrole: Select **Rings**, then select **Cyclopentyl** from the **Rings** menu, and *click* anywhere on screen. Replace one ring C with an sp^3 N fragment. Save the model as "tetrahydropyrrole".

Dihydropyrrole: Select **Save As** from the **File** menu in the main window and save the tetrahydropyrrole model as "dihydropyrrole". Then, select **Edit Structure** from the **Build** menu. A CC double bond can be formed between two atoms that are already joined by a single bond by using the **Make Bond** tool. *Click* on the **Make Bond** button, and *click* on two unfilled valences as shown by "•" above. The valences will be replaced by a bond, but the molecule's geometry will not be affected in any other way. Since the alkene C atoms clearly have an unreasonable geometry (they are nonplanar), it is necessary to "clean them up" by using **Minimize** to perform a fast, but relatively inaccurate geometry optimization (*click* on **Minimize**, then *click* on **OK** in the window that opens). The result is a much more reasonable looking model.

Pyrrole: Save the dihydropyrrole model as "pyrrole" (**File: Save As**). Select **Build: Edit Structure**, use **Make Bond** and **Minimize** to generate the remaining CC double bond (see above).

which is also unsaturated. The properties of pyrrole can, in fact, be explained fairly simply, but not by resonance theory.[6] More to the point, this example illustrates how molecular modeling can produce unexpected insights into what, at first, seems to be a mundane molecule.

OPTIONAL. Does delocalization stabilize pyrrole? Molecules that possess unusual geometries and/or unusual electron distributions often possess unusual energies. In the case of pyrrole, it might be assumed that its "extra" delocalization stabilizes the molecule. This can be investigated by comparing ΔH_{rxn} for the hydrogenation of pyrrole and ΔH_{rxn} for the hydrogenation of dihydropyrrole, i.e.,

If pyrrole enjoys special stabilization then its hydrogenation will be less favorable energetically, i.e., $\Delta H_{rxn}(1) > \Delta H_{rxn}(2)$. The AM1 ΔH_f for H_2 is -5.18 kcal/mol. Combine this value with your previously calculated AM1 ΔH_f values to obtain $\Delta H_{rxn}(1)$ and $\Delta H_{rxn}(2)$. To what degree is pyrrole (de)stabilized? Given this result, do you expect pyrrole to be a stronger or weaker base than its dihydro and tetrahydro derivatives? You should base your answer strictly on energy arguments (note that the conjugate acids of pyrrole and its derivatives all have localized charges, i.e., protonation prevents electron delocalization and any of the energetic effects associated with delocalization).

When you finish, close all of your models.

---

6.   Pyrrole is an example of a heteroaromatic molecule, and delocalization stabilizes it in much the
     same way delocalization stabilizes benzene.

Tutorial 3: Conformation

Many organic molecules are flexible and exist as a mixture of rapidly interconverting conformers. Each "stable" conformer corresponds to a distinct energy minimum, and it is necessary to model each conformer in order to understand how the molecule will behave. (Sometimes it is also helpful to model non-minimum energy structures, such as transition structures, since these can provide information about the rate and mechanism of conformational change.)

Modeling conformationally flexible molecules may be accomplished in one of two ways. For very simple systems, it is possible to examine conformers one at a time "by hand". Such an approach is practical only when the total number of "important conformers" is small. If this is not the case, there is little recourse other than to use some form of "automatic conformer searching". Unfortunately, even automatic techniques, which extensively survey conformation space, may completely miss important conformers, and the chemist needs to be on guard.

This tutorial illustrates both manual and automatic conformation searching techniques. In the latter case, it also shows how to perform operations on the lists of molecules resulting from conformer searches.

Manual Conformer Search: Methylcyclohexane

Cyclohexane exhibits a "chair-like" structure, in which all six carbons are equivalent, but where there are two distinct sets of hydrogens, so-called *equatorial* and *axial* hydrogens. This suggests that substituted cyclohexanes will exist as mixtures of two structures, one in which the substituent is *equatorial* and the other in which it is *axial*, e.g., in methylcyclohexane.

equatorial   axial

Each molecule interconverts rapidly between these two forms, but on average spends more time in the lower-energy form.[1] To identify the lower-energy conformer, and establish the ratio of conformers in a sample at a particular temperature it is necessary to calculate the energies of both forms. You will do this using the AM1 semi-empirical method, This technique is simple enough to be readily applied to organic molecules, and accurate enough to allow you to draw quantitative conclusions regarding the relative abundances of the two conformers.

---

1. The average %time each molecule spends as a conformer is equivalent to the concentration of the conformer in the mixture, i.e., [*axial*]/[*equatorial*] = [%time *axial*]/[%time *equatorial*].

Build and optimize models of *equatorial* **and** *axial* **methylcyclohexane:** To build a model of *equatorial* methylcyclohexane, select **File: New**, *click* on **Cyclohexyl** ring fragment (**Rings: Cyclohexyl**), and then *click* anywhere on screen. Next, *click* on the sp³ C fragment and *click* on any *equatorial* unfilled valence of the ring (use **Edit: Undo Add** or **Delete Atom** if you make a mistake). Use **File: Quit** to exit build mode and save your model ("eq_methylcyclohexane"). Next, optimize the geometry using the AM1 semi-empirical molecular orbital method (**Setup** menu, **Task: Geometry Optimization, Level: AM1**). Pay careful attention to the various dialog windows to make sure that no error messages are reported. Finally, record the model's heat of formation (**Display: Properties: Energy**).

Now repeat the entire process for the *axial* conformer (note: you must connect the sp³ C fragment to an *axial* unfilled valence of the ring); save the model as "ax_methylcyclohexane"). Record the heat of formation of the optimized model. What is the difference in the conformer energies?[2] (Experimental measurements give ΔG(eq → ax) = 1.7 kcal/mol at room temperature.) Which conformer do you predict to be the preferred form?

Finally, compare the optimized geometries of the *equatorial* and *axial* conformers. Do you see any significant differences between the two? If so, which conformer do you think is "distorted"?[3] Do the magnitudes of the structural distortions make sense? It's worth spending a little time looking at these structures and trying to understand what makes their geometry different (beyond the obvious placement of the methyl group), since you will often come across molecules for which structures are known, but not energies.

Close all the molecules on screen.

Automatic Conformer Search: 1-Butanol[4]

All of the preferred conformers of 1-butanol are staggered.

ω(CCCC) = 60°, 180°, 300°   ω(CCCO) = 60°, 180°, 300°

---

2. Your values should agree with the following values to within ±0.1 kcal/mol: *equatorial, -43.7; axial, -42.3.*

3. One way to look for distortions is to check the distances between pairs of *axial* hydrogens located on the same side of the ring (distances should differ by more than 0.001Å for the difference to be significant). In cyclohexane, all of the distances are the same. Other geometrical definitions of "distortion" are possible as well.

4. This section and the next section can only be completed using SPARTAN or other programs equipped to perform automatic conformer searching. It cannot be completed using MacSPARTAN, MacSPARTAN *Plus* or PC SPARTAN.

Interconversion among conformers is rapid, and it is to be expected that several staggered conformers will actually be present in a sample of 1-butanol. Assuming that rotation around two of the single bonds (C_1C_2 and C_2C_3) gives rise to different conformers (rotation around C_3C_4 does not give rise to different conformers, and for the purpose of this tutorial, it will be assumed that the OH bond is *anti* to the C_1C_2 bond), and further assuming that 3 staggered conformations are possible for each bond, then there are 3^2 (or 9) possible conformers. Some conformers will be symmetry related, like the two *gauche* conformers of butane, so the total number of unique conformers will be less than 9. You can manually generate (and name) each of these conformers by building models with each possible combination of the two dihedral (torsion) angles, $\omega(C_4C_3C_2C_1) = 60°, 180°, 300°$ and $\omega(C_3C_2C_1O) = 60°, 180°, 300°$, but this is a tedious task and prone to error. Automatic conformer generation and evaluation is preferred, and this procedure is described below.

Build a single conformer of 1-butanol: First, build a reasonable starting conformation for 1-butanol. Go into build mode (**File: New**), and assemble 1-butanol from one atom fragments. All of the atoms will automatically be placed in staggered positions.

Internal rotation: Now check the orientation of the OH bond. For this exercise, you want the OH bond to be *anti* to the C_1C_2 bond, i.e., $\omega(C_2C_1OH) = 180°$. If the orientation is correct, then exit build mode (save as "butanol"). Otherwise, reorient the OH bond by doing an "internal rotation" about the C_1O bond (note: internal rotation can be done about any non-ring bond). *Click* on the C_1O bond (not the atoms; use **Edit: Undo** to recover from any mistakes). The bond should switch from a solid line to a dashed line. Then press the **space bar** and simultaneously move the mouse up and down (press the middle button). This combined operation produces an internal rotation about the desired bond. Continue this until you have properly oriented the OH bond, and then exit build mode (**File: Quit**).

Define a conformer search: The next step is to define the two bonds to be rotated during the conformer search, and the strategy for carrying out these rotations. Select **Conformer Search** from the **Build** menu. A conformer panel appears which lists several methods for carrying out conformer searches; use the simplest one: **Systematic**.[5] If necessary, select **Systematic**.

Next, move the cursor to the C_1C_2 bond and *double click* on it. A small gold cylinder appears on the bond. This indicates that internal rotation will occur about this bond during the conformer search (the number of rotations to be done, 3, is listed in the **Fold Rotation:** text box in the conformer panel). Now move the cursor to the C_2C_3 bond and *double click* on it. A small gold cylinder appears on this bond too. Save your changes and return to the main window by selecting **Quit** from the **File** menu.

---

5. Any method can be used for conformational searches involving non-ring bonds only inside of *SPARTAN*. Only the systematic (Osawa) method is presently available for searches involving bonds incorporated into rings.

Setup and submit the calculation: Any method for calculating energy and optimizing geometry can be used to drive a conformer search. However, since most searches need to examine a large number of possible conformers, quantum mechanical optimization methods are generally too slow to be practical. Molecular mechanics methods, which calculate a molecule's "strain energy" as a function of its geometry, are practical. Optimizations based on this technique produce geometries with minimal strain energies. Since different conformers correspond to distinct energy minima, differences in conformer energies are equal to the difference in their strain energies.[6] Molecular mechanics does not take a molecule's electron distribution into account. Therefore, molecular mechanics is not a good method for studying conformer relationships when one (or more) conformers is stabilized by a "special" interaction, such as hydrogen bonding or electron delocalization (resonance stabilization), missing in another conformer. And, molecular mechanics energies should never be used to compare the stability of molecules with different bonding patterns.

You will perform the conformation search for 1-butanol using the MM3 molecular mechanics method. Setup and submit an MM3 optimization (**Setup: Mechanics, Task: Geometry Optimization, Force Field: MM3**, then **Setup: Submit**). Pay careful attention to the various dialog windows to make sure that no error messages are reported. Notice that after the calculation is completed, another dialog window appears: "Promoting molecule to list of molecules". *Click* on **OK**.

Each of the lowest energy conformers found during the conformer search is listed in a small window. The names of these conformers ("butanol0001", "butanol0002", etc.) reflect the order in which they were "discovered" during the search. Display their MM3 strain energies by *clicking* on **Show Energy** in the **View** menu. Next, sort the conformers by energy by *clicking* on **by Energy** in the **Sort** menu. The scroll bars can be used to examine any conformers that overflow the window. Examine the different structures by *clicking* on each name in the window; the corresponding conformer will be displayed in the main window and the conformer name ("butanol:butanol000x") will appear in the title bar.

Note that the all *anti* conformer is the most stable (strain energy= 4.2); bringing either OH or CH$_3$ into a *gauche* relationship with the rest of the C chain raises the energy (which group raises the energy more?).[7] Also, note that there are two conformers in which the OH and CH$_3$ groups are both *gauche*, and these have different strain energies. The strain energy of the more stable *gauche-gauche* conformer could also have been predicted by simple addition: 4.2 + 0.4 (effect

---

6.   This assumption is usually valid because conformers usually have identical bonding patterns.
7.   Your MM3 strain energy values should agree with the following values to within ±0.1 kcal/mol:

283

of *gauche* OH, 4.6-4.2) + 0.8 (effect of *gauche* CH_3, 5.0-4.2) = 5.4. The less stable conformer (strain energy = 6.0) is destabilized by an additional nonbonding interaction, namely, interaction of the OH and CH_3 groups, and its energy cannot be predicted by simple addition. In any case, the predicted energy gap separating the five conformers is relatively small, and rapid interconversion between them is to be expected. You can get a feel for this motion by *clicking* on **Animate**. This motion is not meant to be realistic, but it does give an interesting sense of how flexible these molecules truly are.

Finally, note that the "butanol" model is now a list of five structures. You cannot enter build mode with a "list", and if you want to edit a conformer, or define a new conformer search starting with a particular conformer, you first need to save the conformer of interest using the **Extract As** option from the **Member** menu. This would create a new model which could then be manipulated in all of the normal ways.

When you are finished looking at your conformers, close the model.

Working with Conformer Lists: 3-Hydroxypropylammonium Ion

Once a list of conformers has been generated it is easy to examine their electronic structure and energy using quantum mechanical methods. The overall sequence is: 1) build a single conformer, 2) generate a conformer list using molecular mechanics, and 3) perform quantum mechanical calculations on the entire list. This last step can be accomplished with a single "setup/submit" operation, just the way one would work with a normal model. The full procedure is demonstrated for the 3-hydroxypropylammonium ion, a molecule which closely resembles 1-butanol in shape.

Generate a conformer list using the SYBYL molecular mechanics conformer search: Build 3-hydroxypropylammonium ion; connect together 3 sp^3 C fragments and an sp^3 O fragment (entry model kit), then add a tetrahedral, four-coordinate N fragment (expert model kit). Exit build mode and save your model.

Select **Build: Conformer Search** and select both of the CC bonds (**Systematic** strategy) for 3-fold rotation. The MM3 molecular mechanics method employed previously cannot be used to study ammonium ions, and the SYBYL force field will be used instead.[8] Setup and submit a SYBYL optimization (**Setup: Mechanics**,

---

8. Chemists have developed a large number of force fields (MM3, MM2, SYBYL, etc.) to treat various types of molecules. Structures generated by different force fields can be compared, but strain energies calculated using different force fields should never be compared.

Task: Geometry Optimization, Force Field: SYBYL, then **Setup: Submit**). Pay careful attention to the various dialog windows to make sure that no error messages appear. Select **View: Show Energy** and **Sort: by Energy**. What is the structure of the lowest energy conformer? Can you explain the energy ordering?

Optimize the conformers using the AM1 method: As a rule, molecular mechanics methods do not give a good description of charged molecules. A better idea about the relative energies of different conformers is obtained by reoptimizing the structure of each SYBYL generated structure using the AM1 method.

Setup and submit an AM1 geometry optimization (**Setup: Semi-Empirical, Task: Geometry Optimization, Model: AM1, Solvent: None, Charge: 1, Multiplicity: 1, Global**[9], then **Setup: Submit**). As long as the **Global** pushbutton is selected during the setup, all five geometry optimizations will be automatically conducted in sequence using the same setup instructions (the progress of the optimizations will be reflected in the conformer list window). Pay careful attention to the various dialog windows to make sure that no error messages appear. (One error message that might appear is "Optimization failed...". This indicates that one or more of the conformers was not fully optimized. If this occurs, select **Setup: Semi-Empirical**, and *click* on **Restart Using: Wavefunction** and **Restart Using: Hessian** pushbuttons. Then submit the optimization again.)

Sort the conformers by energy. What is the structure of the lowest-energy conformer? Can you explain the energy ordering?[10] Unlike 1-butanol which prefers the all-*anti* conformation, this ion strongly prefers a conformation in which both the OH and NH$_3$ groups are *gauche* to the C chain; this allows the formation of an intramolecular O••H-N$^+$ hydrogen bond. The ion will still be flexible to some degree, but the vast majority of molecules present in a sample at any given moment are predicted to have the same "*gauche-gauche*" conformation.

Calculate the AM1-SM2 energies: Intramolecular hydrogen-bonds play a very important role in determining the conformation of molecules in the gas-phase, molecules dissolved in organic solvents, and molecules in the solid state. The conformational preferences of molecules dissolved in water may be different, however, because water, itself, is such a strong hydrogen-bonding medium. You can test this hypothesis by recalculating the energies of the five conformers using the AM1-SM2 method (use the AM1 geometries for this calculation, i.e., perform

---

9. The **Global** pushbutton is located in the upper right corner of the **Setup** dialog window.
10. Your AM1 ΔH$_f$ values should agree roughly with the following values (kcal/mol) (note the number of conformers detected may vary from three to five):

88   85   88   78

a single point calculation). Setup and submit a single-point AM1-SM2 calculation (**Setup**: **Semi-Empirical**, **Task**: **Single Point Energy**, **Solvent**: **Water (C-T)**, **Restart Using**: **Wavefunction**, **Global**, then **Setup**: **Submit**). Once again, as long as the **Global** pushbutton is selected, all five conformers on the list will be recalculated automatically.

A very different picture of this ion's conformational properties emerge from these calculations. The "aqueous" heats of formation predicted by the SM2 method are all very similar.[11] This makes sense. No price has to be paid for breaking the intramolecular hydrogen bond because the ammonium group and the hydroxyl group can both hydrogen bond to solvent (water) molecules. Therefore, it is to be expected that this ion will be much more flexible in water than in less polar organic solvents (or the gas phase).

When you are finished, close all of your models.

---

11. Your AM1-SM2 ΔH_f values should agree roughly with the following values (kcal/mol) (note the number of conformers detected may vary from two to five):

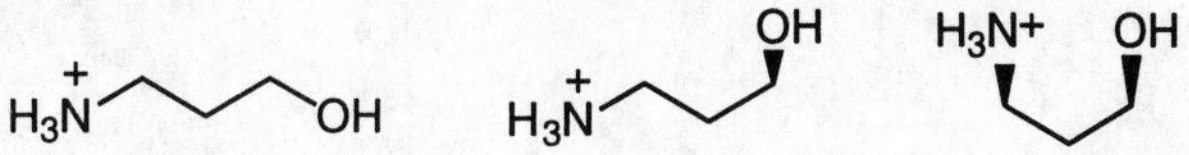

| 15 | 15 | 14 |

Appendix B

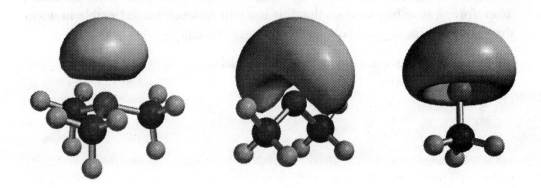

Glossary

Following are brief definitions of a number of terms that are commonly used by computational chemists:

3-21G. A basis set in which each inner-shell atomic orbital is written in terms of three Gaussian functions, and each valence-shell atomic orbital is split into two parts, written in terms of two and one Gaussians, respectively.

3-21G$^{(*)}$. A **3-21G** basis set that has been supplemented by a set of six d-type Gaussian functions for second-row and heavier main-group elements only.

6-31G*. A basis set in which each inner-shell atomic orbital is written in terms of six Gaussian functions, and each valence-shell atomic orbital is split into two parts, written in terms of three and one Gaussians, respectively. Non-hydrogen atoms are also supplemented with a set of six d-type Gaussian functions.

Ab Initio. "From the beginning". The general term used to describe methods seeking approximate solutions to the many-electron **Schrödinger Equation**, but which do not involve empirical parameters.

AM1. Austin Method 1. A semi-empirical molecular orbital method.

AM1-SM2. A modified form of the AM1 semi-empirical method that is specifically parameterized to reproduce aqueous-phase thermochemistry at infinite dilution.

Antibonding Molecular Orbital. A **Molecular Orbital** that is antibonding between particular atomic centers. The opposite is a **Bonding Molecular Orbital**.

Atomic Orbital. A **Basis Function** centered on an atom. Atomic orbitals typically take on the form of the solutions to the hydrogen atom (s, p, d, f....type orbitals).

Basis Functions. Functions usually centered on atoms (but not restricted as such), linear combinations of which make up the set of **Molecular Orbitals**.

Basis Set. The entire collection of **Basis Functions** from which delocalized **Molecular Orbitals** are constructed.

Boltzmann Equation. The equation which relates the composition of equilibrium mixtures to relative thermochemical stabilities and temperature.

Bonding Molecular Orbital. A **Molecular Orbital** that is bonding between particular atom centers. The opposite is an **Antibonding Molecular Orbital**.

Bond Separation Reaction. A special **Isodesmic Reaction** in which a molecule comprising three or more heavy (non-hydrogen) atoms, and described in terms of a conventional valence structure, is broken down into the simplest (two-heavy-atom) molecules containing the same component bonds. The energies of bond separation reactions are well described using simple *ab initio* molecular orbital models.

Density; see **Electron Density**.

Electron Density. A function that gives the number of electrons per unit volume at a point in space. Summed over all space, this gives the total number of electrons.

Electrostatic Charges. Atomic charges chosen to best match the electrostatic potential at points surrounding a molecule, subject to overall charge balance.

Electrostatic Potential. A function describing the energy of interaction of a point positive charge with the nuclei and fixed electron distribution of a molecule.

Electrostatic Potential Map. A graph that shows the value of the **Electrostatic Potential** on an **Electron Density Isosurface** corresponding to a van der Waals contact surface.

Equilibrium Geometry. The geometry corresponding to a **Local Minimum** on the potential energy surface. It cannot actually be directly measured experimentally, for even at 0K, molecules do not reside at the bottom of the potential energy well, but rather at a higher energy level (the **Zero Point Energy** level). In practice, equilibrium geometry can be inferred accurately from experimental measurements on different vibrational states.

Equilibrium Isotope Effect. The change in equilibrium composition caused by isotopic substitution.

Global Minimum. The lowest energy **Local Minimum** for a system of given stereochemistry.

Hartree-Fock Wavefunction. The simplest quantum-mechanically correct representation of the many-electron wavefunction. Electrons are treated as independent particles, and are assigned in pairs to functions termed **Molecular Orbitals**.

Hessian. The matrix of second derivatives of the energy with respect to the 3N-6 (N atoms) geometrical coordinates.

HOMO. Highest Occupied Molecular Orbital.

HOMO Map. A graph that shows the absolute value of the **HOMO** on an **Electron Density Isosurface** corresponding to a van der Waals contact surface.

Imaginary Frequency. A frequency that results from a negative force constant (in the diagonal form of the **Hessian**). An imaginary frequency characterizes the **Reaction Coordinate** in a **Transition State**.

Isodensity Surface. A surface of constant **Electron Density** defined by setting the value of the electron density function to a constant.

Isodesmic **Reaction**. A balanced chemical reaction in which the number of formal chemical bonds of each type is conserved.

Isopotential Surface. A surface of constant **Electrostatic Potential** defined by setting the value of the electrostatic potential to a constant.

Isosurface. A three-dimensional surface defined by the set of points in space where the value of the function is constant.

Isovalue Surface; see **Isosurface**

Kinetically Controlled Reaction. A reaction the product ratio for which is determined solely by the rate at which different products form (product formation must be irreversible).

Kinetic Isotope Effect. The change in reaction rate caused by isotopic substitution.

Linear Synchronous Transit. A procedure that estimates the geometries of **Transition States** based on "averages" of the geometries of reactants and products, sometimes weighted by the overall thermodynamics of reaction.

Local Minimum. Any **Stationary Point** for which all elements in the diagonal representation of the Hessian are positive. Chemically, a local minimum corresponds to an isomer.

Lone Pair. A pair of electrons contained in a **Non-Bonded Molecular Orbital**.

LUMO. Lowest Unoccupied Molecular Orbital.

LUMO Map. A graph that shows the absolute value of the **LUMO** on an **Electron Density Isosurface** corresponding to a van der Waals contact surface.

Molecular Mechanics. Any of the methods for structure, conformation and strain energy calculation based on empirical bond stretching, angle bending and torsional energy formulas as well as formulas accounting for non-bonding interactions. The energy formulas are all parameterized to fit experimental data.

Molecular Orbital. A one-electron function made of contributions of **Basis Functions** at the individual atomic centers (**Atomic Orbitals**) and delocalized throughout the entire molecular skeleton.

Mulliken Charge. Atom charge obtained from a **Mulliken Population Analysis**.

Mulliken Population Analysis. A charge partitioning scheme in which electrons are shared equally between different **Basis Functions**.

Multiplicity. The number of magnetically-equivalent spin states associated with a particular electronic state. Closed-shell molecules are singlets (multiplicity = 1). Radicals are doublets (spin up or down; multiplicity = 2). Diradicals can be singlets or triplets (multiplicity = 3).

Natural Population Analysis. A charge partitioning scheme.

Nonbonded Molecular Orbital. A molecular orbital that does not show any significant bonding or antibonding characteristics. Nonbonded molecular orbitals often correspond to **Lone Pairs**.

Normal Coordinates. The set of coordinates that lead to a diagonal **Hessian**.

Normal Mode Analysis. The process for calculating **Normal Coordinates**.

Overlap Population. The number of electrons shared by two atoms.

Potential Energy Surface. A many-dimensional function of the energy of a molecule in terms of the geometrical coordinates of the atoms.

Reaction Coordinate. The **Normal Coordinate** that connects the **Local Minima** corresponding to the reactant and product. At the **Transition State**, the reaction coordinate corresponds to the normal coordinate with an **Imaginary Frequency**.

Schrödinger Equation. The quantum mechanical equation which accounts for the motions of nuclei and electrons in atomic and molecular systems.

Semi-Empirical. Quantum mechanical methods that seek approximate solutions to the many electron **Schrödinger Equation**, but which involve empirical parameters.

Spin Density. The difference in the number of electrons per unit volume of "up" spin and "down" spin at a point in space. The function is defined over all space, and summed over all space give the difference in the total number of electrons of "up" spin and "down" spin.

Spin Density Map. A graph that shows the value of the **Spin Density** on an **Electron Density Isosurface** corresponding to a van der Waals contact surface.

Stationary Point. Any point on a **Potential Energy Surface** for which all energy first derivatives with respect to coordinate changes are zero.

SYBYL. A very simple molecular mechanics force field developed by Tripos, Inc.

Thermodynamically Controlled Reaction. A reaction the product ratio for which is determined solely by the relative thermochemical stabilities of the different products (product formation must be reversible, or separate low-energy pathways interconnecting the products must exist).

Total Electron Density; see **Electron Density**

Transition State; see **Transition State Geometry**.

Transition State Geometry. The geometry corresponding to a **Stationary Point** on the **Potential Energy Surface** in which all but one of the elements in the diagonal representation of the **Hessian** are positive, and one of the elements is negative. The transition state geometry is usually thought of as corresponding to the highest-energy point on the **Reaction Coordinate**.

Zero Point Energy. The energy of molecular vibration at 0K.

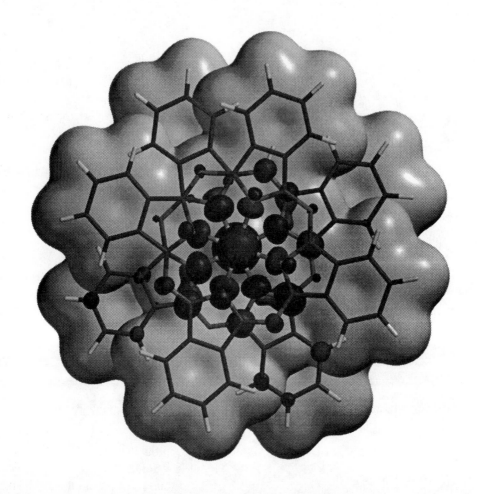

Index

Note: Bold Roman numerals, **I**, **II**, ..., refer to essays; bold arabic numerals **1**, **2**, ... refer to experiments.

Strain; *See* Ring Strain

T

U

V

W

X

Y

Z

Wavefunction Publications

Practical Strategies for Electronic Structure Calculations $ 25.00
W.J. Hehre, 1995.
(ISBN 0-9643495-1-5)

Chemistry with Computation. An Introduction to *SPARTAN* $ 15.00
W.J. Hehre and W.W. Huang, 1995.
(ISBN 0-9643495-2-3)

A Laboratory Book of Computational Organic Chemistry $ 25.00
W.J. Hehre, A.J. Shusterman and W.W. Huang, 1996.
(ISBN 0-9643495-5-8)

A Short Course in Modern Electronic Structure Theory $ 150.00
(spiral bound lecture notes)
W.J. Hehre, 1993-1996

SPARTAN **User's Guide, version 4.1** $ 35.00
(ISBN 0-9643495-3-1)

*Mac*SPARTAN **Tutorial and User's Guide, version 1.0** $ 20.00
(ISBN 0-9643495-4-X)

*Mac*SPARTAN *Plus* **Tutorial and User's Guide, version 1.0** $ 20.00
(ISBN 0-9643495-7-4)

PC SPARTAN **Tutorial and User's Guide, version 1.0** $ 20.00
(ISBN 0-9643495-6-6)

Forthcoming:

Critical Assessment of Modern Electronic Structure Methods
W.J. Hehre, Winter, 1997.

A Textbook of Computational Organic Chemistry
A.J. Shusterman and W.J. Hehre, Winter, 1997.

Wavefunction, Inc.
18401 Von Karman Avenue, Suite 370, Irvine, CA 92612 USA
(714) 955-2120 • Fax: (714) 955-2118
books@wavefun.com
http://www.wavefun.com

NOTES

NOTES